INDUSTRIAL ECOLOGY AND
SUSTAINABLE ENGINEERING

INDUSTRIAL ECOLOGY AND SUSTAINABLE ENGINEERING

T.E. Graedel
Yale University

B.R. Allenby
Arizona State University

Prentice Hall
Boston   Columbus   Indianapolis   New York   San Francisco   Upper Saddle River
Amsterdam   Cape Town   Dubai   London   Madrid   Milan   Munich   Paris   Montreal   Toronto
Delhi   Mexico City   Sao Paulo   Sydney   Hong Kong   Seoul   Singapore   Taipei   Tokyo

Vice President and Editorial Director, ECS:
  Marcia J. Horton
Senior Editor: *Holly Stark*
Associate Editor: *Dee Bernhard*
Editorial Assistant: *William Opaluch*
Marketing Manager: *Tim Galligan*
Production Manager: *Kathy Sleys*
Art Director: *Jayne Conte*

Cover Designer: *Bruce Kenselaar*
Cover Art: Images of Thorncrown
  Chapel/Photographer: *Whit Slemmons*
Full-Service Project Management/Composition:
  Shiji Sashi/Integra Software Services
Printer/Binder: *King Printing Co., Inc.*
Cover Printer: *King Printing Co., Inc.*
Text Font: *10/12, Times Ten Roman*

Credits and acknowledgments borrowed from other sources and reproduced, with permission, in this textbook appear on appropriate page within text.

Microsoft® and Windows® are registered trademarks of the Microsoft Corporation in the U.S.A. and other countries. Screen shots and icons reprinted with permission from the Microsoft Corporation. This book is not sponsored or endorsed by or affiliated with the Microsoft Corporation.

Library of Congress Cataloging-in-Publication Data
Graedel, T. E.
  Industrial ecology and sustainable engineering / T.E. Graedel, B.R.
Allenby.
     p. cm.
Includes bibliographical references and index.
ISBN-13: 978-0-13-600806-4 (alk. paper)
ISBN-10: 0-13-600806-2 (alk. paper)
1. Industrial ecology 2. Sustainable engineering. I. Allenby, Braden R.
II. Title.
TS161.G7425 2010
658.4'08—dc22

                                        2009029798

10 9 8 7

Prentice Hall
is an imprint of

www.pearsonhighered.com

ISBN 10:   0-13-600806-2
ISBN 13: 978-0-13-600806-4

To Our children and grandchildren
Laura, Kendra, Martha, Richard, and Steven
in anticipation of a sustainable world

Contents

Preface xvii

PART I INTRODUCING THE FIELD   1

Chapter 1 Humanity and Technology   1
   1.1 An integrated system 1
   1.2 The tragedy of the commons 2
   1.3 Technology at work 4
   1.4 The master equation 5
   1.5 Technological evolution 7
   1.6 Addressing the challenge 11
   Further Reading 11
   Exercises 11

Chapter 2 The Concept of Sustainability   13
   2.1 Is humanity's path unsustainable? 13
   2.2 Components of a sustainability transition 15
   2.3 Quantifying sustainability 17
      2.3.1 Example 1: Sustainable supplies of zinc 18
      2.3.2 Example 2: Sustainable supplies of germanium 19
      2.3.3 Example 3: Sustainable production of greenhouse gases 20
      2.3.4 Issues in quantifying sustainability 21

2.4 Linking industrial ecology activities to sustainability 22
  2.4.1 *The grand environmental objectives* 22
  2.4.2 *Linking the grand objectives to environmental science* 23
  2.4.3 *Targeted activities of technological societies* 26
  2.4.4 *Actions for an industrialized society* 28
Further Reading 29
Exercises 29

Chapter 3 Industrial Ecology and Sustainable Engineering Concepts   **30**
3.1 From contemporaneous thinking to forward thinking 30
3.2 The greening of engineering 33
3.3 Linking industrial activity with environmental and social sciences 34
3.4 The challenge of quantification and rigor 35
3.5 Key questions of industrial ecology and sustainable engineering 36
3.6 An overview of this book 37
Further Reading 39
Exercises 40

PART II   FRAMEWORK TOPICS   **41**

Chapter 4 The Relevance of Biological Ecology to Industrial Ecology   **41**
4.1 Considering the analogy 41
4.2 Biological and industrial organisms 42
4.3 Biological and industrial ecosystems 44
4.4 Engineering by biological and industrial organisms 47
4.5 Evolution 49
4.6 The utility of the ecological approach 51
Further Reading 53
Exercises 53

Chapter 5 Metabolic Analysis   **55**
5.1 The concept of metabolism 55
5.2 Metabolisms of biological organisms 55
5.3 Metabolisms of industrial organisms 57
5.4 The utility of metabolic analysis in industrial ecology 64
Further Reading 65
Exercises 66

Chapter 6 Technology and Risk   **67**
6.1 Historical patterns in technological evolution 67
6.2 Approaches to risk 71
6.3 Risk assessment 75

6.4   Risk communication   77
6.5   Risk management   78
Further Reading   80
Exercises   80

Chapter 7   The Social Dimensions of Industrial Ecology   **82**
7.1   Framing industrial ecology and sustainable engineering
        within society   82
7.2   Cultural constructs and temporal scales   83
7.3   Social ecology   86
7.4   Consumption   88
7.5   Government and governance   89
7.6   Legal and ethical concerns in industrial ecology   91
7.7   Economics and industrial ecology   93
        7.7.1   The Private Firm   94
        7.7.2   Valuation   94
        7.7.3   Discount Rates   95
        7.7.4   Green Accounting   96
7.8   Integrating the themes   97
Further Reading   99
Exercises   99

PART III   IMPLEMENTATION   **101**

Chapter 8   Sustainable Engineering   **101**
8.1   Engineering and the industrial sequence   101
8.2   Green chemistry   103
8.3   Green engineering   105
8.4   The process design challenge   107
8.5   Pollution prevention   107
8.6   The process life cycle   110
        8.6.1   Resource provisioning   110
        8.6.2   Process implementation   111
        8.6.3   Primary process operation   111
        8.6.4   Complementary process operation   111
        8.6.5   Refurbishment, recycling, disposal   112
8.7   Green technology and sustainability   112
Further Reading   114
Exercises   114

Chapter 9   Technological Product Development   **115**
9.1   The product development challenge   115
9.2   Conceptual tools for product designers   117
        9.2.1   The Pugh Selection Matrix   117
        9.2.2   The House of quality   118

9.3   Design for *X*   118
9.4   Product design teams   121
9.5   The product realization process   122
Further Reading   125
Exercises   125

**Chapter 10   Design for Environment and Sustainability:
Customer Products**   **126**
10.1   Introduction   126
10.2   Choosing materials   126
10.3   Combining materials   129
10.4   Product delivery   131
10.5   The product use phase   134
10.6   Design for reuse and recycling   135
    10.6.1   The comet diagram   135
    *10.6.2   Approaches to design
            for recycling   137*
    10.6.3   Recycling complexities   138
10.7   Guidelines for DfES   143
Further Reading   144
Exercises   144

**Chapter 11   Design for Environment and Sustainability:
Buildings and Infrastructure**   **146**
11.1   The (infra)structures of society   146
11.2   Electric power infrastructure   148
11.3   Water infrastructure   149
11.4   Transportation infrastructure   150
11.5   Telecommunications infrastructure   151
11.6   Green buildings   152
11.7   Infrastructure and building materials
        recycling   153
11.8   Green design guidelines   157
Further reading   158
Exercises   159

Chapter 12   An Introduction to Life Cycle Assessment   **161**
12.1   The concept of the life cycle   161
12.2   The LCA framework   162
12.3   Goal setting and scope determination   164
12.4   Defining boundaries   164
    12.4.1   Level of detail boundaries   165
    12.4.2   The natural ecosystem boundary   165
    12.4.3   Boundaries in space and time   165
    12.4.4   Choosing boundaries   167
12.5   Approaches to data acquisition   167

12.6 The life cycle of industrial products   170
12.7 The utility of life cycle inventory analysis   173
Further Reading   173
Exercises   174

Chapter 13   The LCA Impact and Interpretation Stages   **175**
13.1 LCA impact analysis   175
13.2 Interpretation   181
  13.2.1   Identify significant issues in the results   181
  13.2.2   Evaluate the data used in the LCA   182
  13.2.3   Draw conclusions and recommendations   182
13.3 LCA software   182
13.4 Prioritizing recommendations   183
  13.4.1   Approaches to prioritization   183
  13.4.2   The action-agent prioritization diagram   185
  13.4.3   The life-stage prioritization diagram   187
13.5 The limitations of life cycle assessment   188
Further Reading   189
Exercises   190

Chapter 14   Streamlining the LCA Process   **191**
14.1 Needs of the LCA user community   191
14.2 The assessment continuum   192
14.3 Preserving perspective while streamlining   193
14.4 The SLCA matrix   194
14.5 Target plots   195
14.6 Assessing generic automobiles of yesterday
     and today   197
14.7 Weighting in SLCA   201
14.8 SLCA assets and liabilities   207
14.9 The LCA/SLCA family   208
Further Reading   209
Exercises   210

PART IV   ANALYSIS OF TECHNOLOGICAL SYSTEMS   **211**

Chapter 15   Systems Analysis   **211**
15.1 The systems concept   211
15.2 The adaptive cycle   213
15.3 Holarchies   215
15.4 The phenomenon of emergent behavior   218
15.5 Adaptive management of technological
     holarchies   219
Further Reading   221
Exercises   222

Chapter 16   Industrial Ecosystems   **223**
   16.1   Ecosystems and food chains   223
   16.2   Food webs   226
   16.3   Industrial symbiosis   232
   16.4   Designing and developing symbiotic industrial ecosystems   233
   16.5   Uncovering and stimulating industrial ecosystems   235
   16.6   Island biogeography and island industrogeography   236
   Further Reading   237
   Exercises   239

Chapter 17   Material Flow Analysis   **240**
   17.1   Budgets and cycles   240
   17.2   Resource analyses in industrial ecology   244
         17.2.1   Elemental substance analyses   245
         17.2.2   Molecular analyses   248
   17.3   The balance between natural and anthropogenic mobilization
          of resources   249
   17.4   The utility of material flow analysis   251
   Further Reading   252
   Exercises   252

Chapter 18   National Material Accounts   **254**
   18.1   National-level accounting   254
   18.2   Country-level metabolisms   255
   18.3   Embodiments in trade   260
   18.4   Resource productivity   261
   18.5   Input–output tables   262
   18.6   The utility of metabolic and resource analyses   267
   Further Reading   267
   Exercises   268

Chapter 19   Energy and Industrial Ecology   **269**
   19.1   Energy and organisms   269
   19.2   Energy and the product life cycle   273
   19.3   The energy cycle for a substance   275
   19.4   National and global energy analyses   277
   19.5   Energy and mineral resources   278
   19.6   Energy and industrial ecology   279
   Further Reading   280
   Exercises   280

Chapter 20   Water and Industrial Ecology   **282**
   20.1   Water: An overview   282
   20.2   Water and organisms   283
   20.3   Water and products   285
   20.4   The water footprint   288

20.5   Water quality   290
20.6   Industrial ecology and water futures   292
Further Reading   292
Exercises   293

Chapter 21   Urban Industrial Ecology   **294**
21.1   The city as organism   294
21.2   Urban metabolic flows   296
21.3   Urban metabolic stocks   297
21.4   Urban metabolic histories   299
21.5   Urban mining   301
21.6   Potential benefits of urban metabolic studies   302
Further Reading   303
Exercises   303

Chapter 22   Modeling in Industrial Ecology   **304**
22.1   What is an industrial ecology model?   304
22.2   Building the conceptual model   306
   22.2.1   The Class 1 industrial ecology model   306
   22.2.2   The Class 2 industrial ecology model   309
   22.2.3   The Class 3 industrial ecology model   309
22.3   Running and evaluating industrial
         ecology models   310
   22.3.1   Implementing the model   310
   22.3.2   Model validation   310
22.4   Examples of industrial ecology models   311
22.5   The status of industrial ecology models   314
Further Reading   316
Exercises   318

PART V   THINKING AHEAD   **319**

Chapter 23   Industrial Ecology Scenarios   **319**
23.1   What is an industrial ecology scenario?   319
23.2   Building the scenario   320
23.3   Examples of industrial ecology scenarios   321
23.4   The status of industrial ecology scenarios   325
Further Reading   326
Exercises   327

Chapter 24   The Status of Resources   **328**
24.1   Introduction   328
24.2   Mineral resources scarcity   329
24.3   Cumulative supply curves   333
24.4   Energy resources   335

24.5   Water resources   338
24.6   Summary   338
Further Reading   340
Exercises   340

**Chapter 25   Industrial Ecology and Sustainable Engineering
in Developing Economies**   **341**
25.1   The three groupings   341
25.2   RDC/SDC dynamics and perspectives   343
25.3   Industrial ecology and sustainable engineering
       practice in LDCs   349
25.4   Thoughts on development in LDCs   350
Further Reading   351
Exercises   352

Chapter 26   Industrial Ecology and Sustainability in the Corporation   **353**
26.1   The manufacturing sector, industrial ecology,
       and sustainability   353
26.2   The service sector, industrial ecology, and sustainability   354
26.3   Environment and sustainability as strategic   358
26.4   The corporate economic benefits of environment
       and sustainability   359
26.5   Implementing industrial ecology in the corporation   360
Further Reading   363
Exercises   363

Chapter 27   Sustainable Engineering in Government and Society   **365**
27.1   Ecological engineering   365
27.2   Earth systems engineering and management   366
27.3   Regional scale ESEM: The Florida Everglades   367
27.4   Global scale ESEM: Stratospheric ozone and CFCs   369
27.5   Global scale ESEM: Combating global warming   370
       27.5.1   *Capturing Carbon dioxide   370*
       27.5.2   *Sequestering Carbon in Vegetation   370*
       27.5.3   *Sequestering Carbon in Marine Organisms   372*
       27.5.4   *Scattering Solar Radiation with Sulfur Particles   372*
       27.5.5   *Reflecting Solar Radiation with Mirrors in Space   373*
       27.5.6   *Global Warming ESEM   374*
27.6   The principles of ESEM   374
       27.6.1   *Theoretical principles of ESEM   375*
       27.6.2   *Governance principles of ESEM   375*
       27.6.3   *Design and engineering principles of ESEM   376*
27.7   Facing the ESEM question   376
27.8   Proactive Industrial Ecology   377
Further Reading   379
Exercises   380

Chapter 28   Looking to the Future   **381**
      28.1   A status report   381
      28.2   No simple answers   382
      28.3   Foci for research   383
      28.4   Themes and transitions   383
      Further Reading   384
      Exercise   384

Appendix   Units of Measurement in Industrial Ecology   **385**

Glossary   **387**

Index   **397**

Preface

It has been a maxim of many years' standing that the goals of industry are incompatible with the preservation and enhancement of the environment. It is unclear whether that maxim was ever true in the past, but there is certainly no question that it is untrue today. The more forward-looking corporations and the more forward-looking nations recognize that providing a suitable quality of life for Earth's citizens will involve not less industrial activity but more, not less reliance on new technologies but more, and not less interaction of technology with society and the environment but more; they also recognize that providing a sustainable world will require close attention to integrated industry–society–environment interactions. This awareness of corporations, citizens, and governments promises to ensure that corporations that adopt responsible approaches to industrial activities will not only avoid problems but will benefit from their foresight.

There are three time frames of significance in examining the interactions of industry and environment. The first is that of the past and concerns itself almost entirely with strategies for dealing with inappropriate disposal of industrial wastes. The second time frame is that of the present, and it deals largely with complying with regulations, preventing the obvious mistakes of the past, and conducting responsible operations as those are understood in the context of existing regulatory, social, and environmental knowledge. Hence, it emphasizes waste minimization, avoidance of known toxic chemicals, and "end of pipe" control of emissions to air, water, and soil. Corporate environment and safety personnel are often involved, as are manufacturing personnel, in making small to modest changes to processes that have proved their worth over the years. On neither of these time frames do today's industrial process designers and engineers play a significant role.

The third time frame is that of the future. The industrial products, processes, and services that are being designed and developed today will dictate a large fraction of the industry–environment interactions over the next few decades. Thus, process and product design engineers hold much of the future of industry–environment interactions in their hands, and nearly all are favorably disposed toward doing their jobs with corporate social responsibility and the environment in mind. The same is true of corporate managers, business planners, and service industry employees. Their problem is that doing so requires knowledge and perspective never given to them during their college or professional educations and not readily available in their current positions. Remedying this situation is one of the primary aims of this book.

Industrial ecology and sustainable engineering are emerging fields, still in the process of defining territory and boundaries, but it is clear that they touch on a wide variety of specialties, including sociology, business, biological ecology, economic and technological history, environmental policy, and many more. Specialists and students in those fields of study may therefore find this book useful. Nonetheless, we view our primary audience as technological interdisciplinarians: those industrial designers, engineers, scientists, managers, and planners who in their work are addressing issues of modern technology in all its complexity. Such specialists are as yet few, so the training of students is a central challenge, and the reason for this book's existence. The educational level of the book is at the advanced undergraduate or beginning graduate student stage.

Some of the topics that we cover can be highly mathematical in their full implementation. Many of the most useful tools are mathematically simple, however: life cycle assessment and material flow analysis, to cite two of the most prominent. As a result, this book is not primarily a mathematical treatise on industrial ecology and sustainable engineering. Rather, it is at its core a book about the concepts—new ones and those drawn from other fields—that link modern technology and modern society to social and environmental sustainability.

The book is divided into five sections. The first provides a brief definition of the topic and outlines the approach of the book. The second, "Framework Topics," describes the playing field on which industrial ecology and sustainable engineering operate, and the opportunities and constraints that each of these areas provides. In the third section, "Implementation," topics central to the designers and analyzers of modern technological products are discussed. The central concept of life cycle assessment is introduced here in some detail. The fourth section, "Analysis of Technological Systems," addresses industrial ecology and sustainable engineering on national, regional, and global levels, discussing issues of industrial ecosystems, resource use, and the use of computer models of various types and for various purposes. The final section, "Thinking Ahead," deals with the long term: the sustainability of resources, the use of scenarios as planning tools, and the implementation of industrial ecology and sustainable engineering in corporations, developing countries, governments, and societies.

As industrial ecology has developed, ideas that were embryonic a decade ago have matured, and new ways of approaching technology–environment interactions have emerged. Accordingly, we spend significant parts of this book on topics largely or completely unaddressed in the two editions of the textbook *Industrial Ecology* (Prentice Hall, 1995, 2003). In particular, the following transitions have struck us as particularly notable: the progression of the field from being completely qualitative to increasingly quantitative, an enlarged vision that moves from a focus on products to consideration of systems, a transition from making use of a biological ecology metaphor to the use of the tools of modern ecology, and a progression from concern for environmental effects to the focus on challenges of long-term sustainability, explicitly including social dimensions. In the process, industrial ecology and sustainable engineering are becoming increasingly interdisciplinary (as evidenced by the fact that our far-from-exhaustive Further Reading lists cite some 80 different scholarly journals) and ever more broadly relevant.

The following instructor materials are available from the publisher:

Sample Syllabus for a Course in Industrial Ecology

Annotated Presentation Slides for Each Class

SLCA Guidelines

Sample Examination Questions

Contact your Prentice Hall Sales Representative for more information.

We are grateful to many people for their help during the preparation of this book, especially our many professional colleagues whose work we have attempted to summarize and convey. Probably the most important guidance has come from our students over the years; they have related strongly to some of the topics and presentations, and with less enthusiasm to others, thus inspiring revisions and rewritings. Several anonymous external reviews were also very helpful in determining the appropriate contents for the work. In making our ideas a reality, we appreciate our interactions with the staff at Prentice Hall, especially Holly Stark, who has helped us produce what we feel is a more attractive book. We would also like to thank the following reviewers: Ann Wittig, CUNY City College of New York; Brian Thorn, Rochester Institute of Technology; David Allen, University of Texas, Austin; Angela Casler, California State University, Chico; Zbigniew Bochniarz, University of Minnesota; and Andrea Larson, University of Virginia.

Finally, we thank AT&T, the AT&T and Lucent Foundations, the U.S. National Science Foundation, and the U.S. National Academy of Engineering for their support of industrial ecology initiatives over more than a decade; their help has been essential in the development of this new field.

THOMAS E. GRAEDEL

BRADEN R. ALLENBY
JANUARY 3, 2009

INDUSTRIAL ECOLOGY AND SUSTAINABLE ENGINEERING

PART I   Introducing the Field

C H A P T E R   1

Humanity and Technology

1.1   AN INTEGRATED SYSTEM

Whether we know it or not, whether we like it or not, technology strongly interacts with almost every facet of our lives. It has also come to interact with almost every facet of the natural world. It is this fundamental interdependence that creates the strong linkages between the studies of sustainable engineering, industrial ecology, and the more specific methodologies such as life cycle assessment that will be discussed throughout this book. Our discussion will frequently focus on the engineering or industrial aspects of particular projects or materials, but it should always be remembered that industrial ecology requires an awareness of the broader systems within which the projects or materials are embedded.

Furthermore, the integration of technology with social and environmental systems, a key aspect of sustainability, creates another important dynamic. Technology as a human competence is undergoing a rapid, unprecedented, and accelerating period of evolutionary growth, especially in the key foundational areas of nanotechnology, biotechnology, robotics, information and communications technology, and applied cognitive science. The implications for sustainability, and for industrial ecologists, are profound, and we will discuss them in more detail below. For now, it suffices to note that the effect is to undermine most of the assumptions underlying current engineering disciplines and policy frameworks. Indeed, industrial ecology itself is only beginning to fully engage with all the implications of technological evolution it purports to address.

1.2 THE TRAGEDY OF THE COMMONS

In 1968, Garrett Hardin of the University of California, Santa Barbara, published an article in *Science* magazine that has become more famous with each passing year. Hardin titled his article "The Tragedy of the Commons"; its principal argument was that a society that permitted perfect freedom of action in activities that adversely influenced common properties was eventually doomed to failure. Hardin cited as an example a community pasture area, used by any local herdsman who chooses to do so. Each herdsman, seeking to maximize his financial well-being, concludes independently that he should add animals to his herd. In doing so, he derives additional income from his larger herd but is only weakly influenced by the effects of overgrazing, at least in the short term. At some point, however, depending on the size and lushness of the common pasture and the increasing population of animals, the overgrazing destroys the pasture and disaster strikes all.

A more modern version of the tragedy of the commons has been discussed by Harvey Brooks of Harvard University. Brooks pointed out that the convenience, privacy, and safety of travel by private automobile encourages each individual to drive to work, school, or stores. At low levels of traffic density, this is a perfectly logical approach to the demands of modern life. At some critical density, however, the road network commons is incapable of dealing with the traffic, and the smallest disruption (a stalled vehicle, a delivery truck, a minor accident) dooms drivers to minutes or hours of idleness, the exact opposite of what they had in mind. Examples of frequent collapse of road network commons systems are now legendary in places such as Los Angeles, Tokyo, Naples, Bangkok, and Mexico City.

The common pasture and the common road network are examples of societal systems that are basically local in extent and can be addressed by local societal action if desired. In some cases, the same is true of portions of the *environmental* commons: improper trash disposal or soot emissions from a combustion process are basically local problems, for example. Perturbations to water and air do not follow this pattern, however. The hydrosphere and the atmosphere are examples not of a "local commons" but of a "global commons"—a system that can be altered by individuals the world over for their own gain, but, if abused, can injure all. Much of society's functions are embodied in industrial activity (where the word "industrial" should be interpreted broadly to mean any human action involving the transformation of materials or energy), and it is the relationships between industry, the environment, and society, especially the global commons, that are the topic of this book.

In the 40 years since Hardin's seminal paper, much additional effort has gone into generating a better conceptual picture of the use of what are now termed "common-pool resources." An example is provided by the fish stocks example of Figure 1.1. A feature of this example is that the linear relationship between fishing effort and cost contrasts sharply with the curvilinear relationship between fishing effort and revenue (i.e., number of fish caught). The yield (and the revenue) increases as fishing effort is increased until the maximum sustainable yield (MSY) is reached. After that point, increased effort draws down the stock of fish. However, the best return on effort (the maximum economic yield, MEY) occurs well before the maximum sustainable yield. As with the common pasture, if everyone fishes as an independent actor, neither the MEY nor the MSY is likely to be realized.

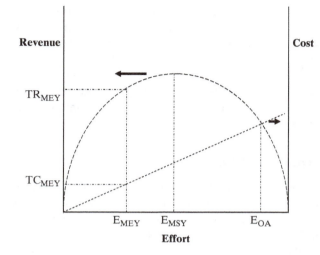

Figure 1.1

The relationships among fishing effort, cost, and revenue. The
symbols are: TR: total revenue; TC: total cost; E: level of fishing
effort; MEY: maximum economic yield; MSY: maximum
sustainable yield; OA: open access. Adapted from R. Townsend,
and J.E. Wilson, An economic view of the commons, in *The
Question of the Commons*, B.J. McCay and J.M. Acheson, Eds.,
Tucson: University of Arizona Press, pp. 311–326, 1987.

Thus, the overuse of common-pool resources (as in the roadway example of
Harvey Brooks) and the challenge of "free riders" who use a resource while doing
nothing to sustain it create a need for institutions to oversee these resources. Municipal
water authorities who meter water use are part of a common example of an overseeing
institution. Institutions can also be established for the whole planet: The Montreal
Protocol on Stratospheric Ozone Depletion demonstrates common-pool resource
institutional functioning on the largest possible scale. Nonetheless, studies of a large
variety of such institutions demonstrate that the success of common-pool institutions is
generally related strongly to small geographical sizes, well-defined boundaries, shared
cultural norms, and responsible monitoring and enforcement, that is, to systems where
participants tend to know each other and where there is respect and trust in the
institution overseeing the resource.

Environmentally related common-pool issues such as protection of coral reefs, land
needed for ecosystem services such as hurricane protection, or emissions restrictions to
protect the global atmosphere are often difficult institutional challenges. Climate change
and resource depletion may often be only dimly visible, if at all, to individual actors.
Scientific uncertainty complicates the process, as does the long timescale often existing
between human action and impact.

Common-pool resources are of concern in industrial ecology and sustainable
engineering because restricted availability of those resources could hamper progress
due to modern technology. Conversely, thoughtless employment of technology could

threaten the availability of energy, water, and the products of technology on which we depend — heating units, medical equipment, electronics, and so forth. The image of Hardin's cows on the village green can thus provide inspiration for a more intelligent technology, one based on minimal and careful resource use, high degrees of recyclability and reuse, and a perspective on life cycles — those of products, of infrastructure, and of colocated ecosystems. Because we have only one planet to work with, sustainability requires that technology work in harmony with the environment, not in opposition to it.

1.3   TECHNOLOGY AT WORK

It is undeniable that modern technology has provided enormous benefits to the world's peoples: a longer life span, increased mobility, decreased manual labor, and widespread literacy, to name a few. Nonetheless, there are growing concerns about the relationships between industrial activity and Earth's environment, nowhere better captured than in the pathbreaking report *Our Common Future*, produced by the World Commission on Environment and Development in 1987. The concerns raised in that report gather credence as we place some of the impacts in perspective. Since 1700, the volume of goods traded internationally has increased some 800 times. In the last 100 years, the world's industrial production has increased more than 100-fold. In the early 1900s, production of synthetic organic chemicals was minimal; today, it is over 225 billion pounds per year in the United States alone. Since 1900, the rate of global consumption of fossil fuel has increased by a factor of 50. What is important is not just the numbers themselves, but their magnitude and the relatively short historical time they represent.

Together with these obvious pressures on the Earth system, several underlying trends deserve attention. The first is the diminution of regional and global capacities to deal with anthropogenic emissions. For example, carbon dioxide production associated with human economic activity has grown dramatically, largely because of extremely rapid growth in energy consumption. This pattern is in keeping with the evolution of the human economy to a more complex state, the increasing growth in materials use and consumption, and an increased use of capital. The societal evolution has been accompanied by a shift in the form of energy consumed, which is increasingly electrical (secondary) as opposed to biomass or direct fossil fuel use (primary), the result being the now familiar exponential increase in atmospheric carbon dioxide that has occurred since the beginning of the industrial revolution. This trend is evidence that human activities are rapidly compromising the ability of the atmosphere to act as a sink for the by-products of our economic practices.

Human population growth is, of course, a major factor in this explosive industrial growth and expanded use and consumption of materials. Since 1700, human population has grown tenfold: It now exceeds six and one-half billion, and is anticipated to peak at nine billion or thereabouts in the twenty-first century. While this growth is generally recognized, it is less widely appreciated how human population growth patterns are tied to technological and cultural evolution. Indeed, the four great historical jumps in human population are synchronous with the initial development of tool use around 100,000 BCE, the agricultural revolution of about 10,000–3000 BCE, the industrial revolution of the eighteenth century, and the public health evolution that began in

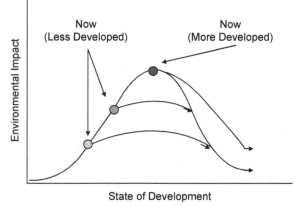

Figure 1.2

A schematic diagram of the typical life cycle of the relationship between the state of technological development of society and its resulting environmental impact.

about the mid-twentieth century. The industrial revolution actually consisted of both a technological revolution and a "neo-agricultural" revolution (the advent of modern agricultural practices), which created what appeared to be unlimited resources for population growth. Our current population levels, patterns of urbanization, economies, and cultures are now inextricably linked to how we use, process, dispose of, and recover or recycle natural and synthetic materials and energy, and the innumerable products made from them.

The above discussion suggests that the planet and its population are far from a steady state and may indeed be on an unsustainable path. Three illustrative routes toward long-term stability have been postulated: (1) managed growth until a long-term sustainable population/technology/cultural dynamic state (which we will call "carrying capacity") is achieved, (2) a managed reduction of population to a lower level sustainable with less technological activity, or (3) an unmanaged crash of one or more of the parameters (population, culture, technology) until stability at some undesirable low level is approached. Figure 1.2 suggests such possibilities. Note, however, that real trajectories will be much more complicated and dynamic; in particular, because population size, technology state, and culture are integrated and mutually dependent, as technology evolves, so might the potential carrying capacity of the planet.

This perspective has significant implications. When we objectively view the recent past—and 200 years is recent even in terms of human cultural evolution, and certainly in terms of our biological evolution—one fact becomes clear: The industrial revolution as we now know it is not sustainable over time. We cannot keep using materials and resources the way we do now, especially in the more developed countries. But what is the alternative?

1.4 THE MASTER EQUATION

A useful way to focus thinking on the most efficient response that society can make to environmental and related health and social stresses is to examine the predominant factors involved in generating those stresses. As is obvious, the stresses on many aspects of the Earth system are strongly influenced by the needs of the population that must be provided for and by the standard of living that population desires. One of the more famous expressions of these driving forces is provided by the "master equation":

$$\text{Environmental Impact} = \text{Population} \times \frac{\text{GDP}}{\text{Person}} \times \frac{\text{Environmental impact}}{\text{Unit of GDP}} \qquad (1.1)$$

where GDP is a country's gross domestic product, a measure of industrial and economic activity. This equation has traditionally been called the IPAT Equation, where A = affluence (GDP/person) and T = technology (impact/unit of GDP). Let us examine the three terms in this equation and their probable change with time.

Earth's population is, of course, increasing rapidly. For a specific geographical region (e.g., city, country, or continent), the rate of population change is given by

$$R = [R_b - R_d] + [R_i - R_e] \qquad (1.2)$$

where the subscripts refer to birth, death, immigration, and emigration. Different factors can dominate the equation during periods of high birthrates, war, enhanced migration, plague, and the like. For the world as a whole, of course, $R_i = R_e = 0$. Given the rate of change, the population at a future time can be predicted by

$$P = P_0 e^{Rt} \qquad (1.3)$$

where P_0 is the present population, t is the number of years in the projection, and R is expressed as a fraction. If R remains constant, the equation predicts an infinite population if one looks far enough into the future. Such a scenario is obviously impossible; at some point in the future, R will have to approach zero or go negative and the population growth will thus be adjusted accordingly.

In practice, demographers predict changes in R on the basis of the age structure of populations, cultural evolution, and other factors. Countries differ on these factors, of course, and the timing and magnitude of Earth's eventual human population peak remain quite uncertain. Even in the mildest reasonable scenario, however, a global population much larger than the present level is anticipated.

The second term in Equation 1.1, the per capita GDP, varies substantially among different countries and regions, responding to the forces of local and global economic conditions, the stage of historical and technological development, governmental factors, weather, and so forth. The general trend, however, is positive, as seen in Table 1.1. This table reflects the aspirations of humans for a better life. Although GDP and quality of life may not be fully connected, we can expect GDP growth to continue, particularly in developing countries.

The third term in the master equation, environmental impact per unit of GDP, is an expression of the degree to which technology is available to permit development without serious environmental consequences and the degree to which that available technology is deployed. The typical pattern followed by nations participating in the industrial revolution of the eighteenth and nineteenth centuries is shown in Figure 1.2. The abscissa can be divided into three segments: the unconstrained industrial revolution, during which

TABLE 1.1 Growth of Real Per Capita Income in Developed and Developing Countries, 1960–2000

Country group	1960–1970	1970–1980	1980–1990	1990–2000
Developed countries	4.1	2.4	2.4	2.1
Sub-Saharan Africa	0.6	0.9	−0.9	0.3
East Asia	3.6	4.6	6.3	5.7
Latin America	2.5	3.1	−0.5	2.2
Eastern Europe	5.2	5.4	0.9	1.6
Developing countries	3.9	3.7	2.2	3.6

Note: The figures are average annual percentage changes, and for the "Developing countries" entry are weighted by population. Figures for 1990–2000 are estimated. (Data from The World Bank, *World Development Report 1992*, Oxford University Press, Oxford, UK, 1992.)

the levels of resource use and waste increased very rapidly; the period of immediate remedial action, in which the most egregious examples of excess were addressed; and the period of the longer-term vision (not yet fully implemented) in which one hopes that environmental impacts will be reduced to small or even negligible proportions while a reasonably high quality of life is maintained.

Although the master equation should be viewed as conceptual rather than mathematically rigorous, it can be used to suggest goals for technology and society. If our aim is to constrain the environmental impact of humanity to its present level (and one could make arguments that we need to do even better than that), we need to look at the probable trends in the three terms of the equation. The first, as discussed above, will likely increase by a factor of about 1.5 over the next half-century. The second is predicted to likely increase over the same time period by a factor of between two and three. Accordingly, to merely hold our environmental impact where it is today, the third term must decrease by something between 65 and 80 percent. This is the inspiration for calls for "Factor Four" or "Factor 10" reductions in environmental impact per unit of economic activity.

Of the trends for the three terms of the master equation, the one which perhaps has the greatest degree of support for its continuation is the second, the gradual improvement of the human standard of living, defined in the broadest of terms. The first term, population growth, is not primarily a technological issue but a social issue. Although countries and cultures approach the issue differently, the upward trend is clearly strong. The third term, the amount of environmental impact per unit of output, is primarily a technological term, though societal and economic issues provide strong constraints to changing it rapidly and dramatically. It is this third term in the equation that appears to offer the greatest hope for a transition to sustainable development, especially in the short term, and it is modifying this term that is among the central tenets of industrial ecology and sustainable engineering.

1.5 TECHNOLOGICAL EVOLUTION

Technological evolution generally proceeds in one of two ways. Most of the time, technological evolution is incremental, marked by small improvements or changes in existing products or systems that, taken together, improve the quality of life but do not significantly change economic, cultural, or natural systems. In some periods, however, so-called "transformative technologies" change the technological landscape so profoundly that change in the related systems is significant and often difficult. Indeed, economists have identified stages in economic development that can be associated with particular enabling technologies (Figure 1.3). For example, the introduction of the railroad, the automobile, and electricity changed not just economic and related technological systems, but also culture, national competitiveness, political systems, and most people's way of life at the individual level. It is indeed accurate to say that the railroad was a necessary, and enabling, technology for the rise of Britain as a world economic and political power. It also necessitated other technologies, such as national communications systems using Morse Code, and required accurate

Technology System Waves

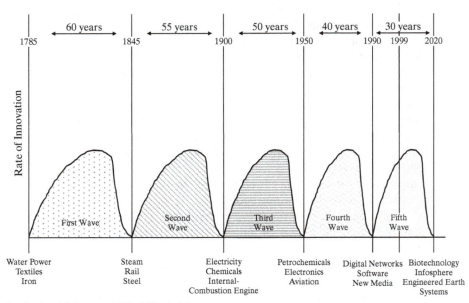

Based on Joseph Schumpeter and Nikolai Kondratieff

Figure 1.3

Economists have identified a relationship between enabling technologies and patterns of economic growth. This relationship hints at the changes in employment, family and personal life, and cultural systems which such technologies frequently entail (consider the effect of the automobile on society, for example). Given this historical view, it is possible to anticipate significant changes as several enabling technologies—nanotechnology, biotechnology, ICT, robotics, and cognitive science—achieve critical mass together, but not to predict with any certainty specific outcomes of that process. Note also that the rate of substitution of enabling technologies has itself accelerated, suggesting an auto-catalytic effect: the more technology humans have, the more rapidly they build on that to create more. Based on Joseph Schumpeter and Nikolai Kondratieff.

timekeeping, thus changing the way time was perceived and measured around the world. Among the other effects of extensive railroad infrastructure was helping to make the American Midwest a viable agricultural enterprise, feeding products to the American East Coast and from there to global markets (and leading to the development of Chicago in the process). That a single technology so structured vast areas of land and affected the economic well-being and personal lives of so many helps make clear the relationship between sustainability and technology.

Given this example, what are we to make of the confluence and accelerating evolution of an entire suite of foundational technologies: nanotechnology, biotechnology, information and communications technologies (ICT), robotics, and cognitive science (sometimes termed the "NBIRC" technologies)? Each by itself has profound implications; taken together, they pose a daunting challenge in at least several ways. First, as they evolve, they greatly increase the complexity of the systems with which industrial ecology deals. This is illustrated by, for example, the integration of enhanced ICT capabilities in urban systems, as discussed in Box 1.1.

Urban Systems: An Industrial Ecology Focus

Perhaps the most obvious demographic trend in modern societies around the world is urbanization. Demographers estimate that about 80 million people a year are moving into cities. Looking forward, the United Nations estimates that the urban populations of Africa, Latin America, and Asia will double over the next 30 years, growing from 1.9 billion in 2000 to 3.9 billion in 2030. By that time, over 60 percent of the world's population will live in cities. Moreover, cities are critical to the social side of sustainability, because they are usually the economic and cultural centers of their regions and have significant roles as consolidation points for energy, water, and material flows. Nonetheless, cities are not well understood at the systems level, an especially worrisome gap given the importance of urbanization as a fundamental demographic trend, the concerns regarding fragility of urban systems in case of natural disaster or deliberate attack, the complexity of the interactions among different infrastructures, and urban systems and their environmental contexts.

Given their central role in human society, urban systems will be significantly affected by a number of technological trends. For example, should people in developed countries routinely begin living past 100, the implications for design of urban buildings, infrastructure, and transportation networks will be significant. But a more challenging example arises when two separate trends, one in urban system design and the other in ICT, are considered together. The nature of urban systems is changing profoundly as ICT capability is increasingly integrated into all levels of urban functionality: smart materials, smart buildings, smart infrastructures, and the like. Sensor systems, sometimes coupled to computers and control devices, are becoming increasingly common. Cities are becoming much more information intensive at all levels. At the same time, ICT systems are changing fundamentally: being

(continued)

(*continued*)

redesigned to be what engineers call "autonomic"—that is, virtualized—and made self-defining, self-monitoring, self-healing, and learning-capable. Moreover, the concomitant introduction of grid computing networks and wireless communications at scales from local to regional to global, combined with the increasing role of urban systems as nodes in energy and financial networks, adds many layers of information complexity to built urban environments.

The implications of these trends for urban systems design, performance, and behavior are not well understood. Some idea of the possibilities and the need for concern can be obtained, however, from the experience of Black Monday, October 19, 1987. The Dow Jones Industrial Average fell 22.6 percent on that day, even though there were no significant changes in underlying financial conditions that warranted such a collapse. Rather, many analysts attributed the occurrence to systems dynamics. What had happened was that many stock trading firms had begun to computerize their trading activity, including building "floors" into their pricing models. Thus, when a certain fall in the market had occurred, the computers, which were not electronically or operationally coupled together, were instructed to sell to minimize losses. When the market declined beyond a certain point, increasingly sophisticated computerized program trading systems began selling into a declining market, creating an unforeseen negative feedback loop that emerged not from individual machines, but from the behavior of the market as a whole. The result was a market crash. Given that these trading systems were far, far simpler than the autonomic structures now being integrated into urban systems in all sorts of formal and informal ways, the potential systems dynamics of urban ICT systems cannot be assumed to be benign.

Second, such rapid (and accelerating) technological evolution undermines and makes contingent many societal assumptions which, because they have changed relatively slowly in the past, are generally assumed to be stable. An obvious example is the span of human life. Many governmental approaches assume that the existing life spans, which vary between 70 and 80 years for many developed countries, are essentially stable through time. But many who work in the medical field believe that within a few decades the expected life span of an individual born in a developed country will be well over 100 years. Under such a scenario, demand and consumption patterns would shift significantly. And this is only one, relatively foreseeable and trivial, implication of the integrated technological evolution that is inevitable at this point.

Technological evolution is thus a major part of the context within which sustainable engineering and industrial ecology studies are conducted. At smaller scales, it may be adequate to simply explicate and revalidate assumptions about technology that underlie the methodology and particular project. At larger scales, however, technological change in itself must be part of industrial ecology studies; indeed, the transdisciplinary nature of industrial ecology is a critical framework for such studies.

1.6   ADDRESSING THE CHALLENGE

The twentieth century was a period of enormous progress, achieved in part by ignoring the possible consequences of the ways in which that progress was being made to happen. The conjunction of inadequately thought-out technological approaches with rapidly rising populations and an increasing culture of consumption is now producing stresses obvious to all.

There are roles for many players in addressing the need to transform the technology–society–environment relationship. Social scientists need to understand consumption and how it may evolve and be modified. Environmental scientists and material specialists need to understand the limits imposed by a planet with limited resources and limited assimilative capacity for industrial emissions. Technologists need to develop design and manufacturing approaches that are more environmentally sound. Industrialists need to understand all these frameworks for action and develop ways to integrate the concepts within today's corporate structures. Policy makers need to provide the proper mix of regulations and incentives to promote the long-term health of the planet rather than short-term fixes.

These are great challenges. They are the ones to which this book is addressed. We cannot treat all of them in detail, nor are all of them sufficiently developed to permit doing so even if we wished. Nonetheless, we can see many approaches that will take us in the right direction. It is time to get started.

FURTHER READING

Chertow, M., The IPAT equation and its variants: Changing views of technology and environmental impacts, *Journal of Industrial Ecology, 4* (4), 13–29, 2001.

Cohen, J.E., *How Many People Can the Earth Support?* New York: W.W. Norton, 1995.

Cohen, J.E., Human population: The next half-century, *Science, 302,* 1172–1175, 2003.

Garreau, J., *Radical Evolution.* New York: Doubleday, 2004.

Hardin, G., The tragedy of the commons, *Science, 162,* 1243–1248, 1968.

Ostrom, E., T. Dietz, N. Dolšak, P.C. Stern, S. Stonich, and E.U. Weber, Eds., *The Drama of the Commons,* Washington, DC: National Academy Press, 2002.

Stern, D.I., Progress on the environmental Kuznets curve? *Environment and Development Economics, 3,* 173–196, 1998.

World Commission on Environment and Development, *Our Common Future,* Oxford, UK: Oxford University Press, 1987.

EXERCISES

1.1 In 1983, the birthrate in Ireland was 19.0 per 1000 population per year and the death rate, immigration rate, and emigration rate (same units) were 9.3, 2.7, and 11.5, respectively. Compute the overall rate of population change.

1.2 If the rate of population change for Ireland were to be stable from 1990 to 2005 at the rate computed in the above problem, compute the 2020 population. (The 1990 population was 3.72 million.)

1.3 Using the "master equation," the "Units of Measurement" section in Appendix, and the following data, compute the 2007 GDP/capita and equivalent CO_2 emissions per equivalent U.S. dollar of GDP for each country shown in the following table.

2007 Master Equation Data for Five Countries

Country	Population (millions)	GDP (billion U.S. dollars)	CO_2 emissions (Tg C/yr)
Brazil	188	621	106
China	1,314	2,512	1,665
India	1,095	796	338
Nigeria	132	83	114
United States	298	13,220	1,709

Source: Data for this table were drawn primarily from J.T. Houghton, B.A. Callander, and S.K. Varney, *Climate Change 1992*, Cambridge, UK: Cambridge University Press, 1992.

1.4 Trends in population, GNP, and technology are estimated periodically by many institutions. Using the typical trend predictions below, compute the equivalent CO_2 anticipated for the years 2010 and 2025 for the five countries in the following table. Graph the answers, together with information from 2007 (previous problem), on an ECO_2 vs. year plot. Comment on the results.

Master Equation–Predicted Data for Five Countries

Country	Population (millions)		GNP growth (%/yr)		Decrease in eCO_2/GNP(%/yr)
	2000	2025	1990–2000	2000–2025	
Brazil	175	240	3.6	2.8	0.5
China	1,290	1,600	5.5	4.0	1.0
India	990	1,425	4.7	3.7	0.2
Nigeria	148	250	3.2	2.4	0.1
United States	270	307	2.4	1.7	0.7

Source: Data for this table were drawn primarily from J.T. Houghton, B.A. Callander, and S.K. Varney, *Climate Change 1992*, Cambridge, UK: Cambridge University Press, 1992.

C H A P T E R   2

The Concept of Sustainability

2.1   IS HUMANITY'S PATH UNSUSTAINABLE?

Unpredicted societal collapse is an occasional feature of the history of humanity. The classic case is that of Easter Island in the southeastern Pacific Ocean. It is very remote, and was not settled until about 800 CE. When the Polynesians arrived, they began to cut trees to create farmland and to make canoes. Soon they began to erect the large statues for which the island is famous, and trees were used to transport the statues and erect them. Over time, the island's trees were all cut for these purposes.

The lack of trees meant that Easter Island had no firewood, mulch, or canoes. Without the ability to catch dolphins from canoes, and with the depletion of nesting birds, the population came under severe pressure, and the island was too remote for help to come. There were no alternatives to a severe and ultimately permanent population collapse.

Easter Island is a special case, certainly, but it is not hard to find other cases in which the misuse of technology has forever changed part of the planet—the acid mine drainage and heavy metal pollution around Butte, Montana, in the western United States is an example.

The discussion of collapse can be generalized by examining the alternative behavioral patterns for complex systems shown in Figure 2.1. The exponential path (Figure 2.1a) traces the path of social progress for some 200 years. This pattern occurs when there are no constraints to growth or when innovation causes apparent limits to recede. The s-shaped curve (Figure 2.1b) is characteristic of the system with fixed constraints in which action is controlled by feedback based on a sense of the

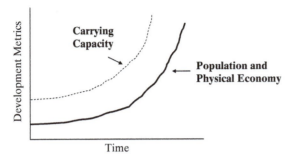

(a) Continuous growth if physical bounds are distant or growing

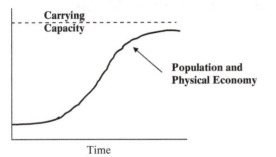

(b) Sigmoidal path occurs when approaches to bounds are seen

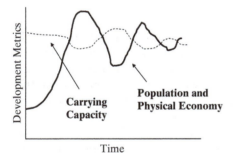

(c) Oscillations occur with delayed signals but robust bounds

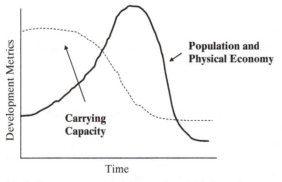

(d) Collapse occurs with delays and erodable bounds

Figure 2.1

Four typical behavior patterns for complex systems. (Adapted from D.H. Meadows, D.L. Meadows, and J. Randers, *Beyond the Limits*, White River Junction, VT: Chelsea Green, 1992.)

distance to the limits. To manage the smooth approach, the system must respond without significant lags and with accurate knowledge of the distance yet to go.

The curve showing oscillatory behavior (Figure 2.1c) is typical of systems where feedback mechanisms are inaccurate and responses are slow. At the point that awareness of some limit becomes sufficient to produce action, it is too late to avoid overstepping the limits and the system continues to move beyond what appears to be some long-run sustainable state. If the stress produced by the overshoot does not completely degrade the system, subsequent corrections can enable the system to oscillate about and approach the limit. Curve (d) depicts initial behavior somewhat like the third curve, but with a critical difference. Here the system is insufficiently robust, corrections are insufficient, and collapse occurs. This is the Easter Island trajectory.

It is important to note that the initial stages of these curves are quite similar. We imagine that we are close to the origin and further imagine intuitively that we are on the exponential growth pattern. If we are not, we must look sustainability in the face and think hard about the robustness and stability of our technological society.

It is useful at this point to define what we mean by sustainability. Many have tried to formulate succinct definitions, two of which we feel have particular merit. John Ehrenfeld's is conceptual: "*Sustainability* is the possibility that human and other forms of life will flourish on the planet forever." The International Institute of Environment and Development defines *sustainable development* (often used as a synonym for *sustainability*) as "A development path that can be maintained indefinitely because it is socially desirable, economically viable, and ecologically sustainable." The words have resonance, but offer minimal guidance to engineers, scientists, political leaders, and citizens. What specific actions will move us in the direction of sustainability?

2.2   COMPONENTS OF A SUSTAINABILITY TRANSITION

In the short term, virtually everything can be sustained. On the longest of timescales, nothing can. In between, there are choices to make. Therefore, to begin the process of operationalizing sustainability, we need to determine for ourselves exactly what it is that we wish to sustain, who are we sustaining it for, and for how long. A variety of positions have been taken on these topics; they are grouped in Figure 2.2. The most common position (not necessarily the "correct" one) on what is to be sustained is *life support systems*, especially for humans. The goal tends also to be human-centered, with economic growth and human development as the central themes. The Board on Sustainable Development also identified linkages that connect sustainability and development to varying degrees.

Choosing the timescale is important, because it enables potential actions to be quantified, as will be shown below. "Now and in the future" and "forever" are naïve and unworkable choices. In a practical sense, making policy for more than a human adult lifetime is not realistic, so most operational planning durations for sustainability fall into the 25–50 year range. This creates obvious discontinuities when working with systems such as the carbon and climate cycles, which involve timescales of many hundreds of years.

Given the general agreement that society's current path is not sustainable, getting closer to the sustainability goal will require a significant and multifaceted transition. This could well involve monitoring and influencing a number of long-term trends, some

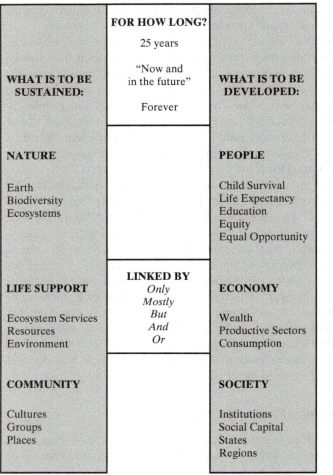

Figure 2.2

Sustainable development: common concerns, differing emphases. (Reproduced with permission from Board on Sustainable Development, *Our Common Journey*, Washington, DC: National Academy Press, 1999.)

of which may not be intuitively associated with sustainability. Industrial ecologists may have minimal influence on some of these, such as health or urbanization, but are clearly central players in changing the rates and modes of use of resources, of energy, and of water, and in minimizing the impacts of technology on the environment.

A feature of the sustainability discussion that is highly relevant to industrial ecology is the distinction between "weak" and "strong" sustainability. The former is the more optimistic position: It holds that sustainability is equivalent to nondecreasing total capital stock, that is, the sum of natural capital and human-made capital. Adherents of strong sustainability take a different position, arguing that natural capital provides certain important functions for which human-made capital cannot substitute. Robert Ayres (2007) lists free oxygen, freshwater, phosphorus, and scarce but very useful heavy elements such as thallium or rhenium in this group and argues that "those who espouse the notion of strong sustainability appear to be closer to the truth than the optimists who believe in more or less unlimited substitution possibilities."

2.3 QUANTIFYING SUSTAINABILITY

How is one to approach the challenge of providing sustainability guidance in such a way that it can be implemented? We explore in this section of the chapter a few examples, all potentially contentious, of how such guidance might be established and provided in such a way that the "journey toward sustainability" might begin.

Realistic and defensible goals for sustainability and their implementation will not be easy to establish in practice, but the principles by which one could proceed are reasonably straightforward. They are:

- Establish the limiting rate of use of the environmental, economic, or equity component.
- Allocate the allowable limit by some appropriate method to those who are influenced by that limit.
- Compare the current situation with the permitted allocation.
- Consider potential corrective actions.

In a number of cases, it will be necessary to select a time horizon over which sustainability is to be evaluated. In accordance with the Board on Sustainable Development (1999), we regard 50 years (i.e., roughly two human generations) as a reasonable period for assessment. By most accounts, the next 50 years will be crucial in determining the long-term sustainability of ecological and human systems. Population is likely to increase dramatically over the next 50 years from 6 billion to 9 billion. A discernable human-induced climate change on the order of 2–5°C could well occur, and commonly used industrial minerals and both oil and natural gas could become increasingly scarce over the next 50 years. We further assume that resource consumption should be planned so that existing resources will last for 50 years at current rates. This allows time for substitution of other resources or the development of alternative ways of meeting the needs that are served by resource consumption.

Once a resource of interest is chosen, then we see four basic steps for beginning a preliminary measure of sustainability.

1. **Establish the virgin material supply limit** by calculating the amount of a resource that can be used per year if that resource is to last for 50 years. To do so, you must first establish the known quantity of the resource available within the region of interest (the globe, a country, a state, etc.). For a nonrenewable resource, the amount often used is the "reserve base," defined as those resources that can be extracted at a profit plus some resources that are known but are not presently economically viable.

2. **Allocate the virgin material supply** according to a reasonable formula (such as dividing it equally among the global population, perhaps). Throughout the following examples, we assume that the *average* global population over the next 50 years will be 7.5 billion people.

3. **Establish the regional "re-captureable" resource base,** which is the known quantity in stockpiles, landfills, and so on, where it might reasonably be accessed. Assume that this resource can be replenished from existing stock in use at the current regional rate of recycling of the chosen material.

4. **Compare the current consumption rate to the sustainable limiting rate** for that resource within the region being assessed.

Once this basic measurement is done, one can begin to speak intelligently and realistically about any necessary policy actions to respond to excess consumption. As the following examples will show, this basic calculation clearly shows the unsustainable nature of current "Western" consumption patterns, especially when available technological resources are equitably distributed across the globe. Whether or not this is a reasonable allocation of resources is a question that will be briefly addressed herein. As is patently obvious, equity in allocation of the global common resources will be a topic for serious and intense political debate for many years to come. However, even the simple sustainability calculations for zinc, germanium, and greenhouse gasses below can serve as starting points for these policy debates.

2.3.1   Example 1: Sustainable Supplies of Zinc

To illustrate the approach, let us proceed to derive a sustainability limit for the use of zinc. Zinc, a fairly representative industrial mineral, is widely used by our modern technological society, yet it is in a relatively short supply from known reserves. About half of zinc production goes to make galvanized steel, in which a thin coating of zinc protects the underlying steel from rusting. Our industrial society also uses zinc in brass, bronze, and other alloys, die casting, and tire manufacture. Applying the four basic steps above, we determine a sustainable limiting rate of zinc consumption in Table 2.1.

Is 1.5 kg of zinc per person per year sufficient for an average person's technological and social requirements? Consider the following exploration of a common technological appliance, the automobile. Automobiles are one of the major technological goods that

---

TABLE 2.1   Calculation of a Global Sustainable Limiting Rate of Zinc Consumption

1. **Virgin material supply limit:** For zinc, the reserve base as of 1999 was 430 Tg, so the virgin material supply limit over the next 50 years is 430 Tg/50 years = 8.6 Tg/yr.

2. **Allocation of virgin material:** Allocating the available zinc equally among all the world's population gives approximately (8.6 Tg/yr)/(7.5 billion people) = 1.15 kg/(person-yr).

3. **Regional "re-captureable" resource base:** For illustrative purposes, assume a 30% zinc recycling rate (the current best-available zinc recycling rate). If 30 % of the 1.15 kg/(person-yr) is recycled, then each person in the region actually has 1.15 + (0.3)(1.15) kg/(person-yr) of zinc available, or 1.5 kg/(person-yr).

4. **Current consumption rate vs. sustainable limiting rate:** This step raises the fundamental question: How we are doing currently? For instance, the United States' zinc consumption in 1999 was 1.6 Tg for a population of 260 million people. This translates to a U.S. per capita zinc use of 6.2 kg per year. The Netherlands zinc use in 1990 was 28.5 Gg for a population of 15 million people. This translates to a Netherlands' per capita zinc use of 1.9 kg per year. The United States clearly exceeds its global sustainable allocation of zinc per person, while the Netherlands is much closer to its global sustainable allocation.

contain zinc, in the form of galvanized steel chassis and body parts. The zinc content of an average car is about 3–4 percent of the total weight of the automobile. Assuming that an average automobile weighs about 2000 pounds (or about 900 kg), then the average automobile contains about 900 kg × 0.035 = 32 kg of zinc. From a "zinc-perspective," the sustainable rate of automobile purchase is thus 32/1.5 = 21 years. Thus, your annual sustainable zinc allotment lets you buy a new car every 21 years. Furthermore, if during those 21 years you want anything else containing zinc, like a brass doorknob or some galvanized fencing, your new-car-purchase cycle will be lengthened.

An argument against this simple calculation would be that substitutes are widely available for galvanized steel (such as aluminum or composite materials for automobile manufacture) that would reduce the need for mined or recycled zinc. Substitution raises a few important questions, such as the following: Is a substitute technologically feasible? Is the substitute economically feasible? Is the substitute sustainable? Is the substitute of equal quality? Is the substitute socially (morally, ethically, etc.) acceptable? Will the substitute be accepted by consumers (due to aesthetics, "feel," or style)? These questions range from the scientific to the political and social. For the case of aluminum or composite materials substituting for zinc-galvanized steel in automobile manufacture, all of these questions could be satisfactorily answered (and probably will have to be in the near future). Answering these and other questions (such as, how do you increase zinc reuse and recycling?) will help policy makers devise the necessary corrective actions and incentives to achieve sustainable zinc use. As the following example for germanium will elucidate, what happens when the material in question has no readily apparent substitute?

2.3.2 Example 2: Sustainable Supplies of Germanium

One potential example of an "unsubstitutable" industrial mineral is germanium, 75 percent of which is used in optical fiber systems, infrared optics, solar electrical applications, and other specialty glass uses. Germanium plays a key role in giving these glasses their desired optical properties. Its most common use is as a dopant in the cylindrical core of glass fibers slightly increasing the refractive index of the core glass compared to the cladding. Lightwaves impinging on the core–cladding interface are trapped inside the core and can transmit the light signal for distances up to 30–60 km. Germanium is more reliable and outperforms all currently known substitutes in these applications. Furthermore, germanium use will likely increase in the future as fiber optic cables replace traditional copper wires and as solar-electric power becomes more widely available. We compute a U. S. sustainable limiting rate of germanium consumption in Table 2.2.

Is the sustainable allocation of 2 mg of germanium per person per year sufficient for an average person's technological and social requirements? Consider the following exploration of fiber optic cable replacing U.S. telephone wire needs. The germanium content of an average telecommunications fiber core is typically 5 mole percent GeO_2. This translates to roughly 14 mg of germanium per kilometer. Assuming that an average road has at least one telephone cable along its length, then replacing the copper telephone cable with fiber optic cable along the 3,830,000 km of paved roads in the United States would require approximately 2 Mg of germanium. Dividing by the 260 million people in the United States, each person would have to contribute

---

TABLE 2.2   Calculation of a U.S. Sustainable Limiting Rate of Germanium Use

---

1. **Virgin material supply limit:** Similar to the above analysis for zinc, the first step is to establish the amount available. For germanium, the reserve base in the United States as of 1999 was 500 Mg (world reserves are unknown at this time), so the virgin material supply limit over the next 50 years is 500 Mg/50 years = 10 Mg/yr.

2. **Allocation of virgin material:** Allocating the available germanium equally among all the world's population gives approximately 10 Mg/yr/7.5 billion people = 1.3 mg/(person-yr).

3. **Regional "re-captureable" resource base:** Worldwide, approximately 25% of the total germanium consumed is produced from recycled materials. If 25% of the 1.3 mg/(person-yr) is recycled, then each person actually has 1.3 + (0.25)(1.3) mg/(person-yr) of germanium available, or 1.6 mg/(person-yr).

4. **Current consumption rate vs. sustainable limiting rate:** The United States' germanium consumption in 1999 was 20 Mg for a population of 260 million people. This translates to a current U.S. per capita germanium use of 77 mg per person per year, compared to the sustainable limiting rate of 1.6 mg per person per year. The United States clearly exceeds its global sustainable allocation of germanium per person.

---

a one-time donation of about 8 mg of germanium to rewire the country. From a "germanium-perspective," replacing copper wire with fiber optic cable for telephone service in the streets of the United States would most likely be sustainable.

Obviously, most countries of the world do not currently have the demand for fiber optic cable suggested above. In a developing country, such an analysis could begin a discussion on whether wiring the streets with fiber optic cable from the beginning, instead of copper wire, would be a more sustainable choice. A third alternative, now apparently going forward but not from a sustainability perspective, is to communicate entirely by cellular telephone and thus avoid much of the necessary cabling completely. Such a choice should, of course, initiate a discussion around the sustainability of all the constituent materials in cellular telephones and base stations.

2.3.3   Example 3: Sustainable Production of Greenhouse Gases

As stated above, two of the major Earth system conditions we wish to maintain are a Holocene-style climate and functioning planetary engineering systems (forests, wetlands, etc.). The sustainability of each is closely linked to global climate change. Perhaps one sustainability threshold for climate change would be to limit human disruption of climate below that which significantly alters ocean circulation patterns, such as the North Atlantic thermohaline circulation. According to some climate change models, a doubling of atmospheric CO_2 (i.e., to approximately 550 ppmv) would most likely not permanently alter Atlantic Ocean circulation (although the circulation would weaken significantly and would take hundreds of years to recover in that case). It would be easy to debate that a doubling of CO_2 would still have some nonzero effects on maintaining climate conditions and ensuring the viability of ecosystem function. Yet, doubling of atmospheric CO_2 has emerged as a political target and a focal point for scientific analysis in most climate change models. Therefore, using the basic steps described above, we calculate in Table 2.3 a sustainable level of CO_2 addition to the atmosphere, if we make the controversial assumption that CO_2 doubling will be reasonably sustainable.

Is a sustainable allocation of 1 metric ton of carbon per person per year reasonable? Consider the following data on automobile usage and carbon production.

TABLE 2.3   Calculation of a Global Sustainable Limiting Rate of Carbon Dioxide Production

1. **Virgin material supply limit:** The IPCC indicates that in order to level off atmospheric CO_2 concentration (the major greenhouse gas of concern for our technological society) below a doubling from the preindustrial level (i.e., below approximately 550 ppmv by the year 2100), global anthropogenic emissions must be limited to ~7–8 Pg of carbon per year.

2. **Allocation of virgin material:** Again, following the simple examples above, each of the average 7.5 billion people on the planet over the next 50 years is allocated an equal share of CO_2 emissions. This translates to roughly 1 metric ton of carbon per person per year.

3. **Regional "recaptureable" resource base:** "Recycling" of carbon in the form of permanent or semipermanent sequestration may eventually be possible through controversial techniques such as deep well injection of carbon dioxide. However, this is still largely theoretical or in the very early stages of experimentation. Future sustainability measures could incorporate carbon recycling if it is eventually accepted as part of the carbon "management" alternatives.

4. **Current consumption rate vs. sustainable limiting rate:** The United States on average produces 6.6 metric tons of carbon equivalents per person, which is clearly well beyond the global sustainable rate of 1 metric ton of carbon per person per year. Inhabitants of Switzerland produce approximately 2.0 metric tons of carbon equivalents per person, which is still approximately twice our calculated sustainable limit.

Driving an automobile produces approximately 100 g of carbon per vehicle mile traveled. Drivers in the United States average 12,500 miles per person per year, which translates to 1.25 metric tons of carbon produced per year by driving. A driver would have to reduce his or her yearly driving miles by 2,500 miles in order to achieve the 1 metric ton of carbon per person sustainability goal. Regardless, a person could use all of his or her sustainable carbon credit on driving, but this would leave nothing for home heating, electricity for a computer, or a personal share in the larger industrial-technological systems that support the economy. Alternative energy sources, carbon sequestration possibilities, less-carbon-intensive production systems, personal driving habits, vehicle technology, public transportation systems, or some combination thereof must all be incorporated into the public discourse. However, as is the theme of this exercise, this public discourse would be well served by having a sustainable target toward which to aim.

These examples and the assumptions made herein raise contentious issues, two of which we address below.

2.3.4   Issues in Quantifying Sustainability

The Simplicity vs. Complexity Issue.   Our analysis here is necessarily simplified, and the simple metrics do not yet handle the inherent complexity of our global environmental system. It is easy to punch holes in the measurements and data, and cumulative and unintended effects are particularly troublesome. We say little about the methods of production for resources, which can involve enormous energy use, serious habitat disruption, environmental degradation, and so forth. For instance, even something seemingly positive such as increased zinc recycling to address sustainability issues may have negative effects on energy consumption and greenhouse gas production through transportation of recyclable material. Whether the sustainability-enhancing aspects of the one outweighs the unsustainability of the other two is cause for serious debate and more in-depth analysis that is beyond the scope of the simple metrics outlined herein.

There is a point where complexity for complexity's sake offers only marginal benefits. Each of our calculations raises the possibility that our current technological systems operate at at least twice the sustainable rate. On the order of magnitude scale, even the simple sustainability measurements herein offer a perspective on the challenges ahead.

The Property Rights Issue.   In calculating preliminary values for sustainable rates of use of various resources on an individual basis, we have allocated resources in the simplest possible way—an equivalent amount to each human being. This choice is comfortable from a global equity standpoint (although in reality far from current social norms), but immediately raises potential legal issues surrounding property rights. Resources are not equally distributed on a geographical basis and they are owned by a variety of entities, including nations, corporations, and individuals. To allocate resources on a global basis is to dictate at least to whom those resources must be sold, and doubtless to have at least some influence on price. Some alternative approaches, all problematic to varying degrees, are as follows:

- The global total extraction rate could be dictated, but allocation left to market forces.
- Regional total extraction rates could be dictated, and residents of resource-rich regions allocated more of the local resource than nonresidents.
- Regional allocations could be based on both local virgin and secondary resources.

Regardless of what choice is adopted, sustainability from a resource standpoint may require the establishment of an upper extraction limit followed by some method for allocating each year's virgin material supply. This would redefine current notions of private property, by imposing limits on extraction, and would transfer these property rights to disenfranchised masses. These are questions of policy, politics, diplomacy, and law. Nevertheless, the few simple calculations presented above can greatly inform the debate that rages among the developed and developing world, the rich and the poor, the large nations and small islands, the North and the South, and Republicans and Democrats.

2.4   LINKING INDUSTRIAL ECOLOGY ACTIVITIES TO SUSTAINABILITY

2.4.1   The Grand Environmental Objectives

Many of the sustainability dialogs involve environmental perturbations, in part because much of sustainability arose from previous environmental movements. Accordingly, it is useful to consider how such issues might be prioritized, without forgetting that sustainability itself requires consideration of a number of dimensions in addition to the environmental. Here, there has been significant progress: Although a number of environmental issues are worth attention, there is indisputable evidence that some environmental concerns are regarded generally or even universally as more important than others. For example, a major global decrease in biodiversity is clearly of more concern than the emission of hydrocarbon molecules from residential heating, and the Montreal Protocol and the Rio Treaty demonstrate that at least most of the countries of the world feel that understanding and minimizing the prospects for ozone depletion and global climate change are issues of universal importance. If one accepts that there are indeed such issues that have general acceptance by human society, one may then postulate the existence of a small number of "Grand Objectives" having to do with life on Earth, its maintenance, and its enjoyment. Determining these objectives requires societal consensus, which may or may not be achievable. For purposes of discussing the concept, a reasonable exposition of the Grand Objectives is the following:

The Ω_1 Objective: Maintaining the existence of the human species

The Ω_2 Objective: Maintaining the capacity for sustainable development and the stability of human systems

The Ω_3 Objective: Maintaining the diversity of life

The Ω_4 Objective: Maintaining the aesthetic richness of the planet

If it is granted that these objectives are universal, there are certain basic societal requirements that must be satisfied if the objectives are to be met. In the case of Ω_1, these are the minimization of environmental toxicity and the provision of basic needs: food, water, shelter, as well as the development of social and environmental resiliency adequate to maintain the species in light of low probability or unanticipated challenges (e.g., nuclear winter). For Ω_2, the requirements are a dependable energy supply, the availability of suitable material resources, the existence of workable political structures, and minimizing cultural conflict. For Ω_3, it is necessary to maintain a suitable amount of natural areas and to maximize biological diversity on disturbed areas, through, for example, the avoidance of monocultural vegetation. Perturbations due to rapid shifts in fundamental natural systems such as climate or oceanic circulation must also be addressed under this objective. Ω_4 requires control of wastes of various kinds: minimizing emissions that result in smog, discouraging dumping and other activities leading to degradation of the visible world, encouraging farming and agricultural practices that avoid land overuse and erosion, and the preservation of commonly held undeveloped land.

The Ω framework is an important prerequisite to determining what societal activities would be desirable, but the framework does not ensure progress toward achieving the objectives, especially when social consensus is involved. That progress results when desirable actions encouraged by the framework occur over and over again. In an industrialized society, a number of those actions are decisions made by product designers and manufacturing engineers. Thus, technological recommendations informed by the Grand Objectives are one means by which favorable decisions can be made.

2.4.2 Linking the Grand Objectives to Environmental Science

The Grand Objectives are, of course, too general to provide direct guidance to the product designer, who deals with specific actions relating to environmental concerns. The objectives and concerns can readily be related (see Table 2.4), but industrial decisions often require, in addition, a ranking of the relative importance of those concerns. This requirement is, in fact, a throwback to the philosophy that societal actions should be taken so as to produce the maximization of the good, and it produces in turn the question, "How does society determine the best actions?"

TABLE 2.4 Relating Environmental Concerns to the Grand Objectives

Grand objective	Environmental concern
Ω_1: Human species existence	
	1. Global climate change
	2. Human organism damage
	3. Water availability and quality
	4. Resource depletion: fossil fuels
	5. Radionuclides
Ω_2: Sustainable development	
	3. Water availability and quality
	4. Resource depletion: fossil fuels
	6. Resource depletion: non–fossil fuels
	7. Landfill exhaustion
Ω_3: Biodiversity	
	3. Water availability and quality
	8. Loss of biodiversity
	9. Stratospheric ozone depletion
	10. Acid deposition
	11. Thermal pollution
	12. Land use patterns
Ω_4: Aesthetic richness	
	13. Smog
	14. Aesthetic degradation
	15. Oil spills
	16. Odor

Note: The numbers in the right column are for later reference purposes.

The particular difficulty of identifying the best actions of society in this instance is that societal activities related to the environment inevitably involve trade-offs: wetland preservation versus job creation, the lack of greenhouse gas emissions of nuclear power reactors versus the chance of nuclear accident, or the preservation and reuse of clothing versus the energy costs required for cleaning, to name but a few. To enable choices to be made, many have proposed that environmental resources (raw materials, plant species, the oceans, etc.) be assigned economic value so that decisions could be market driven. The concept, though potentially quite useful, has proven difficult to put into practice, and this has been further confounded by the fact that the scientific understanding of many of the issues to be valued is itself evolving and thus would require that valuation be continuously performed.

Given this uncertain and shifting foundation for relative ranking, how might specific environmental concerns, several of which are responsive to one or more of the Grand Objectives, be grouped and prioritized in an organized manner? We begin by realizing that sustainability ultimately requires

- Not using renewable resources faster than they are replenished
- Not using nonrenewable, nonabundant resources faster than renewable substitutes can be found for them

- Not significantly depleting the diversity of life on the planet
- Not releasing pollutants faster than the planet can assimilate them

The relative significance of specific impacts can then be established by consideration of those goals in accordance with the following guidelines for prioritization:

- The spatial scale of the impact (large scales being worse than small)
- The severity and/or persistence of the hazard (highly toxic and/or persistent substances being of more concern than less highly toxic and/or persistent substances)
- The degree of exposure (well-sequestered substances being of less concern than readily mobilized substances)
- The degree of irreversability (easily reversed perturbations being of less concern than permanent impacts)
- The penalty for being wrong (longer remediation times being of more concern than shorter times)

These general criteria are perhaps too anthropocentric as stated, and are, of course, subject to change as scientific knowledge evolves, but are nonetheless a reasonable starting point for distinguishing highly important concerns from those less important. Using the criteria and the Grand Objectives, local, regional, and global environmental concerns can be grouped as shown in Table 2.5. The exact wording and relative positioning of these concerns are not critical for the present purpose; what is important is that most actions of industrial society that have potentially significant environmental implications relate in some way to the list.

Of the seven "crucial environmental concerns," three are global in scope and have very long timescales for amelioration: global climate change, loss of biodiversity, and ozone depletion. The fourth critical concern relates to damage to the human organism by toxic, carcinogenic, or mutagenic agents. The fifth critical concern is the availability and quality of water, a concern that embraces the magnitude of water use as well as discharges of harmful residues to surface or ocean waters. The sixth is the rate of loss of fossil fuel resources, vital to many human activities over the next century, at least. The seventh addresses humanity's use of land, a factor of broad influence on many of the other concerns.

Four additional concerns are regarded as highly important, but not as crucial as the first six. The first two of these, acid deposition and smog, are regional-scale impacts occurring in many parts of the world and closely related to fossil fuel combustion and other industrial activities. Aesthetic degradation, the third highly important concern, incorporates "quality of life" issues such as visibility, the action of airborne gases on statuary and buildings, and the dispersal of solid and liquid residues. The final concern, depletion of non–fossil fuel resources, is one of the motivations for current efforts to recycle materials and minimize their use.

Finally, five concerns are rated as less important than those in the first two groupings, but still worthy of being called out for attention: oil spills, radionuclides, odor, thermal pollution, and depletion of landfill space. The justification for this grouping is that the

---

TABLE 2.5 Significant Environmental Concerns

Crucial environmental concerns

  1. Global climate change
  2. Human organism damage
  3. Water availability and quality
  4. Depletion of fossil fuel resources
  8. Loss of biodiversity
  9. Stratospheric ozone depletion
 12. Land use patterns

Highly important environmental concerns

  6. Depletion of non–fossil fuel resources
 10. Acid deposition
 12. Smog
 13. Aesthetic degradation

Less important environmental concerns

  5. Radionuclides
  7. Landfill exhaustion
 11. Thermal pollution
 15. Oil spills
 16. Odor

Note: The numbers are those of Table 2.4. Within the groupings, the numbers are for reference purposes and do not indicate order of importance.

---

effects, while sometimes quite serious, tend to be local or of short time duration or both, when compared with the concerns in the first two groups.

2.4.3 Targeted Activities of Technological Societies

The mitigation of the environmental impacts of human activities follows, at least in principle, a logical sequence. First is the recognition of an environmental concern related to one or more of the Grand Objectives. Global climate change, for example, is related to two: Ω_1 and Ω_3. Next, once that concern is identified, industrial ecologists study the activities of humanity that are related to it. For global climate change, the activities include (though are not be limited to) those that result in emissions of greenhouse gases, especially CO_2, CH_4, N_2O, and CFCs; Table 2.6 lists a number of examples. The concept is that societal activities from agriculture to manufacturing to transportation to services can be evaluated with respect to their impacts on the Grand Objectives and that the link between activities and objectives is the purpose for the environmental evaluation of products, processes, and facilities.

A characteristic of many of the activities of a technological society is that they produce stresses on more than one of the environmental concerns. Similarly, most environmental concerns are related to a spectrum of societal activities. This analytical complexity does not, however, invalidate the framework being developed here.

TABLE 2.6   Targeted Activities in Connection with Crucial Environmental Concerns

Environmental concern	Targeted activity for examination
1. Global climate change	1.1 Fossil fuel combustion 1.2 Cement manufacture 1.3 Rice cultivation 1.4 Coal mining 1.5 Ruminant populations 1.6 Waste treatment 1.7 Biomass burning 1.8 Emission of CFCs, HFCs, N_2O
2. Loss of biodiversity	2.1 Loss of habitat 2.2 Fragmentation of habitat 2.3 Herbicide, pesticide use 2.4 Discharge of toxins to surface waters 2.5 Reduction of dissolved oxygen in surface waters 2.6 Oil spills 2.7 Depletion of water resources 2.8 Industrial development in fragile ecosystems
3. Stratospheric ozone depletion	3.1 Emission of CFCs 3.2 Emission of HCFCs 3.3 Emission of halons 3.4 Emission of nitrous oxide
4. Human organism damage	4.1 Emission of toxins to air 4.2 Emission of toxins to water 4.3 Emission of carcinogens to air 4.4 Emission of carcinogens to water 4.5 Emission of mutagens to air 4.6 Emission of mutagens to water 4.7 Emission of radioactive materials to air 4.8 Emission of radioactive materials to water 4.9 Disposition of toxins in landfills 4.10 Disposition of carcinogens in landfills 4.11 Disposition of mutagens in landfills 4.12 Disposition of radioactive materials in landfills 4.13 Depletion of water resources
5. Water availability and quality	5.1 Use of herbicides and pesticides 5.2 Use of agricultural fertilizers 5.3 Discharge of toxins to surface waters 5.4 Discharge of carcinogens to surface waters 5.5 Discharge of mutagens to surface waters 5.6 Discharge of radioactive materials to surface waters 5.7 Discharge of toxins to groundwaters 5.8 Discharge of carcinogens to groundwaters 5.9 Discharge of mutagens to groundwaters 5.10 Discharge of radioactive materials to groundwaters 5.11 Depletion of water resources
6. Resource depletion: fossil fuels	6.1 Use of fossil fuels for energy 6.2 Use of fossil fuels as feedstocks
7. Land use patterns	7.1 Urban sprawl 7.2 Agricultural disruption of sensitive ecosystems

2.4.4 Actions for an Industrialized Society

The final step in the structured assessment process we are describing is that, given activities for examination, analysts can generate specific design recommendations to improve the environmental and social responsibility of their products.

Thus, the overall process described here occurs in four stages: (1) the definition by society of its Grand Objectives for life on Earth, (2) the identification by environmental scientists of environmental, societal, and sustainability concerns related to one or more of those objectives, (3) the identification by technologists and social scientists of activities of society related to those concerns, and (4) the appropriate modification of those activities. Note that implementing the fourth step in this sequence depends on accepting the definition of step one, believing the validity of step two, and acknowledging the correct attribution in step three, but not necessarily in knowing the magnitude of the impact of step four on improving the environment, that is, the information that is needed tends to be qualitative, not quantitative. From the standpoint of the industrial manager or the product design engineer, what is important is knowing that if step four is taken the corporation's environment and sustainability performance will be improved to at least some degree, and, perhaps at least as important, knowing that customers and policy makers will regard the action as a positive and thoughtful one.

In overview, the four steps of the process are schematically displayed in Figure 2.3, in which it is clear that their influence on the Grand Objectives determines the importance of each of the environmental concerns and that each of the environmental concerns leads to a group of activities for examination, each of which in turn leads to a set of product design recommendations. Note that each of the Grand Objectives and most environmental concerns relate to a number of recommendations, rather than to only one or two, and, conversely, many recommendations respond to more than one environmental concern and

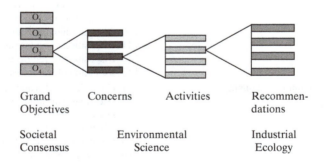

Grand Objectives   Concerns   Activities   Recommendations

Societal Consensus   Environmental Science   Industrial Ecology

Figure 2.3

A schematic representation of the conceptual sequence in life cycle assessment. Each of the four grand challenges is related to a number of concerns, such as climate change (each concern suggested be a horizontal bar); similarly, each of the concerns is related to a number of activities, such as fossil fuel combustion (again, horizontal bars indicate different activities); and each activity is related to a number of recommendations, such as higher-efficiency combustion. As noted at the bottom, different specialist fields treat different stages in the sequence.

perhaps more than one Grand Objective. As shown at the bottom of the diagram, the relationships among the Grand Objectives and recommendations provide logical interconnections among societal consensus, environmental science, and industrial ecology.

FURTHER READING

Ayres, R.U., On the practical limits to substitution, *Ecological Economics, 61*, 115–128, 2007.

Board on Sustainable Development, *Our Common Journey: A Transition Toward Sustainability,* Washington, DC: National Academy Press, 1999.

Diamond, J., *Collapse: How Societies Choose to Fail or Succeed*, New York: Viking Press, 2005.

Ehrenfeld, J.R., Would industrial ecology exist without sustainability in the background? *Journal of Industrial Ecology, 11* (1), 73–84, 2007.

Kates, R.W., and T.M. Parris, Long-term trends and a sustainability transition, *Proceedings of the National Academy of Sciences of the United States of America, 100*, 8062–8067, 2003.

Klee, R.J., and T.E. Graedel, Getting serious about sustainability, *Environmental Science & Technology, 36*, 523–529, 2002.

Parris, T.M., and R.W. Kates, Characterizing and measuring sustainable development, *Annual Reviews of Environment and Resources, 28*, 559–586, 2003.

Solow, R.M., Perspectives on growth theory, *Journal of Economic Perspectives, 8* (1), 45–54, 1994.

EXERCISES

2.1 Section 2.2 proposes that 25–50 years is the best choice for a sustainability planning timescale. Do you agree? Explain.

2.2 Use the information in Table 2.1 to estimate how much zinc could be used in a "sustainable" automobile. Clearly show your reasoning.

2.3 Consider the quantitative sustainability example of greenhouse gases in Section 2.3.3. Discuss the options for allocating the allowable CO_2 emissions, together with the benefits and problems of the options.

2.4 Section 2.4.1 defines four "example" Grand Objectives for sustainability. Do you agree that this is the ideal set? If so, why? If not, present and discuss alternatives.

CHAPTER 3

Industrial Ecology and Sustainable Engineering Concepts

3.1 FROM CONTEMPORANEOUS THINKING TO FORWARD THINKING

Since the industrial revolution, the activities of firms both large and small have defined much of the interactions between humanity and the environment and significantly shaped social and institutional structure and dynamics. These interactions have, however, traditionally been outside the topics of major significance for corporate decision makers. Technology's influence, and especially the potential magnitude of that influence across the full spectrum of economic activity, has been underappreciated in the business world.

However, no firm exists in a vacuum. Every industrial activity is linked to thousands of other transactions and activities and to their environmental and social impacts. A large firm manufacturing high-technology products will have tens of thousands of suppliers located all around the world, changing on a daily basis. The firm may manufacture and offer for sale hundreds of thousands of individual products to a myriad of customers, each with her or his own needs and cultural preferences. Each customer, in turn, may treat the product very differently and live in areas with very different environmental characteristics; these are considerations of importance when use and maintenance of the product may be a source of potential environmental impact (e.g., used oil from automobiles). The services and cultural patterns that the technology enables will also differ significantly in different communities and societies. When finally disposed of, the product may end up in almost any country, in a high-technology landfill, an incinerator, beside a road, or in a river that supplies drinking water to local populations.

TABLE 3.1   Relating Current Environmental Problems to Industrial Responses to Yesterday's Needs

Yesterday's need	Yesterday's solution	Today's problem
Nontoxic, nonflammable refrigerants	Chlorofluorocarbons	Ozone hole
Automobile engine knock	Tetraethyl lead	Lead in air and soil
Locusts, malaria	DDT	Adverse effects on birds and mammals
Fertilizer to aid food production	Nitrogen and phosphorus fertilizer	Lake and estuary eutrophication

In such complex circumstances, with complicated and intertwined social and environmental impacts at many scales, how has industry approached its relationships with the outside world? Satisfying the needs of its customers has always been pretty well done, at least in market economies. Industry has, however, been less adept at identifying some of the consequences, especially the long-term consequences, of the ways in which it goes about satisfying needs. Since the beginning of the 1970s, when modern environmentalism began to arise, analysis of environmental interactions has increased; examples of a few of these interactions have been collected by Dr. James Wei of Princeton University (adapted and displayed as Table 3.1). The table indicates some of the environmental difficulties created for society in a world in which industrial operations have been perceived as essentially unrelated to the wider world, as suggested in the left side of Figure 3.1.

It is important to note that the relationships in Table 3.1 were not the result of disdain for the external world by industry. Several of the solutions were, in fact, great improvements over the practices they replaced, and their eventual consequences could not have been forecast with any precision. What was missing, however, was any structured attempt to relate the techniques for satisfying customer needs to possible environmental consequences; similar efforts to understand systemic long-term social implications of industrial activity lag behind even the environmental methodologies. For example, we can speak of Design for Environment (or DfE) with some clarity, but development of Design for Sustainability (DfS) is just beginning. While making such attempts does not insure that no deleterious impacts will result from industrial activity, proactive consideration has the potential to avoid the most egregious of the impacts and to contribute toward incremental improvements in impacts that are now occurring or can be well forecast. How are such attempts best made?

The old idea: There are "natural" areas and there are "industrial" areas

Industrial System / Natural System

The new idea: Industrial system sits within the natural system

Natural System / Industrial System

Figure 3.1

The transformation from natural and industrial systems as essentially independent entities to the realization that the industrial system is embedded within the natural system.

The broad approach to industry–environment–sustainability interactions that is described in this book is termed "industrial ecology (IE)." The overall concept is, in part, technological. As applied in manufacturing, it involves the design of industrial processes, products, and services from the perspectives of product competitiveness, environmental concerns, and society. Industrial ecology is also, in part, sociological. In that regard, it recognizes that human culture, individual choice, and societal institutions play major roles in defining the interactions between our technological society and the environment. It recognizes as well that modern technological and societal systems are fully connected with and embedded within the natural world, as indicated at the right side of Figure 3.1.

In a later chapter, we will present an extensive definition of industrial ecology that uses a biological analogy to describe the perspective from which industrial ecology views an industrial system. For the present, a working definition of the field is as follows:

> Industrial ecology is the means by which humanity can deliberately approach and maintain sustainability, given continued economic, cultural, and technological evolution. The concept requires that an industrial system be viewed not in isolation from its surrounding systems, but in concert with them. It is a systems view in which one seeks to optimize the total materials cycle from virgin material, to finished material, to component, to product, to obsolete product, and to ultimate disposal.

In this definition, the emphasis on *deliberate* differentiates the industrial ecology path from unplanned, precipitous, and perhaps quite costly and disruptive alternatives. By the same token, the definition indicates that IE practices have the potential to support a sustainable world with a high quality of life for all.

Practitioners of IE and its companion, sustainable engineering (SE), interpret the word "industry" very broadly: It is intended to represent the sum total of human activity, encompassing mining, manufacturing, agriculture, construction, energy generation and utilization, transportation, product use by customers and service providers, infrastructure systems, service networks, and waste disposal. IE is not limited to the domain within the factory walls, but extends to all the impacts on the planet resulting from the presence and actions of human beings. IE thus encompasses society's use of resources of all kinds.

In considering manufactured products, for example, IE may focus on the study of individual products and their environmental impacts at different stages in their life cycles, but a complementary focus is the study of a facility where products are made. In such a facility, raw materials, processed materials, and perhaps finished components produced by others are the input streams, along with energy. The emergent streams are the product itself; residues to land, water, and air; and transformed energy residues in the form of heat and noise. The IE approach to such a facility treats the budgets and cycles of the input and output streams, and it seeks to devise ways in which smaller portions of the residues are lost and more are retained and recycled into the facility itself or into the facilities of others. Key concepts include conservation of mass (all material must be accounted for), conservation of energy (all energy must be accounted for), and the technological arrow of time—the realization that as society becomes more technologically advanced, it builds on its past technological base and so cannot sustain or improve itself without strong reliance on technology.

One of the most important concepts of industrial ecology is that, like the biological system, it rejects the concept of waste. Dictionaries define waste as useless or worthless material. In nature, however, nothing is eternally discarded; in various ways all materials are reused, generally with great efficiency. Natural systems have evolved these patterns because acquiring these materials from their reservoirs is costly in terms of energy and resources, and thus something to be avoided whenever possible. In our industrial world, discarding materials wrested from the Earth system at great cost is also generally unwise. Hence, materials and products that are obsolete should be termed "residues" rather than "wastes," and it should be recognized that wastes are merely residues that our economy has not yet learned to use efficiently. We will sometimes use the term "wastes" in this book where the context refers to material that is or has been discarded, but we encourage the use instead of the term "residues," or perhaps the even less pejorative "experienced resources," thereby calling attention to the engineering characteristics and societal value contained in obsolete products of all sizes and types. In doing so, we acknowledge that the law of entropy prohibits complete reuse without loss, but vision is more useful than scientific rigor in establishing this important perspective.

A full consideration of industrial ecology would include the entire scope of economic activity, such as mining, agriculture, forestry, manufacturing, service sectors, and consumer behavior. It is, however, obviously impossible to cover the full scope of industrial ecology in one volume, particularly given the reality that most services, no matter how abstruse, must rely on physical platforms and consume energy. Accordingly, we limit the discussion in most of this book to manufacturing activities, although some of the final chapters explore the subject in more general terms.

3.2 THE GREENING OF ENGINEERING

Engineering has traditionally been regarded as the specialty that employs scientific principles to achieve practical ends. Because engineers are problem solvers in the context of their cultures, at the time when resources and disposal sites were regarded as limitless, their designs made profligate use of resources and unintentionally caused a great deal of environmental damage. Moreover, considerations of the social implications of their professional actions tended to be limited to issues that directly affected the use of their product or design. These approaches are now clearly recognized as outdated, and modern engineers acknowledge the need to do better. But how?

The first step in this increasingly vigorous transformation of the engineering profession is to practice *green engineering*. There are a variety of definitions of this concept, which applies to all the engineering disciplines, but the essence is

> Green engineering is the design, commercialization, and use of engineering solutions, viewed from the perspective of human and environmental health.

The practice centers on minimizing pollution and risk as a consequence of product manufacture and product use, that is, of being more environmentally responsible than had been the case previously.

The second step in this transformation is to move beyond green engineering to *sustainable engineering*. The distinction is that the first step moves in the direction of more responsible technology, but does not ask, "How can social and environmental considerations be fully integrated into the engineering profession?" Nor does green engineering ask, "How far do we need to go?" This latter question is in many ways the province of industrial ecologists, who study the interactions between technology and the wider world. SE draws upon this information to inform itself about limits and goals and then place them within the traditional framework for accomplishing practical ends. Thus, sustainable engineering can legitimately be regarded as the operational arm of industrial ecology, and the essence of IE and SE can be briefly stated:

> Industrial ecology and sustainable engineering provide a template for the environmentally and societally sustainable redesign of the modern world.

A template is particularly useful only if those who use it understand and address the characteristics of the system they are attempting to redesign. It is useful, therefore, to present some of the characteristics of modern technology:

- *Technology is uncertain* (the best solution is never obvious, and experimentation is vital)
- *Technology is progressive* (change occurs by evolution and transformation)
- *Technology is analytical* (measures actions and new ideas)
- *Technology is cumulative* (builds on previous knowledge and existing capabilities)
- *Technology is systemic* (interdependence of technologies is required for progress)
- *Technology is embedded* (technology sits within natural systems)
- *Technology is accelerating* (the waves of technological transitions are ever shorter)

Any attempt to refashion technology must be responsive to these characteristics, and adapt as they change, else the attempt is doomed to failure.

3.3 LINKING INDUSTRIAL ACTIVITY WITH ENVIRONMENTAL AND SOCIAL SCIENCES

The contrast between traditional environmental approaches to industrial activity and those suggested by industrial ecology can be demonstrated by considering several timescales and types of activity, as shown in Table 3.2. The first topic, remediation, deals with such things as removing toxic chemicals from soil. It concerns past mistakes, is very costly, and adds nothing to the productivity of industry. The second topic—treatment,

TABLE 3.2 Aspects of Industry–Environment Interactions

Activity	Time	Focus	Endpoint	Corporate view
Remediation	Past	Local site	Reduce human risk	Overhead
Treatment, disposal	Present	Local site	Reduce human risk	Overhead
Industrial ecology	Future	Global	Sustainability	Strategic

storage, and disposal—deals with the proper handling of residual streams from today's industrial operations. The costs are embedded in the price of doing business, but contribute little or nothing to corporate success except to prevent criminal actions and lawsuits. Neither of these activities is industrial ecology. In contrast, industrial ecology deals with practices that look to the future and seeks to guide industry to cost-effective methods of operation that will render more nearly benign its interactions with the environment and will optimize the entire manufacturing process for the general good (and, we believe, for the financial good of the corporation). Corporate executives are familiar with the liabilities of past and present industry–environment interactions. A challenge to the industrial ecologist is to demonstrate that viewing this interaction from the perspective of the future is a corporate asset, not a liability.

We began Chapter 1 by discussing the "Tragedy of the Commons," in which a large number of individual actions, clearly beneficial over the short term to those making the decisions, eventually overwhelm a common resource and produce tragedy for all. Originally formulated to describe such local commons resources as public grazing lands, the concept was extended to global resources with discoveries such as the Antarctic ozone hole. Industrial processes and products interact with many different commons regimes, and design engineers should interpret the concept of the commons in a very broad way. The concept certainly includes local venues such as city air, local watersheds, and natural habitats. Regional resources are also included, groundwater and precipitation (and their possible chemical alteration) being examples. The global commons calls other regimes to mind: the deep oceans (can oil spills or ocean dumping significantly degrade this resource, or is humanity's influence modest?), the Antarctic continent (50 years of international scientific activity without environmental controls has left a dubious legacy), and, of course, the atmosphere (and its ozone and climate).

In practice, every product and process interacts with at least one commons regime and probably with several, fragile and robust, monitored and ignored, close and distant. Analyses of industrial designs may produce very different results if the same product or process is to be used under the sea, in the Arctic, or on a spacecraft rather than in a factory. Just as industrial ecology extends to all parts of the life cycle, it extends as well to all commons regimes it may affect.

3.4   THE CHALLENGE OF QUANTIFICATION AND RIGOR

Science and engineering are quantitative sciences, drawing much of their value from their exactness. Industrial ecology and sustainable engineering, in contrast, sprang from a few concepts, and rather simplistic ones at that. Here are some examples:

> *Mimic nature.* Nature does many things well, such as minimizing energy use. It also sometimes does things rather badly, such as depositing so many bird droppings on small oceanic islands that the islands become completely hostile environments. Nature's engineering is often a good model to follow, but not always, and only quantification of the consequences can show which path to follow.
>
> *The "Circular Economy."* The idea of never throwing anything away is a good starting point, but with complex modern technology it is often very expensive in

terms of energy and other resources to recover everything. Deciding when to be circular and when not to be is an analytical issue, not a conceptual one.

"Reduce, reuse, recycle." Again, a good general rule, but sometimes reusing an old, inefficient product can result in more harm than good. Is following this rule "committing a little less sin rather than solving the problem," as William McDonough puts it?

The "environmental footprint." It sounds really bad that Amsterdam's footprint is more than 10 times its physical area, but is it? We only know by understanding the assumptions built into the methodology, and the consequences of that footprint, in some detail.

Concepts are important, and they have played a big role in the inauguration of the field described and discussed in this book. As the field moves into its more mature phase, however, it is increasingly quantitative and rigorous. We will present and discuss the important concepts in the chapters that follow, but we will present and discuss even more the ways in which those concepts are now being quantified and made increasingly rigorous.

3.5   KEY QUESTIONS OF INDUSTRIAL ECOLOGY AND SUSTAINABLE ENGINEERING

As in any field, there are key questions in industrial ecology and sustainable engineering. Unlike biological ecology, we are interested not in the functioning of the technological system per se, but on the industrial ecosystem's interactions with and implications for the natural and social systems of the planet. We specifically concentrate on a single species (humans), its relationship with the environment, and the impacts of industrial operations and choices on its social systems. From this broad framework, it is possible to propose a more or less analogous set of key questions to be addressed in this book:

1. How do modern technological cycles operate?
   1.1 How are industrial sectors linked?
   1.2 What are the environmental and social opportunities and threats related to specific technologies or products?
   1.3 How are technological products and processes designed, and how might those approaches be usefully modified?
   1.4 Can cycles from extraction to final disposal be established for the technological materials used by our modern society?
   1.5 How do technological cycles interact with culture and society, and what are the implications inherent in these "second order" effects of technology?
2. How do the resource-related aspects of human cultural systems operate?
   2.1 How do corporations manage their interactions with the environment and society, and how might corporate environmental management evolve?
   2.2 How can the influence of culture/consumption on materials' cycles be modulated?

    2.3 How can engineers appreciate their relationships with environment and society?

    2.4 How might IE systems be better understood?

3. What are the limits to the interactions of technology with the world within which it operates?

    3.1 What limits are imposed by nonrenewable, nonfossil resource availability?

    3.2 What limits are imposed by the availability of energy?

    3.3 What limits are imposed by the availability of water?

    3.4 What limits are imposed by environmental and/or sustainability concerns?

    3.5 What limits are imposed by institutional, social, and cultural systems?

4. What is the future of the technology–environment–society relationship?

    4.1 What scenarios for development over the next several decades form plausible pictures of the future of technology and its relationship to the environment and social systems?

    4.2 Should systems degraded by technological activity, local to global, be restored, and if so how?

These key questions form the intellectual basis for the discussions throughout the rest of this volume.

3.6   AN OVERVIEW OF THIS BOOK

This book, directed toward codifying and explicating ways in which to transform our technological society from what is largely a nonsustainable system to resemble more and more closely a sustainable system, is divided into sections, as shown in Figure 3.2. Part I, "Introducing the Field," is composed of the information in the first several chapters. The descriptions are intended to set the stage by examining trends and patterns of industrial development and societal and environmental impact, presenting in some detail the biological metaphor upon which industrial ecology is based, discussing the interactions of technology with society and culture, and presenting the concept of sustainability.

    Part II introduces and discusses a central concept in industrial ecology, the life cycle of products and how the stresses related to each stage of the life cycle can be assessed. It begins the process of "thinking beyond the factory gate" by considering two simple questions with big implications: "Where did it come from?" and "Where is it going?"

    Green and sustainable engineering, which we regard as the operational arms of industrial ecology, are presented in Part III of the book. This is where product design and development, process design and implementation, and industrial ecosystems are described. Part IV addresses the broader topic of technological systems at levels from the individual firm to the planet. It discusses the use, recycling, and loss of resources, country-level material accounts, the implications of energy and water use by technology, analyzing an urban area by treating it by analogy with a biological organism, and the use of models to better understand the interactions that link technology, human systems, and the environment. The fifth and final part of the book is yet more expansive in scope—it deals with corporations and societies as systems; it also tries to understand

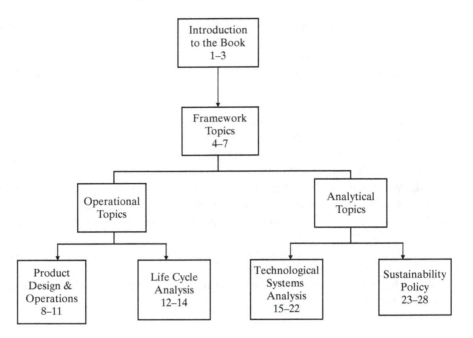

Figure 3.2

The structural outline of this book. Each box refers to the numbered book chapters indicated.

how the human species may address environmental and resource constraints and how it might begin the transition toward more sustainable development. Limits to technology and the ways in which scenario approaches can be useful are included. Finally, we address the contentious issue of whether technology should be employed in the restoration of planetary systems, small and large, that human actions have substantially altered over time. Thus, the practical segment of the book provides guidance for improving our environmental and social performance, while the systems approaches help define in broad terms where we should be going.

Industrial ecology and sustainable engineering by their nature have distinctive characteristics of their own, and these are related to the characteristics of technology itself:

- IE/SE is *adaptive*, not rigid (recall that technology is uncertain). (See Chapters 15 and 26.)
- IE/SE is *quantitative* (as well as qualitative) (recall that technology is analytical). (See Chapters 12–14 and 17–22.)
- IE/SE is *designed in*, not added on (recall that technology is cumulative and inclusive). (See Chapters 8–11.)
- IE/SE is *evolutionary*, not static (recall that technology is progressive). (See Chapters 4 and 5.)
- IE/SE is *encompassing*, not fragmented (recall that technology is systemic). (See Chapters 7 and 15.)

TABLE 3.3   Tools That May Be Brought to Bear upon the Key Questions of Industrial Ecology and Sustainable Engineering

Key question	Applicable chapters in this text	Industrial ecology tools
1.1	GIF book[*]	Industrial sector plots
1.2	GIF book, 12–14	Life cycle assessment
1.3	8–11, 16	Design for environment, sustainable engineering, industrial symbiosis
1.4	17, 18	Material flow analysis, national material accounts
2.1	26	Total environmental management, environmental metrics
2.2	7	Consumption analysis
2.3	15, 16, 21	Sustainable engineering, systems analysis, urban industrial ecology
2.4	22	Industrial ecology models
3.1	24	Criticality matrix
3.2	19	Energy cycle analyses
3.3	20	Water cycle analyses
3.4	2	Sustainability analyses
4.1	23	Industrial ecology scenarios
4.2	27	Earth system engineering and management

[*] T.E. Graedel, and J. Howard-Grenville, *Greening the Industrial Facility*, New York: Springer, 2005.

- IE/SE is *ecologically integrated*, not isolated (recall that technology is embedded). (See Chapters 8, 10, 11, and 16.)
- IE/SE *looks to the future*, not the past (recall that technology is accelerating). (See Chapters 23–25 and 27–28.)

It is possible also to think of the structure of this book from the perspective of the principal research questions posed earlier in this chapter. We list those questions in Table 3.3, together with the chapters that discuss them and the tools used to address them. It is a rich toolbox, if one that is as yet fairly early in its development, especially as regards the societal dimensions of SE and IE. Notwithstanding that limitation, it appears likely that industrial ecology craftspeople and sustainability engineers will be major participants in the construction of a more sustainable society than that which we now have.

FURTHER READING

Allen, D.T., and D.R. Shonnard, *Green Engineering: Environmentally Conscious Design of Chemical Processes*, Upper Saddle River, NJ: Prentice Hall, 2002.

Ayres, R.U., and L.W. Ayres, *A Handbook of Industrial Ecology*, Cheltenham, UK: Edward Elgar, 2001.

Clark, W.C., and R.E. Munn, Eds., *Sustainable Development of the Biosphere*, Cambridge, UK: Cambridge University Press, 1986.

Graedel, T.E., and J.A. Howard-Grenville, *Greening the Industrial Facility*, New York: Springer, 2005.

Grübler, A., *Technology and Global Change*, Cambridge: Cambridge University Press, 1998.

Smil, V., *Cycles of Life: Civilization and the Biosphere*, New York: Scientific American Library, 1997.

EXERCISES

3.1 Choose a room of your apartment, dormitory, or house. Conduct an inventory of the physical items or "artifacts" in the room. Divide them into four categories: (1) The artifact is necessary for survival; (2) The function performed by the article is necessary, but the artifact represents unnecessary environmental impact (e.g., clothes may be necessary, but a fur coat or 10 pairs of shoes may not be); (3) The artifact is unnecessary for survival but is culturally required; (4) The artifact is both physically and culturally unnecessary (albeit probably desirable, or it wouldn't be there). Can you extrapolate this result to your general consumption patterns? Based on these results, what percentage of your consumption represents unnecessary environmental impact?

3.2 Technology brings benefits, such as food, home heating, and medications. It also brings potential problems, such as air pollution and ecosystem disruption. Considering your personal interactions with the products of technology (see Exercise 3.1) and those of others, what sort and scope of technology do you think is appropriate for Earth in the twenty-first century? How should that technology operate?

3.3 Describe a sustainable world in your own words. Include a description of the lifestyle you would expect in such a world, as well as estimates of how large a population could be supported in such a world. What data and analysis would you need to be confident that your vision was, in fact, sustainable?

PART II   Framework Topics

C H A P T E R   4

The Relevance of Biological Ecology to Industrial Ecology

4.1   CONSIDERING THE ANALOGY

Industrial ecology (IE) consciously incorporates the word "ecology," a term originated with reference to biological systems. On the face of it, combining a word associated with the natural world with one associated with its exact opposite seems inappropriate, or at least injudicious. However, the idea of conceptualizing human social systems (including industrial systems) from an organismic point of view is at least a century old. The distinction between previous thinking along this line and the evolving IE approach is that the former concentrated on behavior and social structure whereas the latter focuses especially on physical and chemical parameters: resource flows, energy use, and the like. The vision of industry in all its facets engaged in the cycling of resources rather than in one-time resource use was presented in 1989 by Robert Frosch and Nicholas Gallopoulos of the General Motors Research Laboratories in Michigan. Their paper is regarded by many as the first publication in the field of industrial ecology.

A working definition of biological ecology (BE) is *the study of the distribution and abundance of organisms and their interactions with the physical world.* Along the same lines, IE can be defined as follows:

> Industrial ecology is the study of technological organisms, their use of resources, their potential environmental impacts, and the ways in which their interactions with the natural world could be restructured to enable global sustainability.

IE has been thought to be an appealing analogy to BE because it encourages the idea of the cycling (i.e., the reuse) of materials, but the analogy may be more extensive than that. In this chapter, we investigate whether there are indeed recognizable similarities between BE and IE, and whether they seem natural rather than contrived. The key issue we address is utility: whether any perspectives that result or methods that have been developed are useful for industrial ecology analysis and/or implementation.

4.2   BIOLOGICAL AND INDUSTRIAL ORGANISMS

The elementary unit of study in BE is the organism, which in the dictionary is defined as "an entity internally organized to maintain vital activities." Organisms share several characteristics, broadly defined, and it is instructive to list and comment on some of them.

1. A biological organism is capable of independent activity. Although biological organisms vary greatly in the degree of independence, all can take actions on their own behalf.
2. A biological organism utilizes energy and materials resources. Biological organisms expend energy to transform materials into new forms suitable for use. They also release waste heat and material residues. Excess energy is released by biological organisms into the surroundings, as are material residues (feces, urine, expelled breath, etc.).
3. A biological organism is capable of reproduction. Biological organisms are all able to reproduce their own kind, though the lifetimes and number of offspring vary enormously.
4. A biological organism responds to external stimuli. Biological organisms relate readily to such factors as temperature, humidity, resource availability, potential reproductive partners, and so on.
5. All multicellular organisms originate as one cell and move through stages of growth. This characteristic is commonly recognized in every living being from moths to humans (see, for example, Shakespeare's famous "ages of man" speech in *As You Like It*).
6. A biological organism has a finite lifetime. Unlike some physical systems such as igneous and metamorphic rocks, which for most purposes can be regarded as having unconstrained existence, biological organisms generally have variable but limited lifetimes.

The word "organism" is not used only to refer to living things. A second definition is "anything analogous in structure and function to a living thing." Hence, we speak of "social organisms" and the like. But what about industrial activity; does it have entities that meet the definition? To make such a determination, let us select a candidate organism—the factory (including its equipment and workers)—and look at its characteristics from the perspective of the biological organism.

1. Is an industrial organism capable of independent activity? Factories (through their employees) clearly undertake many essentially independent activities on their own behalf: acquisition of resources, transformation of resources, and so forth.

2. Do industrial organisms use energy and material resources and release waste heat and material residues? Industrial organisms expend energy for the purpose of transforming materials of various kinds into new forms suitable for use. Energy residues are emitted by industrial organisms into the surroundings, as are material residues (solid waste, liquid waste, gaseous emissions, etc).

3. Are industrial organisms capable of reproduction? An industrial organism is designed and constructed not for the purpose of re-creating itself, but to create a nonorganismic product (such as a pencil). Generally speaking, new industrial organisms (factories) are created by contractors whose job is to produce any of a variety of factories to desired specifications rather than to create replicates of existing factories. If reproduction is defined as the generation of essentially exact copies of existing organisms, then industrial organisms do not meet the definition. If substantial modification is allowed for, however, we can recognize that copies or similar organisms are indeed generated. However, industrial organism reproduction is not a function of each individual organism itself, but of specialized external actors.

4. Do industrial organisms respond to external stimuli? Industrial organisms relate readily to such external factors as resource availability, potential customers, prices, and so on.

5. Does an industrial organism move through stages of growth? The answer is generally yes. Few factories are unchanged during their lifetimes, although they may not follow the orderly or predictable progression of life stages of the biological organism.

6. Does an industrial organism have a finite lifetime? This characteristic is certainly true.

A factory thus seems to be an appropriate candidate as an industrial organism, since it utilizes energy to transform materials just as does a biological organism. A candidate that stores materials but does not transform them is a repository, not an organism, just as soil contains the nitrate that organisms have transformed from atmospheric N_2, but it does not itself act on the N_2 or nitrate.

The concept of industrial organisms can be expanded in ways that seem useful, even if all the above conditions are not satisfied. Indeed, the definition of a biological organism seems to require only two conditions: The candidate organism must not be passive (as is a sedimentary rock or a coffee cup) and the organism must make use of resources during its lifetime (as does a flower or a washing machine factory). Thus, organisms can manufacture other organisms (badgers make little badgers; factories make washing machines) and/or nonorganismal products (badgers make fecal pellets; factories make sludge). The key signature of an organism, biological or industrial, is that it is involved in resource utilization after, as well as during, its own manufacture.

Another important characteristic of organisms is their life cycle, which comprises birth, growth, relative stasis, and eventual death. At these different stages, an organism's requirements for resources and energy are quite different, as are its waste products. In much the same way, the life cycle of an industrial product provides fertile ground for analysis, from resource extraction through manufacture, use, and eventual obsolescence and discard. In both cases, there are resources potentially available for use at end of life. The use of life cycle assessment in industrial ecology is discussed in Chapters 12–14.

4.3 BIOLOGICAL AND INDUSTRIAL ECOSYSTEMS

An ecosystem is the sum total of the organisms in a functioning space, together with their physical environment. In this connection, it is instructive to think of the materials' cycles associated with a postulated primitive biological system such as might have existed early in Earth's history. At that time, the potentially usable resources were so large and the amount of life so small that the existence of life forms had essentially no impact on available resources. This process was one in which the flow of material from one stage to the next was essentially independent of any other flows. Schematically, what we might term a "Type I" system takes the form of Figure 4.1a.

As the early life forms multiplied, external constraints on the unlimited sources and sinks of the Type I system began to develop. These constraints led in turn to the development of resource cycling as an alternative to linear material flows. Feedback and cycling loops for resources were developed because scarcity drove the process of change. In such systems, the flows of material within the proximal domain could have been quite large, but the flows into and out of that domain (i.e., from resources and to waste) eventually were quite small. Schematically, such a Type II system might be expressed as in Figure 4.1b. This transition from Type I to Type II, or from one Type II system to another, is termed "succession."

The above discussion of Type I and Type II systems refers to the planet as a whole and regards the ecosystem types as sequential. Individual ecosystems, however, may be of either type in any epoch. A Type I system, known ecologically as an open system, is one in which resource flows into and out of the system are large compared with the flows within it. A Type II system, known as a closed system, is the reverse. Complicating the issue, especially on the small scale, is the fact that an ecosystem can be open with respect to one resource (e.g., water) and closed with respect to another (e.g., nitrogen).

A Type II system is much more efficient than a Type I system, but on a planetary scale it clearly is not sustainable over the long term because the flows are all in one direction; that is, the system is "running down." To be ultimately sustainable, the global biological ecosystem has evolved over the long term to the point where resources and waste are undefined, because waste to one component of the system represents resources to another. Such a Type III system, in which complete cyclicity has been achieved (except for solar energy), may be pictured as in Figure 4.1c.

As suggested above, recycle loops have, as inherent properties, temporal and spatial scales. The ideal temporal scale for a resource recycle loop in a biological ecosystem is short, for two reasons. First, material viewed by a specific organism as a resource may degrade if left unused for long periods and thus become less useful to that organism. Second, resources not utilized promptly must be retained in some sort of storage facility,

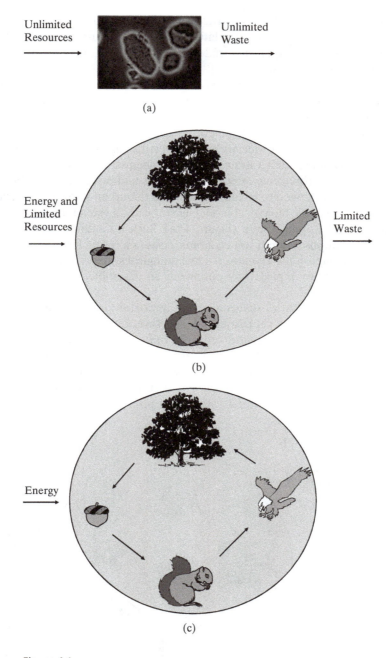

Figure 4.1

(a) Linear materials flows in Type I biological ecology. (b) Quasicyclic materials flows in Type II biological ecology. (c) Cyclic materials flows in Type III biological ecology.

which must first be found or constructed, then defended. Still, examples of resource storage in biological systems are relatively common: Squirrels store nuts, birds store seeds, and so forth; the organisms trade off the costs of storage against the benefits to themselves or their offspring.

Analogously, the ideal spatial scale for most resource recycle loops in a biological ecosystem is small. One reason is that procuring resources from far away has a high energy cost. A second is that it is much more difficult to monitor and thus ensure the availability of spatially distant resources. As with temporal scales, counterexamples come readily to mind, such as eagles that hunt over wide areas. However, the eagles would probably have evolved to hunt close to home if such a strategy provided adequate resources.

The ideal anthropogenic use of the materials and resources available for industrial processes (broadly defined to include agriculture, the urban infrastructure, etc.) would be one similar to the cyclic biological model. Historically, however, human resource use has mimicked a Type I open system (Figure 4.1a). Such a mode of operation is essentially unplanned and imposes significant economic costs as a result. IE, in its implementation, is intended to accomplish the evolution of technological systems from Type I to Type II, and perhaps ultimately to Type III, by optimizing in ensemble all the factors involved, as suggested in Figure 4.2.

Temporal and spatial scales are also important considerations in IE. As with biological systems, temporal resource loops should, in general, be short, lest corrosion or other processes degrade reusable materials. Society's long-term resource storage facilities, many of which are termed "landfills," cannot be recommended from an ecosystem viewpoint: They are expensive to maintain; they mix materials, which makes their recovery and reuse difficult; and they tend to leak. The ideal spatial scale in an industrial ecosystem also mimics its biological analog: Small is

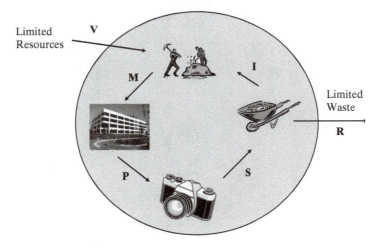

Figure 4.2

The Type II materials flow model of industrial ecology. The letters refer to the following mass flows: V = virgin material; M = processed material; P = product; S = salvaged material; I = impure material; R = uncaptured residues.

best, again because of the energy requirements of long-range resource procurement and the uncertainty of supply continuity in a world where political issues as well as resource stocks may provide constraints. However, industrial organisms can view resources on a global basis (whereas biological systems generally cannot) and can acquire resources on very large spatial scales if the combination of resource attributes and resource cost is satisfactory.

4.4   ENGINEERING BY BIOLOGICAL AND INDUSTRIAL ORGANISMS

Clive Jones and colleagues have identified six classes of material transformation by organisms, as diagrammed in Figure 4.3 and exemplified in Table 4.1. They discuss the classification scheme in detail; it is subject to a number of gray areas that need not be of concern here, but its central tenets are most instructive. The first case—the direct provisioning of resources by one species to another, as in an osprey eating a fish—is not ecosystem engineering, because although materials are utilized (i.e., the fish's parts are transformed), no environmental change occurs.

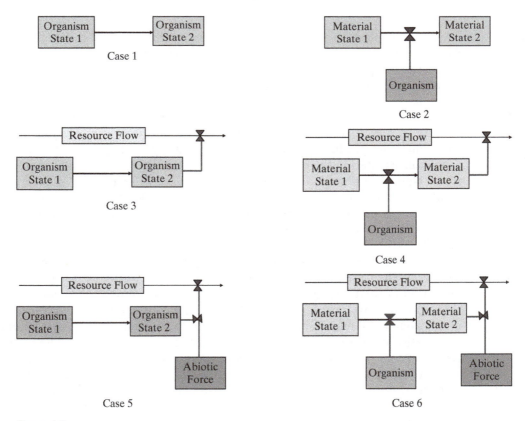

Figure 4.3

Conceptual models of engineering by organisms and engineers. Terms are defined and examples given in Table 4.1. The "bow tie" symbol indicates points of material or resource transformation. (Adapted from C.G. Jones, J.H. Lawton, and M. Shachak, Organisms as ecosystem engineers, *Oikos, 69*, 373–386, 1994.)

TABLE 4.1 Examples of Ecosystem Engineering and Industrial Engineering

Case (Figure 4.3)	Example
1	Ecosystem: An osprey eats a fish
	Industrial: A factory roof is repaired
2	Ecosystem: Rabbits dig a warren
	Industrial: Neolithic humans construct a log road
3	Ecosystem: Plankton blooms warm surface waters
	Industrial: Urban areas alter water flows
4	Ecosystem: Beavers build dams
	Industrial: Humans manufacture portable radios
5	Ecosystem: Mussel beds protect sediments
	Industrial: Urban areas raise regional temperatures
6	Ecosystem: Plankton emit cloud-forming dimethyl sulfide
	Industrial: CFC emissions create ozone hole

From T.E. Graedel, On the concept of industrial ecology, *Annual Reviews of Energy and the Environment, 21*, 69–98, 1996.

The simplest case of true ecosystem engineering (Case 2 in Figure 4.3) occurs when an organism transforms materials into resources useful to other organisms. "Transform" is perhaps too strong a word here; "shape" or "rearrange" might be better, because the process need not involve modifying resource flows. A biological Case 2 example is a bird's nest in the above-water portion of a muskrat lodge, constructed by the muskrat without the slightest thought of providing avian apartments.

In IE, Case 1 activities are also those done to maintain organism function but not to manufacture products for others. Examples include fixing the roof of a factory, or repainting its interior. Examples of Case 2 are much less common today then they once were: Nomadic tribes moved rocks, trees, and brush to provide shelter; Neolithic residents of Britain used tree trunks to make roads 6000 years ago; and so forth. Even now, in certain parts of the world one can find cobblestone and slate walkways and streets being constructed; if the materials are used in their natural state, the activities constitute Case 2 industrial engineering.

In Cases 3 and 4, ecosystem engineering results in an alteration in the flows of - resources. The Case 3 flow alteration arises from the tissues of the engineer, as in the alteration of solar energy to the upper ocean produced by planktonic absorption of solar radiation. In Case 4, the flow alteration is a consequence of the transformation of environmental materials. Case 4 is perhaps the easiest of all ecosystem engineering classifications to appreciate; it is exemplified by the beaver building a dam that results in changes in streamflow, nutrient transport, and wetlands habitat.

A common IE example of Case 3 is the alteration by buildings and their associated infrastructure of roads and parking lots of the water budget of an urban area. Most urban areas were originally settled because they possessed abundant natural water supplies. After construction of the modern city is accomplished, water must generally be transported to the city from outside sources, often quite remote ones.

Case 4 is, of course, exactly the historical concept of industrial engineering. A manufacturer consumes plastic and metal and produces a portable radio, for example, and during its life the radio uses an occasional battery made by another supplier. Or, an automobile plant builds a sedan from metal, glass, plastic, and rubber, and the sedan goes on to consume gasoline, oil, antifreeze, and various replacement parts.

Engineering Cases 5 and 6 are those in which the transformation of materials results in the modulation not of resource flows, but of one or more major abiotic forces of nature. Jones and colleagues had in mind such natural but spatially restricted processes as hurricanes and tornadoes. Their example of Case 5 engineering is mussels that form matted colonies on the seabed, minimizing the sediment disruption that often results from roiled waters caused by hurricanes. In much the same way, a Case 6 example is that of clams that secrete cement to anchor themselves to rocks and subsequently to each other, again resulting in stabilization of the ocean substrate.

Industrial engineering activities appropriate for designation as Class 6 can also be readily identified. A sea wall, for example, is a classic Case 6 example, designed to minimize beach erosion and structural damage from high waves; contour plowing to minimize soil erosion is another example. If we include climate as a major abiotic force, the widely recognized urban heat island (the increase in regional temperature produced by the combination of radiation absorption, power consumption, and heat dissipation in urban areas, with regional increases also in cloudiness and precipitation) is clearly an example of Case 5 industrial engineering.

Case 5 and Case 6 engineering for both BE and IE activities influences major abiotic forces on a global scale as well. The classic BE Case 5 is the carbon dioxide/oxygen photosynthesis and respiration cycle that establishes a natural greenhouse, maintaining the temperature of Earth above the freezing point of water. The widely discussed Gaia theory of Lovelock and collaborators arises from the fact that vegetation alters, in a passive way, the CO_2/O_2 ratio in the atmosphere so that it is optimal for plant growth and sustainability.

Case 6 ecological engineering is applicable to another widely discussed and still contentious proposal: that gaseous emissions of dimethyl sulfide from oceanic phytoplankton control the rate of cloud formation, thus of solar radiation to the surface ocean, thus of ocean temperature, thus (because CO_2 loss to the ocean is a function of temperature) of global climate. We discuss this idea in more detail in Chapter 27.

Case 6 examples of industrial engineering are also easy to point out. One that is clearly established is the Antarctic ozone hole and the correlated decrease in midlatitude stratospheric ozone as a consequence of the emission of chlorofluorocarbon propellants, foams, and refrigerants. A second is enhanced global warming from anthropogenically produced carbon dioxide, methane, and other greenhouse gases.

4.5   EVOLUTION

Evolution is, of course, the central organizing principle of modern biology and biological ecology. Since the time of Charles Darwin, the origin and modification of species has been pictured as the "tree of life" (Figure 4.4a), with the different organisms being regarded as descended from organisms lower on the tree of life than themselves. The evolution is driven by random genetic variation, changes in local ecosystem structure, and changes in environmental pressures or constraints.

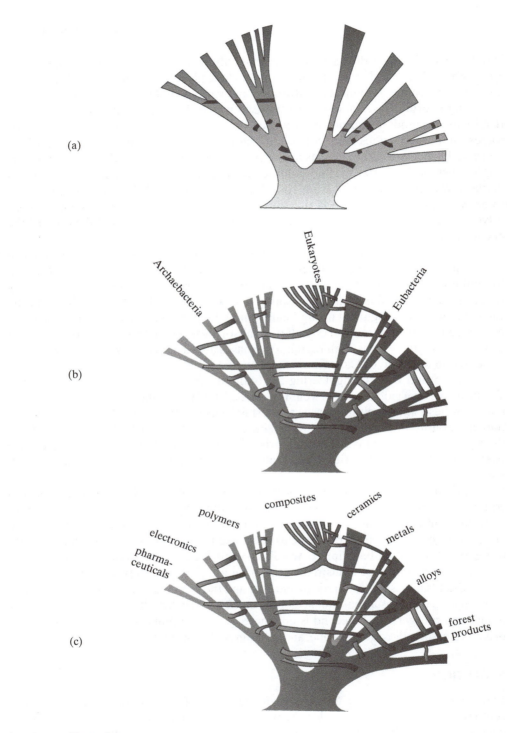

Figure 4.4

(a) The traditional tree of life; (b) The contemporary network of life; (c) The network of technology. (Adapted from McInerney, J.O., J.A. Cotton, and D. Pisani, The prokaryotic tree of life: Past, present . . . and future? *Trends in Ecology and Evolution, 23*, 276–281, 2008.)

With the determination of DNA structure, and particularly with the onset of DNA sequencing, it has been recognized that it is increasingly common for prokaryotes (unicellular organisms without a membrane-bound nucleus) to exchange DNA in what is termed "horizontal gene transfer." This transforms the tree of life into a network (Figure 4.4b), in which the network connections produce symbionts that draw DNA from two or more branches of the tree. As this is written, evolutionary ecologists are hard at work determining how best to represent the phylogeny of life.

Modern technology and its developmental history have much in common with biology's tree of life, and even more so the network of life. Technology developed from its own set of driving factors: tool use, new sources of energy (steam, petroleum, nuclear), new materials (alloys, polymers, etc.), and the information age. Throughout much of the industrial revolution, the emphasis was on making existing materials perform better through minor modifications, structural redesign, and the like. More recently, it has become common to see a product utilizing materials from quite different historical technologies, and so we see machines made largely from metal but employing polymers and being controlled by electronics. It is also increasingly common for engineers to utilize composites, technologically generated multiphase materials in which the constituent phases are chemically dissimilar and separated by a distinct interface. Thus the tree of technology (Figure 4.4c) has come to resemble the network of life, with composites as its symbionts. Just as biological symbionts survive by filling a unique niche, so do composites, whose physical and chemical properties are designed to be superior to those of its parent materials. Modern industrial sectors are so closely integrated, in fact, that their network (Figure 4.5) can be understood as a variant of the network of technology of Figure 4.3c.

4.6   THE UTILITY OF THE ECOLOGICAL APPROACH

This chapter has demonstrated that the concepts and tools of BE can without much difficulty be related to our perspective of industry and its relationships to the natural world. The notion of scale is a central idea, enabling us to consider industrial facilities as organisms acting both individually and collectively. This leads naturally to the idea of cycling and reuse of resources. In this regard, quantitative measures of efficiency in resource use provide useful comparative benchmarking. Also important is the concept of the food chain, which provides a framework for and an analytical approach to the flows of resources in the technological society. We have also discussed life cycles (of animals and industrial products) and how ecosystem succession in biology is mirrored by the effects produced in industry by invention and implementation. All of these concepts, drawn from biological ecology, help to inform IE discussions and analyses.

Perhaps just as important is the symbolic appreciation of both nature and industry as interconnected systems—systems that ideally act to conserve and reuse resources, to be resilient under stress, and to evolve in response to need. Industrial ecology includes such highly focused topics as detailed product design and methods for choosing materials, but at heart it is a systems science.

Beyond the utility of key concepts of biological ecology (the *organism*, its *life cycle*, the *reuse of resources*, *ecosystem succession*) for industrial ecology is the potential

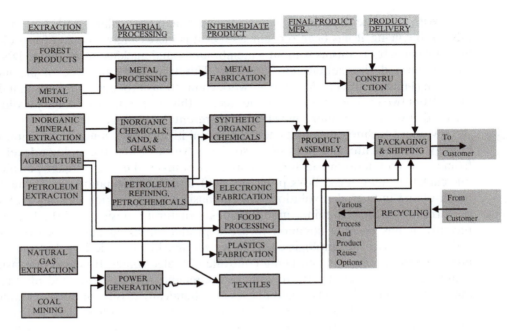

Figure 4.5

The major industrial sectors and the interrelationships among them.

utility of biological ecology's "tool kit," the methods developed by biologists over many decades. In subsequent chapters, we will see a number of examples of industrial ecology's use of some of the tools from this tool kit:

- *Metabolic analysis.* Metabolism is the aggregate of all physical and chemical processes taking place within an organism or group of organisms. This topic and its usefulness in industrial ecology are discussed in Chapter 5.
- *Material flow analysis (MFA).* MFA is an analytical approach to how materials are used by an (industrial) ecosystem. This topic and its usefulness in industrial ecology are discussed in Chapters 17 and 18.
- *Energy analysis.* An organism's metabolism or an ecosystem's material use can be expressed in terms of energy used or contained, an approach that often proves useful. This topic and its usefulness in industrial ecology are discussed in Chapter 19.
- *Symbiosis.* Symbiosis is a close association of two organisms for the benefit of either or both. This topic and its usefulness in industrial ecology are discussed in Chapter 16.
- *Island biogeography.* Biological ecologists often find their analyses easier and more informative if they study the organisms on an island rather than on an equivalent area on a large landmass. This topic and its usefulness in industrial ecology are discussed in Chapter 16.

It is fair to point out that questions have sometimes been raised concerning the validity of the BE/IE analogy. Perhaps the most thoughtful critic has been Robert Ayres (2004), who makes the following points:

- There is no primary producer in the economic system analogous to photosynthesizers.
- In the biosphere there are no products as such.
- In the biosphere there are no markets.
- Evolution in nature is driven by random variations, in industry by directed invention.

Do Ayres's objections render the analogy useless? In our view, they do not. No analogy constitutes perfect symmetry, only correspondence in some respects. This chapter and others to come demonstrate that many features of biological ecology have no counterpart in industrial ecology and vice versa. Nevertheless, the wide acceptance of the term "industrial ecology" has proven that it is more than a convenient label; rather, it is a reflection of the value that has resulted and will continue to result from explorations of the linkages between fields of study once thought to be completely unrelated but increasingly shown to have mutually useful synergies.

FURTHER READING

Archer, K., Regions as social organisms: The Lamarckian characteristics of Vidal de la Blache's regional geography, *Annals of the Association of American Geographers, 83*, 498–514, 1993.

Ayres, R.U., On the life cycle metaphor: Where ecology and economics diverge, *Ecological Economics, 48*, 425–438, 2004.

Frosch, R.A., and N.E. Gallopoulos, Strategies for manufacturing, *Scientific American, 261* (3), 144–1552, 1989.

Graedel, T. E., On the concept of industrial ecology, *Annual Reviews of Energy and the Environment, 21*, 69–98, 1996.

Keeling, P.J., et al., The tree of eukaryotes, *Trends in Ecology and Evolution, 20*, 670–676, 2005.

McInerney, J.O., J.A. Cotton, and D. Pisani, The prokaryotic tree of life: Past, present . . . and future? *Trends in Ecology and Evolution, 23*, 276–281, 2008.

Wright, J.P., and C.G. Jones, The concept of organisms as ecosystem engineers ten years on: Progress, limitations, and challenges, *Bioscience, 56*, 203–209, 2006.

EXERCISES

4.1 A factory is proposed in this chapter as a candidate industrial organism, and some of its characteristics evaluated from that perspective. There are other possible candidate organisms, however. Evaluate the following as industrial organisms, compare their characteristics to those of a factory, and determine the most appropriate organism analog:
   (a) a multinational corporation
   (b) a city of one million people

4.2 Is a fully functioning Type II industrial ecosystem the ultimate? Is it realistic to work toward a Type III industrial ecosystem?

4.3 Ecologists have demonstrated that analyses of ecosystem engineering provide an organizational framework for understanding biological influences on material flows. Under what sorts of conditions or in what sorts of locations might such flows dominate over flows influenced by industrial engineering?

4.4 Do you regard the criticisms of Robert Ayres concerning the BE/IE analogy to be valid and important? Why or why not?

CHAPTER 5

Metabolic Analysis

5.1 THE CONCEPT OF METABOLISM

Organisms, whether biological or industrial, utilize resources of various kinds to perform their functions. The study of these processes is metabolic analysis, where metabolism is defined as

> The aggregate of all physical and chemical processes taking place within an organism or group of organisms.

Unlike the study of organisms, which is centered on *attributes* (size, lifetime, etc.), the study of metabolisms is centered on *processes* taking place within organisms.

As in the previous chapter, where the relevance of biological ecology to industrial ecology was explored, metabolic analysis is pursued in industrial ecology not only because it is interesting but because it has the potential to be quite useful. We begin this chapter by describing classical metabolic analysis in biology so as to set the stage for considering metabolism in its industrial context.

5.2 METABOLISMS OF BIOLOGICAL ORGANISMS

A biological organism takes in a variety of resources (plants, smaller animals, etc.); uses them to enable a variety of bodily functions (breathing, motion, etc.) and to build muscles, bones, and other parts of its body; and then emits or excretes a variety of waste products. The first step in characterizing an organism's metabolism is identifying and quantifying these inputs and outputs. The next step is to identify and understand the *pathways*, or internal chemical transformation sequences, that occur. In principle, every transformation can be described as a reaction, or a series of reactions. As with all sets of chemical reaction, however, complexity is inherent in the process. For example, an intermediate product, termed a "metabolite," may often be formed by several different pathways. The flow of material down these pathways is generally controlled by *enzymes,* which are molecules that act as catalysts. By-products are common, as a reaction often produces both the desired metabolic product (to be built upon) and a less desired product (to be removed by further reaction or discarded).

These features of metabolism are illustrated by the classic tricarboxylic acid cycle shown in Figure 5.1. In this cycle MAL, FUM, OGA, and OAA are the metabolites; NADH and NADPH are the enzymes; and many (but not all) of the reactions can be reversible depending on factors such as enzyme concentrations and temperature.

The first step in constructing the tricarboxylic acid cycle was the realization that a resource *I* gives rise to a product *O* (Figure 5.2a). With better analytical tools, metabolites *A* and *B* were identified (Figure 5.2b). More detailed analysis eventually produced the system shown in Figure 5.2c. This system is certainly complex, but it is only a small part of the total metabolism of a typical organism. The metabolic network of the bacterium *Escherichia coli,* for example, is comprised of several hundred metabolites, perhaps twice as many reactions, and more than hundred pathways of interest. As a result, in the words of bioinformatics specialist Peter Karp,

> The theoretical understanding of a system such as the biochemical network of *E. coli* is too large for a single scientist to grasp . . . As scientific theories reach a certain complexity, it becomes essential to encode those theories in a symbolic form within a computer database.

The computer approach (termed "biological informatics") begins by representing the networks with a matrix *S*, where the rows correspond to the metabolites and the columns to the reactions (Figure 5.2d). The matrix elements are the stoichiometric coefficients of the reactions. If all metabolites are identified, and all fluxes determined,

Figure 5.1

A metabolic example from biology: the tricarboxylic acid (Krebs) cycle. The arrows indicate reversibility or irreversibility of the reaction. f_x indicates the fluxes of the reactions. FUM, fumerate; MAL, malonate; OAA, oxoloacetate; OGA, 2-oxoglutorate; NADH, nicotinamide adeninedinucleotide; NADPH, nicotinamide adeninedinucleotide phosphate. The large arrows indicates a flux to biomass building blocks. (Adapted from M. Emmerling, et al., Metabolic flux responses to pyruvate kinase knockout in *Escherichia coli, Journal of Bacteriology, 184,* 142–164, 2002.)

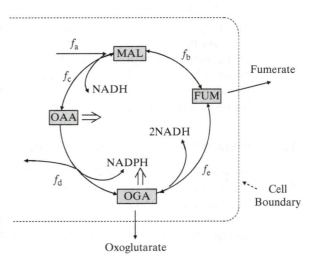

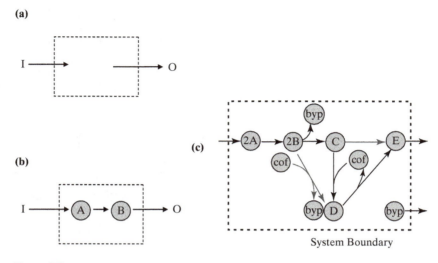

Figure 5.2

The progressive analysis of metabolic pathways in biology, as further discussed in the text. (a)—resource *I* and product *O*, the transformation occurring within the system (such as a cell) indicated by the dotted line; (b)—as with (a), but with two metabolites identified; (c)—as with (b), but with many more metabolites and many potential reaction paths identified; (d)—the *S* matrix that describes the system in (c); (e)—the pathway analysis resulting from (d); (f)—two of the active pathways deduced from the pathway analysis. A, B, C, D, and E are metabolites; v_x, the relative reactions; b_x, fluxes that cross the system boundary; cof, cofactor; and byp, by-product. The solid arrows in (f) correspond to active reactions, dashed arrows to inactive ones. (Adapted from J.A. Papin, et al., Metabolic pathways in the post-genome era, *Trends in Biochemical Sciences, 28,* 240–248, 2003.)

the result then specifies matrix *P*—the potential alternative pathways for the transformation of 2*A* to *E* (Figure 5.2e). Three of these pathways are shown in Figure 5.2f, where the dark lines indicate the deduced reactions and the faint lines inactive reactions; in any particular circumstance the pathway taken may be a function of temperature, cofactor concentration, or some other system parameter. With the availability of computer analysis, very large networks can be analyzed in this way, and the existence and properties of unknown metabolites and unnamed reactions can be predicted.

The pathway analysis approach is data intensive, as befits such a complex system. The result of utilizing this information from biochemistry, physiology, and genomics is the elucidation of the processes within an organism, thereby leading to increased understanding and even the possible design of engineered biological systems.

5.3   METABOLISMS OF INDUSTRIAL ORGANISMS

A human being is a biological organism, of course, and each of us possesses a complex *biological* metabolism as just described. An alternative perspective is provided by characterizing the physical metabolism of the individual. Such an analysis reveals the material inputs and outputs related to the lifestyle of the average human, or of a specific person. A classic analysis of this type was performed by Iddo Wernick and Jesse Ausubel of the

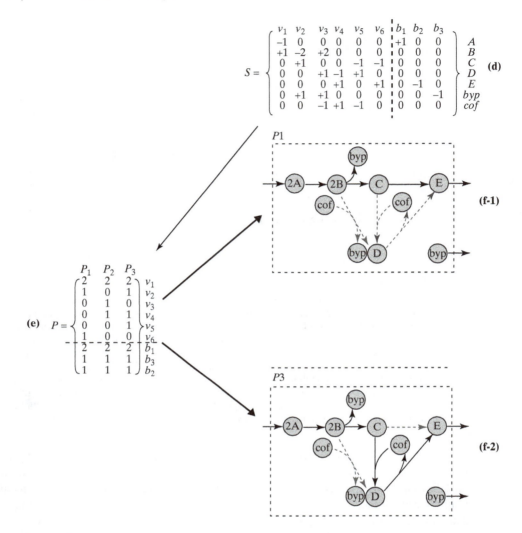

Figure 5.2

Continued

Rockefeller University, who studied per capita resource flows in 1990 for the United States. Some of the results are given in Table 5.1. Disregarding such very large flows as water or mine overburden, construction materials and energy-generating fossil fuels were found to be the largest inputs. The major new aspect of this study was the inclusion of outputs, of which carbon dioxide from fossil fuel combustion was seen to be the largest.

One of the most complete contemporary industrial versions of physical metabolism, that of the Toyota Motor Company, is shown in Figure 5.3. Here the major inputs and outputs are identified—water, energy, carbon dioxide, and so forth. Flows subject to regulations of one sort or another are quantified. Transformations are not indicated, however, and individual materials are not identified. The result is an analysis that parallels the metabolic studies that biologists were developing some three or four decades earlier.

TABLE 5.1   United States Per Capita Resource Flows for 1990

	Flows (kg/day)
Inputs	
Energy materials	21.5
Construction minerals	21.1
Imports	6.9
Agriculture	6.9
Forestry products	2.9
Industrial minerals	2.7
Metals	1.2
Outputs	
Air emissions	19.0
Wastes	6.1
Export	4.5
Dissipation	1.6

Source: From I.K. Wernick, and J.H. Ausubel, National material flows and the environment, *Annual Review of Energy and the Environment, 20*, 463–492, 1995.

Just as for biological organisms, human or otherwise, it is possible and useful to analyze the metabolism of an industrial organism (the factory or corporation) in much more detail than is indicated in Figure 5.3. To do so, we need to revise somewhat the definitions of metabolic terminology:

- *Industrial metabolite*—an intermediate product in the transformation of resources into final products. Industrial metabolites can also be termed "parts," "subassemblies," and so on.
- *Industrial enzyme*—an industrial process or piece of equipment that results in a transformation, also termed "reactor," "milling machine," "lathe," and so on.
- *Industrial pathway*—the sequence of transformations that convert resources into final products.

We further see that an industrial enzyme may enable a physical transformation (e.g., drilling, shaping) as well as a chemical one.

An industrial pathway diagram shares many characteristics with biochemical pathway analysis, as can be seen in Figure 5.4. In overview, at the top of the figure, steel, parts, and chemicals enter the facility, and a packaged product emerges. To construct the diagram for a given material (and there could be concurrent diagrams for other materials), one needs to know which process transforms or transports the material and which final products contain the material. To specify the fluxes, one needs to know the mass fluxes of the material inputs, the material concentrations in the inputs, and the details of the transformation process. This information then generates the industrial metabolic diagram (bottom of Figure 5.4) which indicates the metabolites, enzymes, and by-products. It is straightforward to form the S matrix from this diagram.

It turns out that a form of metabolic analyses in industry has existed for about a quarter of a century under the name *material requirements planning* (MRP). MRPs

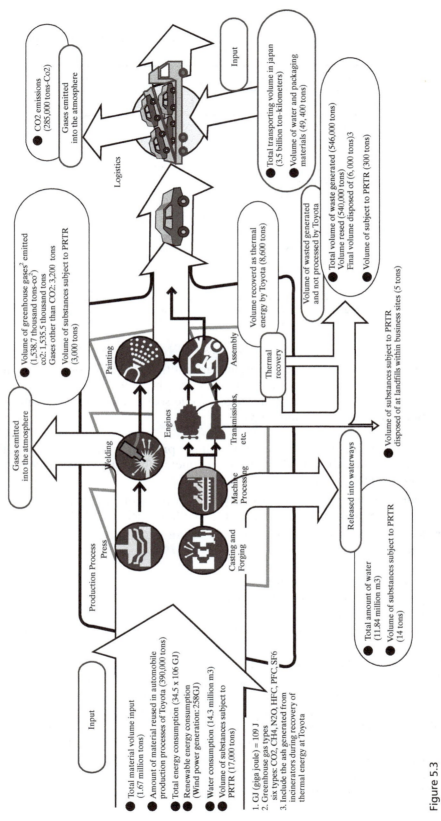

Figure 5.3

The industrial metabolism for Toyota's manufacturing processes in 2005. Only some of the materials and flows are indicated, but clearly there is much underlying detail from which this figure was drafted. (Courtesy of Toyota Motor Company.)

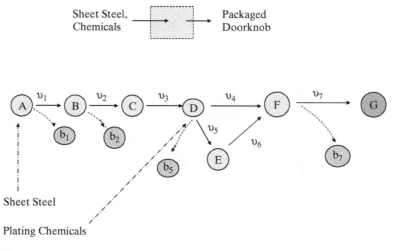

Figure 5.4

Industrial metabolic reaction pathways, from simple to complex: the doorknob cycle. A, sheet steel in receiving bin; B, thin coil of sheet steel; C, disk of sheet steel; D, shaped knob; E, plated knob; F, assembled doorknob; G, packaged doorknob; v_1, slitter; v_2, punch press; v_3, drawing press; v_4, assembly station; v_5, plating bath; v_6, assembly station; v_7, packaging station.

are software programs that link production schedules with "bills of materials" files (sources of materials information on every part and component), inventory status files, and materials requirements files (Figure 5.5a). This approach has enabled businesses to more efficiently manage manufacturing so as to "build to order" rather than "build

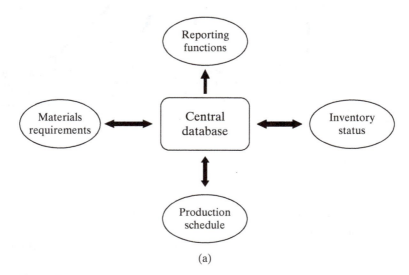

(a)

Figure 5.5

(a) The scope of a materials requirements planning (MRP) system; (b) The scope of an enterprise resource planning (ERP) system; (c) The scope of a hypothetical expanded ERP system.

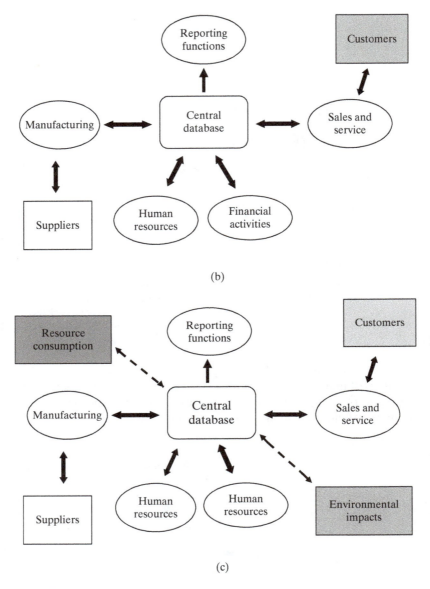

(b)

(c)

Figure 5.5

Continued

to stock." In essence, MRPs are the industrial equivalent of modern biological informatics systems.

There are several distinctions between biological metabolism analysis and industrial metabolism analysis:

- In principle, each industrial metabolite is known, as is each industrial enzyme, because they are specified by the designer rather than having to be deduced.

- The biological approach assumes a steady state for all reactions, thus simplifying the mathematics at the cost of simplifying the problem. Industrial approaches often look at disruptions caused by a perturbation.
- MRP is used not to deduce how a given organism functions, as is generally the case in biology, but to explore the desirability of alternative ways of functioning (although we acknowledge that drug design involves exploring perturbations of biological pathways).

The use of MRP for alternative manufacturing approaches can be pictured with the aid of Figure 5.6, which shows three different sequences for the manufacture of an assembly from six parts P_1–P_6. Figure 5.6a includes the formation of two subassemblies

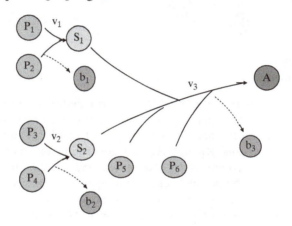

(a)

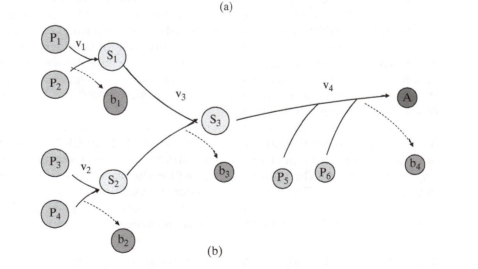

(b)

Figure 5.6

Optional metabolic reaction pathways for an industrial product assembly process. P, part; S, subassembly; A, assembly; b, by-product; R, resource. (Adapted from A. Kusiak, *Intelligent Manufacturing Systems,* Englewood Cliffs, NJ: Prentice Hall, p. 341, 1990.)

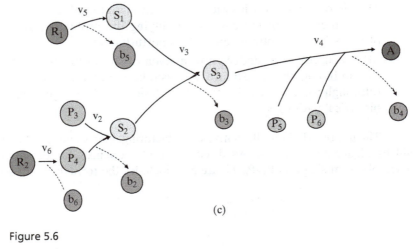

Figure 5.6

Continued

and three reactions. Figure 5.6b adds an intermediate subassembly and thus an additional reaction. In Figure 5.6c, part P_4 and subassembly 1 are manufactured within the facility, thereby eliminating the need to acquire parts P_2 and P_4 from suppliers, but introducing resource requirements R_1 and R_2 and generating two new by-products. The choice between the three sequences depends in large part on the fluxes of the pathways, and the degree and types of by-product generation. As before, the systems are amenable to matrix analysis. In a more realistic example of a factory manufacturing several final products, each containing scores of parts and requiring 10 or 20 reactions or transformations, computer approaches become essential if optimization is to be achieved.

Factors other than resource transformation options enter into decisions about the industrial metabolism: financial systems, labor, management, and so on. Analytical (software) approaches that encompass this entire suite of considerations are termed "enterprise resource planning (ERP) tools" and are in extensive use in industry. A conceptual picture of a modern ERP system is shown in Figure 5.5b. As with MRP systems, a central database unites the various functions. MRP and ERP are examples of what are sometimes termed "expert systems."

From an industrial ecology standpoint, it is of interest that two aspects of an industrial metabolism that are not currently part of MRPs or ERPs are the tracking of individual materials (copper rather than switches or relays) and the environmental impacts arising from the transformations within the facility (solid waste disposal, air emissions, etc.) (see Figure 5.5c). The latter is customarily tracked separately, and sometimes the former as well, but efficiency and information gains are likely if these flows were to be integrated into the software management systems.

5.4   THE UTILITY OF METABOLIC ANALYSIS IN INDUSTRIAL ECOLOGY

The motivation for metabolic analysis in biology and industrial ecology is clearly different. The biological focus is on how substances are changed, what the products of the change may be, and what regulatory mechanisms direct the system. The industrial ecology

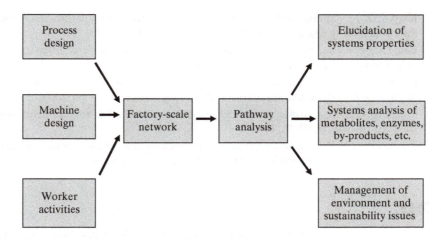

Figure 5.7

The linkage of industrial metabolic data, pathways, and applications. (Adapted from J.A. Papin, et al., Metabolic pathways in the post-genome era, *Trends in Biochemical Sciences, 28*, 240–248, 2003.)

focus is more toward mass flows in the system and understanding how a disruption to one part of the system ripples through the system as a whole. In both cases, however, increased knowledge is at the center of the activity, and physical laws constrain the possibilities.

The industrial metabolic analyses described in this chapter are capable of addressing concerns of the environment or of sustainability as well as manufacturing and efficiency. Industrial ecologists will thus incorporate flux analysis and follow by-product generation. In this way, they integrate the tools of biology and industrial management to serve larger societal goals.

The information flow in metabolic industrial analysis is sketched in Figure 5.7, where the metabolic information is seen to feed the industrial networks and thus to enable pathway analysis. The elucidation of systems properties then permits the active management of those aspects of industrial or societal activity that link to environmental or sustainability concerns.

The discussion in this chapter is centered on individual organisms—the squirrel or the factory. The approach can be scaled up, however: Some biologists study the metabolism of an ecosystem, and some industrial ecologists study the metabolism of a city. We will see aspects of metabolic analyses at different scales later in this book, in discussions of industrial ecosystems, material flow analysis, and urban ecology.

FURTHER READING

Brady, J.A., E.F. Monk, and B.J. Wagner, *Concepts in Enterprise Resource Planning,* Boston: Course Technology, 2001.

Brown, J.H., et al., Toward a metabolic theory of ecology, *Ecology, 84*, 1771–1789, 2004.

Davenport, T.H., Putting the enterprise into the enterprise system, *Harvard Business Review, 76*, 121–131, 1998.

Karp, P.D., M. Krummenacker, S. Paley, and J. Wagg, Integrated pathway-genome databases and their role in drug discovery, *Trends in Biotechnology, 17*, 275–281, 1999.

Lee, J.M., E.P. Gianchandani, and J.A. Papin, Flux balance analysis in the era of metabolomics, *Briefings in Bioinformatics, 7*, 140–150, 2006.

Löfving, E., A. Grimvall, and V. Palm, Data cubes and matrix formulae for convenient handling of physical flow data, *Journal of Industrial Ecology, 10* (1–2), 43–60, 2006.

Miller, J.G., and L.G. Sprague, Behind the growth in materials requirements planning, *Harvard Business Review, 53*, 83–91, 1975.

Papin, J.A., J.L. Reed, and B.O. Palsson, Hierarchical thinking in network biology: the unbiased modularization of biochemical networks, *Trends in Biochemical Sciences, 29*, 641–647, 2004.

EXERCISES

5.1 In Figure 5.4, identify the industrial metabolites, the industrial enzymes, and the pathway for the manufacture of the doorknob.

5.2 Construct the S matrix for the process illustrated in Figure 5.4.

5.3 Construct S matrices for the three optional processes of Figure 5.6.

5.4 Combine the S matrices of Exercise 5.3 into a composite S matrix for the manufacture of assembly A. Construct the P matrix.

5.5 In Figure 5.6b, if $v_3 = 0$, what is the pathway for the manufacture of assembly A?

CHAPTER 6

Technology and Risk

6.1 HISTORICAL PATTERNS IN TECHNOLOGICAL EVOLUTION

Although many people think of technology as physical artifacts, it is a far broader concept than that, particularly in the context of industrial ecology. It cannot be readily separated from the economic, cultural, and social context within which it evolves, nor can technology be separated from the natural systems with which it couples. Technology is the means by which humans and their societies interact with the physical, chemical, and biological world.

It was with the advent of agriculture that humans began to exert, through their technology, significant impact on their surroundings. Human migrations spread technology across broad areas of the globe, impacting many local ecosystems. Early civilizations also caused a noticeable increase in atmospheric carbon: For example, one jump resulted from the deforestation of Europe and North Africa in the eleventh through thirteenth centuries. Greenland ice deposits reflect copper production during the Sung Dynasty in ancient China circa 1000 BCE, and episodes of high lead concentrations in lake sediments in Sweden reflect Greek, Roman, and medieval European production of that metal many centuries earlier.

But the real changes in patterns of human, technological, and environmental interaction date from the industrial revolution, and from its concomitant demographic and economic shifts. People moved from agrarian communities to urban centers, and the economy shifted from agrarian activities to manufacturing. Global transportation and communication infrastructures dramatically increased economic activity, and the industrial revolution created the resource base for a significant population increase as well. The results can be seen in the growth in global GDP; if we take 1500 as a baseline, by 1820 GDP had almost tripled, by 1900 it was up by a factor of 8.2, by 1950 it was up by over 22 times, and by 1998 it had grown over 155 times. The result was a similar accelerating growth in human impacts on natural systems (Figure 6.1).

Regardless of the technology, there is a surprising regularity in technological evolution. At all scales, technology tends to exhibit the familiar logistic growth pattern: It begins in research, invention, and innovation; experiences exponential growth as it is introduced into the market; peaks at market saturation; and is usually replaced by a newer technology as the original becomes obsolete (Figure 6.2). This general pattern, albeit over different periods of time, characterizes electricity, color television, air conditioning, and computers, among many others (Figure 6.3).

It is also apparent that there are regularities in technological evolution at higher levels, particularly in the way that constellations of core technologies tend to define technology clusters. Table 6.1 presents one example of a technology cluster assessment of the industrial revolution, characterized not only by the major technologies constituting the cluster but also by the nature of the environmental impacts. Note the shifts in the geographic center of activity, which can be correlated with the movement of technology from the center to the periphery as indicated in the idealized technology life cycle of Figure 6.2. Another important regularity, already discussed in Chapter 1, is that the rate of innovation and technological change is accelerating, with the information-rich Fifth Wave taking only half the time of the water-powered First Wave (Figure 1.3).

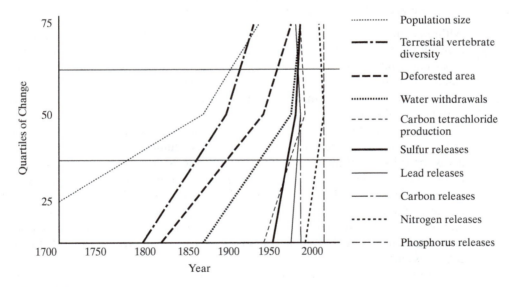

Figure 6.1

Trends in anthropogenic environmental transformation accelerated through the period of the industrial revolution. This figure shows the times required to achieve the second and third quartiles of change for a number of parameters: underlying these shifts are a number of technological transformations. (Adapted from R.W. Kates, B.L. Turner II, and W.C. Clark, The great transformation, in *The Earth as Transformed by Human Action*, Cambridge, UK: Cambridge University Press, pp. 1–17, 1990.)

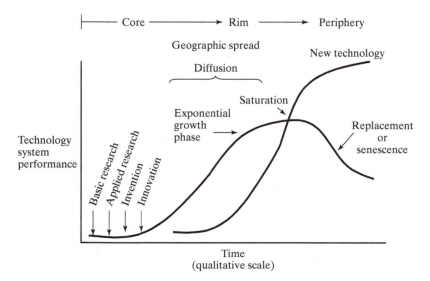

Figure 6.2

The idealized technology lifecycle. Note also the geographic spread of technology, which usually begins in the industrialized core, spreads to rim areas as it becomes more common and reaches the further peripheries only as it is already becoming obsolete in the core. Technology thus spreads in three dimensions: economically, geographically, and temporally. (Adapted from A. Grübler, *Technology and Global Change*, Cambridge, UK: Cambridge University Press, 1998.)

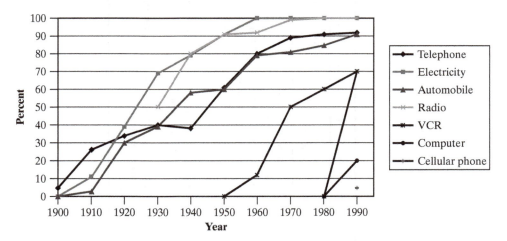

Figure 6.3

Consumer technology penetration rates, based on U.S. data. Such data, and research on the underlying dynamics, lie behind the idealized technology lifecycle model presented in Figure 6.2.

TABLE 6.1 Technology Clusters Characterizing the Industrial Revolution

Cluster	Major technology	Geographic center of activity	Date	Nature of technology	Nature of environmental impacts
Textiles	Cotton mills/coal and iron production	British midlands/ Landshire	1750–1820	Physical infrastructure (materials)	Significant but localized (e.g., British forests)
Steam	Steam engine (pumping to machinery to railroads)	Europe	1800–1870s	Enabling physical infrastructure (energy)	Diffuse; local air impacts
Heavy engineering	Steel and railways	Europe, U.S., Japan	1850–1940	Physical infrastructure (advanced energy and materials sectors)	Significant; less localized (use and disposal throughout developed regions)
Mass production and consumption	Internal combustion engine, automobile	Europe, U.S., Japan	1920s–present	Application of physical infrastructure, mass production	Important contributor to global impacts (scale issues)
Information	Electronics, services and biotechnology	U.S., Pacific Rim	1990s–present	Development of information (nonphysical infrastructure)	Reduction in environmental impact per unit quality of life

Is the history of technology a useful guide to the technology of the future? This is a difficult question to answer, because there are important aspects of technological evolution for which we do not as yet have good explanations. Among the most important are

1. How rapidly do fundamental technologies evolve, and what are the relevant constraints?

2. Are different paths of technological evolution equivalent, or are there ones that are more likely to achieve desired results? What determines the paths that the evolution of a particular technology is likely to take? In particular, can deliberate policy or regulatory decisions affect paths of technological evolution in desirable ways, or are the internal dynamics of such systems resistant to efforts to control them? Do the answers to these questions differ depending on how fundamental a technology is?

3. What are the differential distributions of risks, costs, and benefits of technological evolution across sectors, social groups, and nations, and who is empowered to decide whether technologies proceed or not?

4. Technology is generally the province of private firms. What should the relationship be between private firms, interested stakeholders, the public, and governmental entities, and which groups are responsible for which elements of technological evolution?

This last question is of considerable interest in such ethically and culturally sensitive areas as biotechnology, drug development and marketing, and the information industry, where the potential for a "digital divide" between the information rich and the information poor is a concern. The combination of questions three and four raises another issue: that of risk, and how to think about it, especially in a complex world where not acting is frequently as risky as acting, and where risks, as a form of cost, frequently fall on those who are not represented in the political debates. The future of technology is thus inherently bound up with its risks, real or imagined, and with how those risks are assessed and evaluated.

6.2   APPROACHES TO RISK

Risk, that is, the probability of suffering harm from a hazard, is a reflection of a problematic world in which the future cannot be known with certainty. In environmental circles, the hazards are usually defined as impacts to human health or the environment, but in a broader social sense they include economic, social, cultural, and psychological impacts as well. Sustainable engineering and industrial ecology require consideration of both environmental and sociocultural risk, although both fields reflect to some extent their history, which is primarily environmental.

Risk has two dimensions: objective and subjective. The objective dimension is quantitative and is frequently captured through a series of algorithmic methodologies: It may involve the engineering of complex systems, evaluating the toxicology of a new substance, or assessing the potential economic implications of a course of action.

TABLE 6.2 Annual Mortality Rate Associated with Certain Occurrences and Activities in The Netherlands

Activity/Occurrence	Annual mortality rate	Lifetime mortality rate
Drowning as a result of dike collapse	1×10^{-7} (1 in 10 million)	1 in 133,000
Bee sting	2×10^{-7} (1 in 5.5 million)	1 in 73,000
Being struck by lightning	5×10^{-7} (1 in 2 million)	1 in 27,000
Flying	1×10^{-6} (1 in 814,000)	1 in 11,000
Walking	2×10^{-5} (1 in 54,000)	1 in 720
Cycling	4×10^{-5} (1 in 26,000)	1 in 350
Driving a car	2×10^{-4} (1 in 5,700)	1 in 76
Riding a moped	2×10^{-4} (1 in 5,000)	1 in 67
Riding a motorcycle	1×10^{-3} (1 in 1,000)	1 in 13
Smoking cigarettes (one pack per day)	5×10^{-3} (1 in 200)	1 in 3

Source: *Ministry of Housing, Physical Planning, and Environment*, National Environmental Policy Plan: Premises for Risk Management, p. 7, The Hague, The Netherlands, 1991.

Engineers and scientists are usually trained to work with quantitative risk. The subjective is not capable of being reduced to numbers, but in practice often outweighs more objective approaches. Most stakeholder groups tend to evaluate risk in this more subjective way. It is important to understand that neither perspective is necessarily "right": They both provide different ways of thinking about risk, and both must be respected by the industrial ecologist.

Many risks can be quantified quite easily. For example, Table 6.2 presents annual and lifetime mortality rates associated with common activities in The Netherlands. Note that the activities involving the highest annual mortality rates, such as smoking or driving various vehicles, are all undertaken voluntarily, an important subjective dimension of risk evaluation by individuals. In connection with the information in Table 6.2, it is of interest that standards for environmental cleanups in the United States are frequently based on a 10^{-6} lifetime risk, or one additional fatality in a million over a lifetime. Thus, based on the data in this table, walking, cycling, or driving a car are all far more risky (have a higher probability of resulting in mortality) than the standards the United States chooses to impose on cleanups of contaminated sites. This is a significant efficiency consideration when it is recognized that such standards are the primary determinant of cost for virtually any cleanup.

But with risk evaluations, as with economic analyses, quantitative data must be treated with caution. For example, it is well known that air travel is safer than car travel—or is it? As Table 6.3 shows, this is true on a per kilometer basis but not on a per journey basis—and on a per hour basis, the two are equivalent. Buses, on the other hand, are less risky than any other mode of transportation on either a per journey or a per hour basis, and virtually as safe as air travel on a per kilometer basis.

Professionals tend to think of risk as objective, but risks are regarded by most people as intensely subjective. This difference in approach explains studies that have asked experts and informed laypeople to rank risks of technology, broadly defined,

TABLE 6.3   Using Fatality Rates from Great Britain

Mode of transport	Per 100 m passenger		
	Journeys	Hours	Kilometers
Motorcycle	100	300	9.7
Air	55	15	0.03
Bicycle	12	60	4.3
Foot	5.1	20	5.3
Car	4.5	15	0.4
Van	2.7	6.6	0.2
Rail	2.7	4.8	0.1
Bus or Coach	0.3	0.1	0.04

Source: Based on *The Economist*, January 11, 1997, p. 57.

as shown in Table 6.4. This difference in perception appears to arise because the public integrates a number of subjective factors into its determination of risk, particularly:

1. The extent to which the risk appears to be controllable by the population at risk (note in Table 6.4 that college students, who ride bicycles a lot, and thus feel that they are in control on them, rank that risk much lower than the experts);

2. Whether the risk is feared, even dreaded (the focus of much risk assessment on human cancer arises, in part, because of the dreaded nature of the disease);

3. The extent to which the risk is imposed rather than voluntarily assumed (note from Table 6.4 that the League of Women Voters rank contraceptives as much less risky than the experts, probably reflecting not only voluntary assumption of risk, but also familiarity with contraceptives, and quite possibly a personal understanding of the trade-offs in risk involved in using contraceptives as well);

4. The extent to which the risk is easily observable, especially by the at-risk population, and, if observable, is manageable with current technologies;

5. Whether the victims are especially sympathetic and vulnerable (particularly children);

6. The extent to which the risk is new and unfamiliar, and unquantifiable or previously unknown to science;

7. The extent to which the victims are identifiable as individuals, as opposed to statistical groupings (media coverage of air crashes, for example, tends to focus on individual victims, particularly children, which may explain why the public perception of the risk associated with them differs significantly from the expert assessment);

8. The extent and type of media attention (sensationalist as opposed to factual reporting);

TABLE 6.4 Perception of Risk From Most Risky (1) to Least Risky (30) by Three Target Audiences: Educated and Politically Involved Female Citizens; College Students; and Experts

Activity or technology	League of Women Voters	College students	Experts
Nuclear power	1	1	20
Motor vehicles	2	5	1
Handguns	3	2	4
Smoking	4	3	2
Motorcycles	5	6	6
Alcoholic beverages	6	7	3
General (private aviation)	7	15	12
Police work	8	8	17
Pesticides	9	4	8
Surgery	10	11	5
Fire fighting	11	10	18
Large construction	12	14	13
Hunting	13	18	23
Spray cans	14	13	26
Mountain climbing	15	22	29
Bicycles	16	24	15
Commercial aviation	17	16	26
Electric power (nonnuclear)	18	19	9
Swimming	19	30	10
Contraceptives	20	9	11
Skiing	21	25	30
X-rays	22	17	7
High school and college football	23	26	27
Railroads	24	23	19
Food preservatives	25	12	14
Food coloring	26	20	21
Power mowers	27	28	28
Prescription antibiotics	28	21	24
Home appliances	29	27	22
Vaccinations	30	29	25

Source: From Slovic, P., Perception of risk, *Science, 236*, 280–285, 1987.

9. The perceived equity of the distribution of the risk among different groups (the public is more likely to be concerned where the costs, benefits, and risk of a particular activity are disproportionately allocated among groups—"distributed justice"—than the expert risk assessor, interested primarily in cumulative increases or decreases in absolute risk); and,

10. The type of risk involved (specific risks associated with technology systems can usually be fairly easily quantified; environmental risks are more subjective, but still can be quantified in many cases; and sociocultural risks are frequently difficult to quantify at all—in fact, in the latter case, people may even disagree on whether a particular aspect is a cost or a benefit).

There are three steps in the sequence of risk analysis: assessment, communication, and management. The first is largely objective and attempts to use statistical and laboratory data to quantify the risk, as in Table 6.2. The second step involves the ways in which the results of the assessment are communicated to interested parties. The third deals with the actions taken by organizations or governments to minimize the risk, a step which usually integrated subjective perspectives. We discuss each of these in turn.

6.3   RISK ASSESSMENT

The type of risk assessment commonly used in environmental regulation is highly quantitative, typically focuses on health, especially carcinogenic risks to humans, and generally consists of five stages: (1) hazard identification; (2) delivered dose; (3) probability of an undesirable effect as a result of the delivered dose; (4) determination of the exposed population; and (5) characterization. The last is the calculation of the total risk impact: The number of individuals exposed multiplied by the probability that the delivered dose will cause the undesirable effect. This may be expressed mathematically as:

$$I = NP(d) \tag{6.1}$$

where I is the total risk impact, N is the number of individuals exposed, and $P(d)$ is the probability, P, that the indicated dose, d, will cause the effect.

Determining the probability of an undesirable effect at the delivered dose is the great challenge of risk assessment, particularly where one is dealing with human exposures to low levels of chemicals whose influence may be obvious only years later. Testing for such impacts as carcinogenesis or mutagenesis is not done on humans nor at typical delivered doses, but on laboratory animals and at doses high enough to produce measurable effects in relatively short times. The results must be evaluated to determine whether they are realistic surrogates for human response and then extrapolated to typical delivered dose levels. The methods used to extrapolate the results tend to be problematic, yet the choice of method can determine the outcome of the risk assessment. Figure 6.4, for example, compares four different methods of extrapolating animal data for trichloroethylene ingestion. If one anticipates drinking water concentrations of 10 µg/l and wishes to hold lifetime risk below 10^{-6}, the W and L extrapolations indicate a problem while the M and P extrapolations do not. For many chemicals, the probability determination is more certain than in this example, but in general quantification becomes less reliable as probability becomes smaller.

Comprehensive risk assessment (CRA) models are based on the recognition that there are qualitatively different categories of risk associated with environmental concerns. Most models use a taxonomy adopted by the Government of The Netherlands that establishes three categories of risk. The first concerns damage to biological systems in general and humans in particular. The second category includes risks that aesthetically degrade the environment but may or may not damage biological systems. The final category is risks involving damage to fundamental planetary systems.

This risk categorization can be used to derive an illustrative CRA methodology. To do so, first consider the generic risk equation (6.1). For the first category, damage to biological systems, the equation can be written as:

$$B = \beta NP(d_i) \tag{6.2}$$

Figure 6.4

Bioassay data for ingestion of trichloroethylene in water (the asterisks in the upper right corner) and extrapolations by the logit (L), multistage (M), log probit (P), and Weibull (W) models. (Adapted with permission from C.R. Cothern, Uncertainties in quantitative risk assessment — Two examples: Trichloroethylene and radon in drinking water, in Cothern, C.R., M.A. Mehlman, and W.L. Marcus, Eds., *Risk Assessment and Risk Management of Industrial and Environmental Chemicals*, pp. 159–180. Copyright 1988 by Princeton Scientific Publishing Company.)

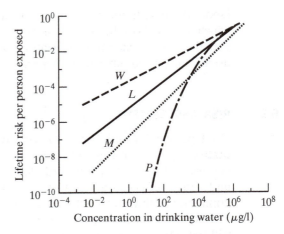

where B is the comprehensive biological risk, i refers to the "i"th source of impact, and β is a weighting factor, agreed to by social consensus, reflecting both the objective and subjective value placed on biological systems by society. If desired, this term can be broken in two, reflecting different weighting for human and nonhuman systems.

Risk associated with aesthetic degradation can be expressed in a similar manner as follows:

$$A = \alpha N P(d_i) \tag{6.3}$$

where A is the aesthetic risk, N is the number of people affected by aesthetic degradation (including those who may not be physically present, but who value the impacted environment), P is the probability of an effect for the dose d of the "i"th source of impact, and α is again a weighting factor, reflecting societal consensus. It is likely that α would be less than β, reflecting the fact that aesthetic degradation is felt by most people to be less serious than damage to biological or human systems.

As with other categories of environmental risk, precise weighting factors and dose–response relationships for damage to planetary systems have not been established, though the direction is clear. Weighting factors should be high since the effects in this category potentially constrain the sustainability of the entire planet. It is necessary to integrate global impacts over time since they may extend for several generations. Thus,

$$G = \gamma \int_{t_0}^{t_1} N(t) P(d_{i,t}) dt \tag{6.4}$$

where "G" is the global risk, γ is the weighting factor, and the integration is performed from the present time "t_0" through the lifetime "t_1" of the substance or insult in question. Note that the dose and the affected population are time-dependent.

From these three equations, the CRA is given by:

$$CRA = B + A + G \tag{6.5}$$

where the comprehensive risk equals the sum of the biological (B), aesthetic (A), and global (G) impacts for any particular subject of assessment.

Note one very important caveat to this process: it assumes that there is underlying social consensus on risk definition. In many cases, especially involving environmental issues, this is a reasonable assumption. For example, people may value species differently, but most agree with the proposition that species extinction is a bad thing. With social issues, however, such a consensus may not exist. For example, an industrial ecology study of solders could not conclude unequivocally that the social implications of stimulating additional mining activity were understood by society to be either good or bad: Different groups have different opinions about that.

6.4 RISK COMMUNICATION

Risk communication follows risk assessment, and it is the stage during which the risk assessment results are made known to interested communities, organizations, and individuals. Risk assessment, in principle at least, is an objective exercise in the scientific interpretation of data. Risk communication may involve corporations, governments, and the news media. Because of these disparate actors, accurate communication of research results can be challenging, especially in emotionally charged situations.

A classic example of failure in risk communication involved the Brent Spar oil storage platform in the North Sea. When this platform became obsolete in the late 1980s, Royal Dutch Shell (co-owner of the platform with Exxon) hired environmental experts to decide how best to decommission the Brent Spar. The options were

- Disassembly or disposal on land
- Sinking in its North Sea location
- Disassembly at the North Sea location
- Deep sea disposal

Shell and its experts chose the final option on environmental, employee safety, and economic grounds. The announcement, however, did not effectively discuss the detailed evaluation that had taken place, nor the environmental disadvantages of the rejected options. This was particularly problematic because some stakeholders had very strong subjective opinions about any disposal activity involving the ocean. As a result, the choice was widely viewed as corporate disregard for the environment. Much business and goodwill were lost as a result, and Shell eventually bowed to public pressure and acquiesced to disassembly on land. To some extent, then, this example also indicates a failure to understand not just the quantitative, but the normative, aspect of risk.

In any risk communication, the potential participants include the originator or discoverer of the risk (frequently a corporation), expert analysts, various special interest groups, and the public at large. It is often difficult to find common ground, but experience shows that engaging all likely interested parties as early as possible, coming to agreement on the validity of the risk assessment, and considering everyone's goals and motivations as much as possible has a much higher probability of communicating risks with suitable accuracy and limited acrimony than is the case if these steps are ignored.

6.5 RISK MANAGEMENT

The final step in a structured basis for risk-related regulatory decisions or policy formulation is risk management. At this stage, the risk has been quantitatively evaluated (risk assessment) and interested parties have been informed (risk communication). The challenge then becomes to decide whether any policy or regulatory action is desirable. These decisions are inherently combinations of the scientific, the economic, and the sociological. Moreover, because they may be intensely political, they may also be the point where intensely subjective perspectives are introduced into the process.

The task is conceptually challenging but potentially achievable. It may be accomplished, for example, by defining a process similar to the CRA for economic, cultural, and other impacts, assuming one is able and willing to express as many results as possible in comparable units (e.g., monetary ones). The quantification of the impacts inherent in the equations can be derived from cost–benefit or other kinds of economic analysis.

For economic impact, for example, one can write:

$$E = \varepsilon NP(d_i) \tag{6.6}$$

where E is economic impact, and ε is a weighting factor reflecting the fact that the monetary value of economic impacts may be subjectively assessed by the public as more or less than one when compared with other values. Also note that economic impacts occur over time; this formulation assumes present discounted value is used, or, in other words, that the integration over time has been performed in the underlying assessment.

A practical approach to risk management has been proposed by Granger Morgan of Carnegie Mellon University. Morgan defines his strategy as follows:

> No individual shall be exposed to a lifetime excess probability of death from this hazard to greater than X. Whether additional resources should be spent to reduce the risks from this hazard to people whose lifetime probability of death falls below X should be determined by a careful benefit-cost calculation.

The method is illustrated by Figure 6.5. It takes the results of the risk assessment, shown as the curve in the figure, and chooses a specific level to be the maximum acceptable risk (MAR). This example is for human mortality risk, and the MAR is set to that for the probability of death by lightning. Any mortality risk above that level must be abated. Possible risk abatement below that level could be determined by benefit–cost analysis. In Morgan's method, risk management is pursued up to the point where marginal benefits equal marginal costs.

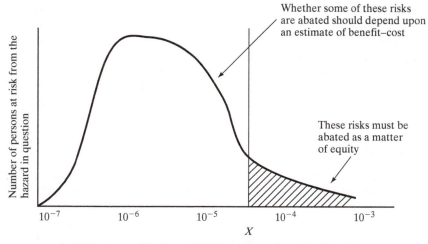

Whether some of these risks are abated should depend upon an estimate of benefit–cost

These risks must be abated as a matter of equity

X

Individual excess lifetime probability of death due to the hazard

Figure 6.5

A possible risk management strategy. X is the maximum acceptable excess lifetime risk from the hazard for any individual in society. In this example, X is set to 3×10^{-5}. (Adapted from M.G. Morgan, Risk management should be about efficiency and equity, *Environmental Science & Technology, 34*, 32A–34A, 2000.)

Not all risk management decisions are, or can be, made by this combination of science and economics. Social and cultural values are also commonly included, either implicitly or explicitly. In the latter case, the expression

$$C = \chi NP(d_i) \qquad (6.7)$$

can be defined, though no general agreement exists on how to determine the cultural impact C and the appropriate weighting factor χ.

A Comprehensive Policy Support Assessment (CPSA) that integrates the CRA with economic and cultural considerations can in principle be defined for each option under consideration. For the jth policy option, for example, one would obtain:

$$\text{CPSA}_j = B_j + A_j + Gj + E_j + C_j \qquad (6.8)$$

or, equivalently:

$$\text{CPSA}_j = \text{CRA}_j + E_j + C_j \qquad (6.9)$$

The above methodology implies a quantitative process. In many cases, this may be unrealistic: Resources may not permit such detailed analysis; options may be difficult to define with sufficient specificity to support a quantitative assessment; or data may be so sparse or uncertain as to render quantification misleading in itself. More fundamentally, there may be values and ideologies involved that defy easy quantification: To many people, for example, driving a species to extinction is not a quantifiable risk, but involves transcendental values. Similarly, many social and cultural effects of decisions about industrial systems or technologies are impossible to quantify in ways that relevant stakeholders accept as valid. Nonetheless, we believe that if properly done, such an approach has value in that it ensures that the full range of implications of the decision under consideration for the system is being considered at some point in the assessment process. It must be used, however, to help explicate the decision, rather than to stifle dissent and those who disagree with the basic qualitative assumptions behind the analysis.

FURTHER READING

Cothern, C.R., M.A. Mehlman, and W.L. Marcus, Eds., *Risk Assessment and Risk Management of Industrial and Environmental Chemicals*, Princeton, Princeton Scientific Publishing Company, 1988.

Finkel, A.M., and D. Golding, Eds., *Worst Things First? The Debate over Risk-Based National Environmental Policies*, Washington, DC: Resources for the Future, 1994.

Gold, L.S., T.H. Slone, B.R. Stern, N.B. Manley, and B.N. Ames, Rodent carcinogens: setting priorities, *Science, 258*, 261–265 (1992).

Grubler, A., Time for a change: On the patterns of diffusion of innovation, in *Technological Trajectories and the Human Environment,* J. Ausubel, Ed., Washington, DC: National Academy Press, pp. 14–32, 1997.

Grubler, A., *Technology and Global Change*, Cambridge, UK: Cambridge University Press, 1998.

Lofstedt, R.E., and O. Renn, The Brent Spar controversy: An example of risk communication gone wrong, *Risk Analysis, 17*, 131–136, 1997.

Masters, G.M., *Introduction to Environmental Engineering and Science,* Chapter 5, Englewood Cliffs, NJ: Prentice Hall, 1991.

van der Voet, E., L. Van Oers, and I. Nikolic, Dematerialization: Not just a matter of weight, *Journal of Industrial Ecology, 8* (4), 121–137, 2005.

EXERCISES

6.1 Using the list of risks in this chapter, create your own risk prioritization list. Explain and defend your choices, in each case differentiating between scientific and technical assessments on the one hand and values and ethical judgments on the other.

6.2 The average concentration of trichloroethylene in the drinking water of a nearby town is 100 μg/l. Using Figure 6.4, estimate the lifetime risk per person exposed for each of the four data extrapolations. What is the ratio of the highest to lowest of your estimates? How would you communicate this result to the town's citizens?

6.3 Rice is the staple crop in much of Asia, so its lack of Vitamin A has significant negative health implications for the poor in that region of the world, especially in terms of child mortality. A number of researchers and companies have developed a rice cultivar that has been genetically engineered to contain Vitamin A and are giving up certain intellectual property rights to allow such rice to be grown in those areas. Some major environmental groups oppose any use of the modified rice regardless of health or mortality benefits, claiming that it is an attempt by corporations to make genetically modified organisms (GMOs), which they equate to "playing God," politically acceptable. As a member of the local government, what do you advocate and why?

6.4 You are an environmental regulator in a developed country faced with a decision as to whether to permit mining in a wetlands wilderness area that probably contains threatened species. You are required to perform an assessment of the desirability of this activity.

   (a) What should your option set be?

   (b) Perform a CPSA for each option (this can be qualitative). What stakeholders should you involve in order to support your CPSAs? What are the major issues that can be resolved by gathering data, and what issues involve value judgments?

   (c) Are there any issues you feel are important, but you cannot fit into a CPSA? Assuming that there are, how would you ensure they are considered as part of the regulatory process?

CHAPTER 7

The Social Dimensions of Industrial Ecology

7.1 FRAMING INDUSTRIAL ECOLOGY AND SUSTAINABLE ENGINEERING WITHIN SOCIETY

The relationship between industrial ecology, sustainable engineering, and sustainability inevitably reflects the history of each of the concepts. In particular, industrial ecology originated from an environmental perspective, and many of the methodologies that were developed displayed a strong environmental focus, for example, tools for life cycle assessment and design for environment tools primarily dealt with environmental considerations and only indirectly and imperfectly with social or cultural concerns. We are now seeing those tools evolving to more directly address the full panoply of sustainability issues. This transition is critical for industrial ecology as a field, because most of the systems that industrial ecologists study have important social and cultural dimensions, and these cannot be ignored if the resulting analysis is to be robust.

The social and cultural dimensions of technology systems have a strong scale dependence. Consider Figure 7.1, showing the concentric levels in a typical industrial ecology system. At the lowest level, that of the component, the environmentally related decisions are those of the design engineer, and the interaction with society and culture is negligible. At successively higher levels, however, the degree of interaction increases. A societal preference for privacy and status, for example, is reflected in sprawling communities far from jobs and shopping. A consequence of that preference is an incentive to produce larger, more powerful, and more comfortable vehicles than would be the case in a more condensed urban pattern, and to substitute personal vehicles for mass transportation, because the latter requires high population density to be effective. Technology and society are thus inherently related, and

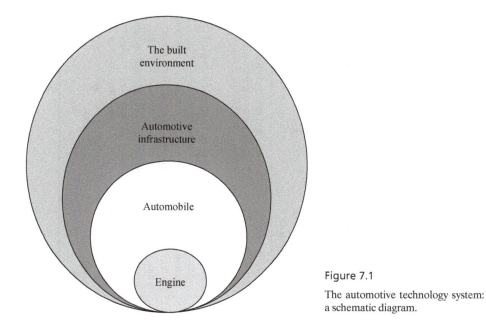

Figure 7.1

The automotive technology system: a schematic diagram.

because environmental impacts result from those relationships the environment is closely linked as well.

7.2 CULTURAL CONSTRUCTS AND TEMPORAL SCALES

Stability is generally assumed to be a desirable property of social and cultural frameworks. Most human thinking and planning relates to the short term, in which stability is generally a useful assumption. Over long timescales, however, this assumption may not be valid and may result in decisions that cause significant environmental, social, or economic harm. Table 7.1 emphasizes that system structures become much more flexible over time, with additional elements becoming variable the further out in time one goes. In the short term, there are many systems components that can legitimately be assumed to be fixed: major industrial systems such as transportation or energy networks, economic structures such as private companies, or many government policies. In the medium term, more elements become flexible; whereas a major product redesign might not be feasible in a five-year framework, it may well be possible in a ten-year framework. Fundamental shifts in consumer demand—away from cigarettes or toward larger automobiles—typically occur in such a timeframe. In the longer term, major technology systems and their acceptance by the public can change completely—perhaps from reliance on fossil fuel to a nuclear-based hydrogen economy, for example. This is one reason why predictions based on extensions of current trends are almost always wrong.

The changeable nature of almost everything in the long term illustrates the importance of what sociologists and psychologists call "cultural constructs." These are ideas that are invented for certain purposes within a society but soon begin to seem-unquestionable, especially as they are embedded in ideological structures. For example,

TABLE 7.1 Social System Structure over Different Timescales

Timescale	Endogenous factors	Exogenous factors	Principal implementation mechanism	Principal R&D component	Integration of natural and artifactual systems
Short term (*ca.* 5 years)	Incremental technology evolution within existing major technology systems	Whims, passions of the moment	Policy	Short-term industrial ecology R&D (e.g., DfES)	Experimental stage involving small systems (e.g., bioreactors, drug production in genetically engineered sheep)
Medium term (*ca.* 5–10 years)	Evolution of product and process technology systems, marginal cultural change	Population level, significant cultural change	Changes in legal structures and disciplinary assumptions	Industrial ecology infrastructure (e.g., environmentally preferable materials database)	Partial integration of biological and engineered systems (e.g., commercial energy from biomass; engineered wetlands for flood control and waste processing)
Long term (*ca.* 10–100 years)	Significant evolution of major technology systems; link between quality of life and material consumption; most aspects of culture	Almost nothing	Metrics, changes in fundamental conservative cultural systems	Industrial ecology systems (e.g., resource and energy maps of communities and regions)	Management of integrated regional and global systems (e.g., water cycles in Yellow River watershed); Earth systems engineering and management

some people are concerned that current environmental policies espoused by developed countries represent a powerful but unrecognized drive to favor Earth's present peoples over those who will populate the planet in the future. In this regard, the global climate change negotiations have been criticized as seeking to stabilize current climatic conditions, thereby removing an important source of variability that has affected the evolution of life on this planet. By the same token, opposition to biotechnology implicitly grants precedence to current genetic structures over what may be evolved in the future. This ideological and ethical statement is unconscious for most participants in these dialogs.

Industrial ecologists have cultural constructs as well. Perhaps the most obvious one is "sustainable development," popularized in the 1987 book *Our Common Future*. As successful cultural constructs tend to do, "sustainable development" and the looser term of "sustainability" have become over the past 15 years a major policy goal for many. In the process, the contingency of the term—although quite explicit in its history—has vanished for many people. The dangers this presents, while common to many cultural constructs, are apparent when one places sustainable development in a framework of basic political values; it then becomes apparent that the term represents a fairly specific culture—basically Northern European social democratic traditions. Conceptualizations resulting from these different approaches form the basis of Figure 7.2.

As William Cronon has pointed out in his book *Uncommon Ground*, the prevalence of cultural constructs is not limited to "sustainable development," but extends

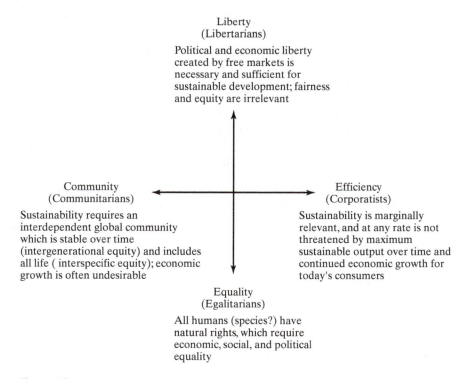

Figure 7.2

Belief systems regarding the economy, individual liberty, the environment, and sustainability.

to many other ideas that are seen as absolutes: "wilderness," "nature," and, indeed, "environment" itself. That a number of the terms used in industrial ecology are cultural constructs is not necessarily good or bad, since cultural constructs are necessary for intellectual order, but it does suggest that a reasonable self-awareness and respect for conflicting values is appropriate for the industrial ecologist or sustainable engineer.

7.3   SOCIAL ECOLOGY

Social ecology is a branch of the social sciences and of industrial ecology. It can be defined as the study of societies and their evolutionary behavior from the perspective of societal use of energy and materials. This approach enlarges traditional definitions of metabolism to treat human communities (often, but not necessarily, countries) and their material and energy flows as constituting a social metabolism. Thus, just as the metabolic analysis of a bird can be expanded to include the nest it builds, so can that of a society address aquaducts and railroads.

Throughout almost all of human existence, agrarian lifestyles have been the norm. Those who live in countries that are technologically advanced tend not to realize that this is essentially still the case for two-thirds of Earth's people. We can expect, however, that the next few decades will witness an unprecedented level of agrarian to industrial transition. What might be the implications of this process?

Some clues are given by Table 7.2, in which properties of agrarian and industrial lifestyles are contrasted. For example, industrial regions customarily feature population densities five to eight times higher than the agrarian norm. The relative use of energy and materials carries a similar multiplier. Further, because the using population is more geographically condensed, energy and material use per unit of land area is ten to twenty times higher. Energy sources also undergo a marked transition—from an almost total dependence on biomass in the agrarian regions to a strong emphasis on fossil fuels in the industrial.

Many opportunities arise as a consequence of these differences. One is that fossil fuel energy permits mechanized farming, which frees much of the labor force for other employment (Figure 7.3). Another is that the increased population density enables public transportation, better schooling, and improved health care. Liabilities come

TABLE 7.2   Attributes of Agrarian and Industrial Economies

Attribute	Agrarian regions	Industrial regions
Population density (cap/km^2)	<40	100–300
Energy use per cap [GJ/(cap-yr)]	50–70	150–400
Energy use per unit area [GJ/(ha-yr)]	20–30	200–600
Biomass (%)	95–100	10–30
Fossil fuels (%)	0–5	60–80
Other (%)	0–5	0–20
Material use per cap [t/(cap-yr)]	2–5	15–25
Material use per unit area [t/(ha-yr)]	1–2	20–50

Abstracted from M. Fischer-Kowalski, H. Haberl, and F. Krausmann, Conclusions: Likely and unlikely pasts, possible and impossible futures, in *Socioecological Transitions and Global Change,* M. Fischer-Kowalski and H. Haberl, Eds., Cheltenham, UK: Edward Elgar, pp. 223–255, 2007.

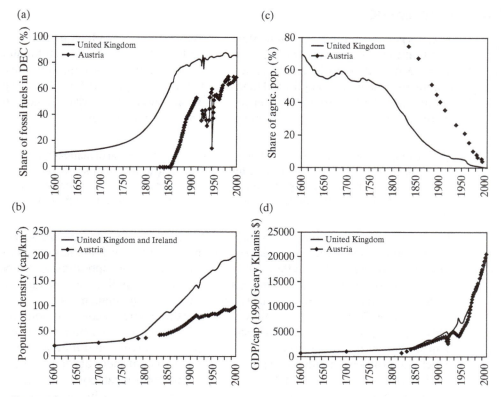

Figure 7.3

A comparison of the historical transitions of the United Kingdom and Austria. (a) Share of fossil fuels in primary energy use; (b) Population density; (c) Share of the population engaged in agriculture; (d) Per capita income. (Reproduced with permission from M. Fischer-Kowalski, H. Haberl, and F. Krausmann, Conclusions: Likely and unlikely pasts, possible and impossible futures, in *Socioecological Transitions and Global Change*, M. Fischer-Kowalski and H. Haberl, Eds., Cheltenham, UK: Edward Elgar, pp. 223–255, 2007.)

along with the transition as well, however: a changed social structure and spatial concentration of pollution sources, to name only two.

Fischer-Kowalski has used a tribe of Brazilian Indians to illustrate the magnitude of the transition. Originally completely self-sufficient, the Indians were drawn into trade relationships with international business enterprises, becoming reliant on others for the manufactured goods on which they had come to depend. From the standpoint on resources and energy, this has resulted in the substitution of a *trade-based* metabolism for a *native* metabolism, and of a *regional–global* metabolism for a *local* one.

The metabolic structure of societies can be studied at various spatial levels. At the local level, studies address villages and small islands (see Grünbühel et al., 2003; Singh and Grünbühel, 2003); such studies tend to focus on societies in the midst of the agrarian–industrial transition. At the country level (see Chapter 18), societal use of resources and energy can be compared among countries differing in their location along the transitional path. Finally, in cities (see Chapter 21), the comprehensive metabolism of modern industrial societies can be examined in detail.

7.4   CONSUMPTION

There is general agreement among those who have studied the phenomenon of consumption that today's intensive consumption in highly developed nations is ultimately unsustainable. As with many issues, the concept and implications of consumption turn out when examined to be more complicated than might at first be supposed. First, let us explore two definitions, adapted from Paul Stern of the U.S. National Research Council and Edgar Hertwich of the Norwegian University of Science and Technology:

> *Consumption* is the human and human-induced transformation of materials and energy. *Sustainable consumption* is a pattern of human activities that satisfy basic needs, offer humans the freedom to develop their potential and are replicable across the planet without significantly compromising Earth's ongoing natural processes.

The key phrase in these definitions, and one that combines the physical and social sciences, is "offer humans the freedom to develop their potential," which implies access to modern electronics, travel, and a plethora of other possessions and opportunities embedded in appropriate social and cultural patterns. How should we think about the juxtaposition of consumption, human potential, social systems, and environmental impacts?

Thomas Princen suggests that we consider three types of consumption: *background consumption* (that which satisfies basic needs), *overconsumption* (that which undermines a species' own life support system for which choice exists), and *misconsumption* (that which undermines an individual's own well-being even if there are no aggregate effects on the population or species). Princen's overconsumption category is relevant to industrial ecology in two ways: (1) if the overconsumption results in serious environmental impact, or (2) if the overconsumption threatens the long-term sustainability of resources. In this context, it is useful to recall the IPAT equation from Chapter 1. Population (P) is an important driver of impact (I) on the environment and sustainability, but not part of industrial ecology. Technology (T) is one of the principal subjects of this book. The affluence (A) term remains and is directly linked to background consumption and overconsumption. There is perhaps no other place where the physical and social sciences meet so directly.

Because not all consumption results in significant environmental impacts, it is important to identify the problematic forms of consumption. This is not straightforward, however, and is often a matter of degree rather than kind—the purchase of a large vehicle instead of a smaller one rather than no vehicle purchase, for example. It is also a matter of scale: a single purchase of a large vehicle has virtually no impact on environmental systems; a large number of such decisions aggregated across society can have substantial resource and emission implications. A key question is why such behavior occurs: At an individual level, decisions may be related to peer pressure, age, gender, or economics; at an aggregate level, they may be related to marketing, peer group choices, and availability of options. A related question is how this behavior can be changed: It may be changed by increased information, social change, regulations, or other factors. Technology plays an important role in determining problematic consumption, but the human element must then choose whether or not to follow the environmentally preferable path.

A consumption issue that often frustrates planners is the "rebound effect," in which a consumption decision that seems beneficial stimulates behavior that neutralizes the decision. For example, a choice is made to replace a vehicle with poor fuel efficiency by one much more fuel efficient. The buyer then discovers his fuel expense has decreased and is able to afford to drive more than he did before.

Other behavioral factors relate to societal change. As household sizes decrease, there is need for more appliances, and more housing stock. As people choose to eat in restaurants with greater frequency, transportation and food waste may increase. As more women enter the labor force, a variety of new consumptive behaviors are stimulated, in part because additional disposable income is usually generated.

In the final judgment, consumption is often intensely personal and conflated with ideas of image, responsibility, and morality, but relates as well to public goods and actions such as extensive highway construction. It demands attention, yet is problematic in the ways it can be addressed. Levels of consumption drive economies in obvious ways and often relate inversely to the environment and sustainability. Understanding consumptive behavior and changing it where desirable remain major challenges for the planet's future.

7.5   GOVERNMENT AND GOVERNANCE

Governance is the process by which societies implement their values, usually through decision making, allocating and monitoring power, and verifying performance, to address issues such as the common-pool resource limitations discussed in Chapter 1. In today's world, the most important governance mechanism remains the nation-state. Increasingly, however, the nation-state is supplemented or replaced by a number of other relatively independent governance mechanisms, including private firms, nongovernmental organizations, political groups, and international communities, many of them increasingly virtual (Internet-based, for example).

Nation-states are particularly relevant for two reasons. First, they are the only entities which current international law recognizes as sovereign, and thus able to represent their people in international forums. Moreover, nation-states have authority over any activities occurring within their borders, and they are often the primary jurisdiction within which remediation and compliance activities are regulated. However, many sustainability and environmental issues are beyond the borders of any single nation-state: acid precipitation, watershed pollution, ozone depletion, global climate change, loss of biodiversity and habitat. There is thus a mismatch in scale between the political bodies with the most authority and legitimacy and the sustainability and environmental perturbations with which they must deal. The global scope of many of the anthropogenically perturbed natural support systems, in combination with an unwieldy international law system, raises questions about the appropriate venue for addressing environmental issues and adds to increasing devolution of nation-state obligations to international organizations, transnational corporations, and political subunits.

Existing international environmental treaties and agreements are negotiated, approved, and enforced at the nation-state level. Reflecting the shift in environmental focus from compliance and enforcement to industrial ecology, these international agreements are becoming increasingly numerous. Strictly speaking, they apply only to signatory countries that specifically agree to be bound by the requirements, but their influence tends to be broader. The most obvious examples include the Montreal Protocol, under which production and consumption of CFCs and other ozone-depleting chemicals are being phased out; the Basel Convention, under which transnational shipment of hazardous residues is controlled; and the Kyoto Protocol on Climate Change, designed to deal with emissions of greenhouse gases.

The complexities that arise as society shifts from an environmental to a sustainability focus can be illustrated by agreements that were designed solely as environmental initiatives. Consider, for example, the complexity of integrating environment and trade policy. Potential conflicts arise because environmental laws generally seek to control the means by which goods are made and to forbid or discriminate against environmentally inappropriate processes, products, or technologies, while trade laws in general seek to liberalize the flow of goods among nations. This dynamic has made the interaction of trade and environment extremely contentious. To complicate matters further, developing countries are very concerned that efforts to inject environmental considerations into trade negotiations are not about environmental protection at all, but simply attempts by developed countries to discriminate against developing country products to protect domestic markets. Yet, trade policy is only a small part of the social dimension that sustainability adds: national and international legal and cultural structures surround questions of human rights, labor rights, corporate governance issues, economic development and local institution building, and national security.

The role of government, especially as it acts through the development and implementation of legal and policy systems, can play important roles in the implementation of environmentally responsible technology. Governments at all levels dramatically influence the behavior of firms by the way in which these options are employed and thus influence the effects of firms on sustainability and the environment. Here policies have shifted over time as environmental laws and regulations have evolved away from a traditional command-and-control approach to a broader sustainability approach centered largely on market instruments and sustainability goals. These strategies reflect increased experience with, and understanding of, the interaction of market systems with legal and regulatory structures on the one hand and impacted environmental systems on the other (Figure 7.4).

An obvious way in which governments influence corporations positively is that they are large purchasers of products for their own use. They may also be thought of as "indirect customers," expressing the demand of consumers as formulated in legislation, regulation, and less formal policies and practices. More than is commonly realized, governments as well as customers shape markets, and they have it within their power to create new ones. Similarly, governments are beginning to appreciate the strong coupling between technology and environmental policy systems. Policies can encourage environmentally preferable technologies and industrial ecology principles: product take-back legislation, if properly implemented, is one example. Alternatively, policies can impede progress; for example, laws that restrict the ability of government entities to purchase products containing refurbished subassemblies or components. In general, government technology policies that assist the diffusion of new technologies will also indirectly increase the use of environmentally preferable technologies, because newer technologies tend to be more efficient in the use of resources and energy.

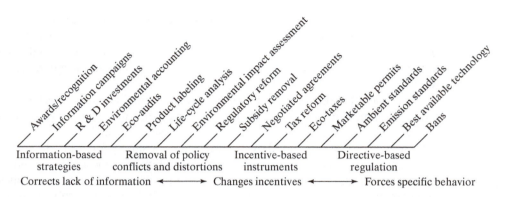

Figure 7.4

Governmental strategies to achieve environmentally responsive behavior in a market economy. (Courtesy of B. Long, *Organization for Economic Cooperation and Development,* Paper presented at ECO '97 International Congress, Paris, February 24–26, 1997.)

Governments also establish incentives through their tax and subsidy systems. Reductions in subsidies that encourage inefficient use of resources and energy, and increases in the taxation of environmentally problematic activities are well-studied examples, and if properly implemented appear to offer both social and environmental benefits. Most countries subsidize energy consumption, which, all things being equal, leads to people using more energy than they otherwise would. "Green taxes" based on carbon emissions from energy consumption are an attempt to reduce such consumption, but they have generally not been widely adopted at levels that actually affect behavior. Broad proposals for tax reform that would tax environmental problems such as pollution and relieve taxes on social benefits such as employment have been widely proposed, but have not yet been politically successful.

Dietz and colleagues (2003) recognize that governance related to issues of environment and sustainability must be adaptive. They offer the following as a checklist of desirable actions:

- *Provide information,* so that issues are approached in a transparent manner.
- *Deal with conflict,* to encourage resolution of contentious issues.
- *Induce rule compliance,* encouraging voluntary and innovative approaches to common problems.
- *Provide infrastructure*—technology, transportation, social systems.
- *Be prepared for change*—the world, its people, and the environment are not static, so governance must be flexible in planning and implementation.

7.6   LEGAL AND ETHICAL CONCERNS IN INDUSTRIAL ECOLOGY

Sustainability arises primarily from the environmental discourse and thus has a long legal history, because environmental laws are not a recent phenomenon. As early as 1306, London adopted an ordinance limiting the burning of coal because of the degradation of local air quality. Such laws became more common as industrialization created substantial point source emissions. For example, the LeBlanc system for producing soda (sodium carbonate), patented in 1791, resulted in substantial emissions of gaseous HCl to the air in the environs of the facilities. As a result, the English Parliament eventually passed the Alkali Act of 1863, requiring manufacturers to absorb the acid in special towers designed by William Gossage.

As in these historical examples, much environmental law has reflected the perception of environmental problems as localized in time, space, and media (i.e., air, water, soil). For example, it was not uncommon in the recent past for groundwater contamination by organic solvents to be eliminated by "air stripping," or simply releasing the solvent to the air, where it contributed in many cases to the formation of tropospheric ozone. Similarly, many hazardous waste sites in the United States have been "cleaned" by simply shipping the contaminated dirt somewhere else, which not only leaves the problem unsolved but creates the danger of incidents during the removal and transportation process. Environmental regulation has thus traditionally focused on specific phenomena and adopted the so-called "command-and-control" approach. This relative simplicity fails as one moves from environmental to sustainability policies and raises a number of fundamental legal issues, as discussed below.

Intragenerational Equity. The distribution of wealth and power both among nation-states and between the elites and the marginalized populations within individual nation-states is one of the principal themes of political science. It is a highly ideological and contentious arena, with the poles in the debate being equality of outcome, called "egalitarianism," and equality of opportunity, called "libertarianism" (recall Figure 7.2). Sustainable development is an egalitarian concept; it contemplates reasonably equal qualities of life for all people and thus implies a substantial shift in resources between rich and poor nations, as well as within nations. Whether and how this should be accomplished are the subjects of extensive debate.

Intergenerational Equity. This concept, which would impose egalitarian ethics across generations, is also part of the sustainable development concept. In addition to being contentious for its egalitarian implications, it is problematic from a practical perspective. It would require, for example, some representation of the interests of future generations in current disputes such as storage of radioactive residues from electric power production, deforestation and reforestation, and the extent of climate change mitigation efforts. Although many legal and philosophical traditions recognize some general requirement for fairness across generations, future generations are not viewed as having enforceable legal rights, in part because it is virtually impossible to identify future individuals or their interests with sufficient specificity to involve them in adjudication. Moreover, the uncertainty of future technological, social, and environmental states makes simple efforts to identify future interests almost assuredly wrong (although, for the same reason, exactly where they are wrong is also impossible to determine).

Flexibility of Legal Tools. Because legal systems in many societies tend to be important components of social structure, they are usually conservative and relatively inflexible. One does not want a situation, for example, where inheritance laws are changed every six months: the passage of property between generations, the assurance of free title, and the preservation of family harmony and expectations all argue for a reasonably stable inheritance system. The price for this inflexibility is paid in terms of inability to adjust to changing situations. Where change is limited and foreseeable—as it may be with inheritance issues—this inflexibility seems appropriate. Where change is rapid and fundamental, however, the result can be substantial inefficiency. This effect can also be seen as environmental regulations are extended from command-and-control requirements—which may be expensive, but don't really impact choices of materials, manufacturing technologies, product design, or customer choice—to pollution prevention and product regulation. For example, an overly conservative requirement for scrubbers might cause manufacturing plants to spend marginally more than they should on such technologies. However, an overly conservative process or product standard skews manufacturing, product design, and consumption patterns for a long time as it becomes embedded in sunk costs, a process known as "lock-in." The initial inflexibility built into such systems is augmented by the well-known tendency of regulation to create and nurture interest groups that benefit from continued regulation and that consequently form a further barrier to subsequent regulatory rationalization. For example, reform of hazardous waste laws tends to be impeded by engineering firms that produce government-approved technologies and waste management services, and, more subtly, the environmental groups that use the fear of hazardous waste as a membership recruitment and fund-raising device.

Regulatory Management Structure.   Traditional command-and-control regulatory mechanisms have been applied with significant success to easily visible environmental problems. Unthinking extension of such simple regulatory tools to far more complex sustainability situations, however, can frequently be both environmentally and economically costly. Rather than continuing to rely primarily on centralized command-and-control, a more sophisticated sustainability management system that recognizes the complex nature of the systems at issue should be used. Note that such a sophisticated management system requires flexible legal tools, implying that political resistance will be strong in many cases.

Determining Appropriate Jurisdictional Level.   Political jurisdictions are creations of human culture and history, and there is no a priori reason why their boundaries should reflect relevant underlying characteristics of social or natural systems. For example, many of Africa's developmental issues arise from discrepancies between traditional tribal boundaries and those imposed by historic overlords; it is no surprise that many problematic environmental perturbations are not coextensive with political boundaries. These distortions can sometimes be reduced by policy harmonization across multiple jurisdictions, which is desirable in imposing substantively different constraints on economic activity at small geographic scales, or to avoid exportation of risk.

7.7   ECONOMICS AND INDUSTRIAL ECOLOGY

Economics is perhaps the most powerful discipline in terms of its capability to shape policy. Its analyses strongly influence most national and international policy formulation, and economic performance is inherent in the concept of sustainable development. In recent years the specialty of *ecological economics* has emerged, as classical economics attempts to encompass relevant aspects of the natural world within which human society exists. In this context, it has become apparent that the study of economics without an understanding of industrial ecology will grow increasingly sterile, and that ecological economics and industrial ecology are natural partners. In the subsections below, we explore some of the aspects of that relationship.

7.7.1 The Private Firm

Private firms are where social organization, economic goals, design decisions, and environmental impacts all come together, and private firms are pivotal economic agents in any modern economy. They also reflect and create the cultures and economies within which they function. Individual firms may be components of self-organizing and loosely linked industrial districts, such as Silicon Valley in the United States or the collection of textile firms near Florence, Italy. These companies compete intensely, while also learning from one another about changing markets and technologies through informal communication and collaborative practices, and forming complicated supply networks among themselves. The functional boundaries within firms are porous in a network system, as are the boundaries among firms themselves and between firms and local institutions such as trade associations and universities.

The private corporate enterprise is such an intrinsic part of the modern capitalist economy that few realize its relative youth. It is possible to trace the antecedents of the corporation back to the medieval merchant guild systems, or, more recently, to trading companies enjoying monopolies granted under royal charter, such as the British East India Company. However, the advent of the truly modern firm awaited the development in the early nineteenth century of laws under which any entity meeting statutorily defined criteria was able to incorporate. The pattern subsequently established in Western economies—a complex network of independent firms, frequently competing on the basis of technological and scientific creativity and with successful innovation rewarded in the marketplace—became the basis for modern, materially successful economies, and a critical part of the innovative engine behind modernity itself. Thus, the modern corporation appeared at a certain stage in the development of the industrial revolution, because such a construct was necessary for the continued evolution of the industrial economies characterizing the modern state. Indeed, it can also be argued that such entities as "virtual firms" represent a continuing evolution of the firm into increasingly complex and flexible entities, a response to the substantially increased complexity of the modern global economy.

7.7.2 Valuation

Traditionally, many economists have viewed economics as an objective, not normative, discipline. This has been strongly challenged by those who claim that the foundation of economic analysis—an assumption that all things can be quantitatively valued in terms of money—is a fundamental statement about morality, and an incorrect one. Those who accept the premise, however, face the pragmatic challenge of determining valuation. A number of tools, or valuation methods, have been developed to quantify such difficult phenomenon as the health effects of pollution. Examples include:

1. The *human capital* method, which measures earnings foregone due to illness or premature death as a result of pollution exposure.
2. The *cost of illness* method, which measures lost workdays plus out-of-pocket medical and associated costs resulting from pollution exposure.
3. The *preventive/mitigative expenditure* method, which measures expenditures on activities to mitigate or reduce the effects of pollution, such as putting in new water delivery systems to avoid exposure to contaminated groundwater.
4. The *wage differential* method, which uses wage differentials between areas differing in pollution exposure as a surrogate for the implicit value of less pollution for people.
5. The *contingent valuation* method, which uses surveys to determine what value people say they put on pollution avoidance.
6. The *surrogate actions* method, which infers an economic value to environmental goods by costing the actions people have taken that are surrogates for the goods themselves. For example, the *travel cost* method examines how much people are willing to travel for a higher quality environmental amenity, such as an uncrowded park.

All these methods offer a means by which dollar values can be assigned to sustainability issues, but all have drawbacks and confounding factors and must therefore be used with caution. In addition, it is particularly difficult to apply them to complicated social domains, where benefits and costs often accrue to different communities and ethical values may not be explicit.

Quantification methods illustrate a common pitfall in economic analyses. Although such an approach simplifies analysis, and can generate more rigorous and understandable results, it also means that factors that cannot be quantified are, in practice, simply not included in the analysis. Even when qualitative impacts are considered, they tend to be assigned lesser weight than the former. In practice, this tends to elevate environmental considerations over the much more difficult and complex social dimensions of sustainability.

7.7.3 Discount Rates

Standard economic analysis asserts that money today is worth more than the same amount of money tomorrow because of inflation and the returns over time that can be anticipated if the money is invested. This is expressed by applying a "discount rate" to future returns as compared to current returns. Technically, this is represented by an equation which gives the present value A of an amount V which will be available t years from now, where i is the discount rate:

$$A = V(1 + i)^{-t} \tag{7.1}$$

The use of discount rates to value resources and plan investments in business and government is ubiquitous and, in many cases, appropriate. Without such an approach, it would be difficult to compare investments that required expenditures and generated streams of returns in differing time periods. Obviously, however, this approach also provides a strong incentive to use resources as soon as possible rather than save them for the future. If the investment's return is itself properly invested so that it provides a stream of benefits over time, the future may indeed benefit more from the economic growth generated than it would have if the resources were conserved. Thus, the issue does not appear to be one of rejecting the concept outright, but of understanding under what conditions such analyses are useful. In particular, discount rates should be applied cautiously in cases where significant social value issues and externalities are present, especially for the longer-term issues so frequently encountered in sustainability.

7.7.4 Green Accounting

Information concerning the economic performance of the firm is generally captured in management accounting systems. Traditionally, such systems have treated environmental costs—even real, quantifiable environmental ones, such as residue disposal costs—as overhead, and have therefore not broken them out by activity, product, process, material, or technology. The result has been that managers, not having access to the environmental cost information concerning their choices, have had neither the incentive nor the data needed to reduce those costs.

The solution, called "green accounting," is conceptually simple: Develop managerial accounting systems that break out such costs, assign them to the causative activity, and thus permit their rational management. In practice this may be a difficult task. For example, in many complex manufacturing operations, developing sensors and systems to provide the physical data on the contributions of different processes and products to a liquid residue stream is a nontrivial task, and one that involves engineering design and capital investment. Moreover, managers tend to resist additional elements of the business process for which they will be made responsible. Also, the assignment of "potential costs," such as estimates of future regulatory liability for present residue disposal practices, may be resisted for fear of creating unnecessary legal liability (it might be argued that a company which foresaw potential future liabilities was thereby admitting its planned behavior was inappropriate or illegal). Nonetheless, it is clear that development of appropriate managerial accounting systems, and their supporting information subsystems, is critical to completing a necessary feedback loop for environmentally appropriate behavior by corporations, and is currently under way in many of them.

Green accounting procedures are also appropriate at the national level. Traditionally, national economic accounts are dominated by a focus on gross national product (GNP). GNP is typically defined as the aggregate money demand for all products, including consumer goods, investments, government expenditures, and export spending. It and similar national account systems are frequently taken as a measure of individual economic welfare, or, more controversially, as a measure of quality of life. As environmental issues have become more important, such metrics have been increasingly criticized for their failure to depreciate so-called "natural capital" as it is used to produce monetarized assets. Technically, systems based on the United Nations System of National Accounts (SNA), the international standard, recognize land, mineral, and timber resources as assets in a nation's capital stock, but do not recognize them in the income and product accounts. Accordingly, if a natural resource is used, the national income and product accounts show no equivalent depreciation. It is increasingly argued, however, that if a forest or a mineral deposit has been depleted, national income accounts should reflect this reduction in value of a natural asset even as they may reflect money income derived from that depletion. Failure to do so in essence values the existing forest as zero until it is destroyed.

Adoption of green accounting methods is still in its infancy. Even more remote is the development of "sustainability accounting," which, because of the difficulty of quantifying social values and the lack of data, has yet to be conceptually fleshed out. While it is true that the market performs at least one test of social value—if a product or service is not purchased, it obviously failed to appeal to potential consumers at the price offered—there are still many social issues that the market does not reflect. Therefore, developing "sustainability accounting" systems is an appropriate but daunting challenge.

7.8   INTEGRATING THE THEMES

Because many industry ecology and sustainable engineering activities relate to long-term issues such as climate change or the sustainability of resources, the field has an inherent interaction with culture and society. This is uncomfortable territory for most scientists and engineers, even more so as culture and cultural constructs evolve over time and place. Every product designer might not need to face this challenge directly, but those developing software tools or analytical methods must do so to avoid generating products that appear inappropriate, insulting, or simply outdated. As a result, sociologists, cultural anthropologists, and international environmental policy specialists are increasingly important partners in creating responsive and adaptable approaches to sustainable development.

Issues of governance and policy have historically been central to industrial ecology and sustainable engineering, and continue to be so today. Restrictions on the emission and sometimes the use of hazardous chemicals have expanded in recent years, and the European Union's REACH regulations now apply across the industrial chemical spectrum (see Chapter 8). Energy and water use restrictions apply to processes and products (see Chapters 9–11), and thereby have a strong influence on design.

Legal issues, and often ethical ones as well, tend to be straightforward at the nation-state level. They are often related to risk in one way or another (see Chapter 6), and have the potential for negotiated settlement under well-understood conventions. In a global economy, however, unexpected differences in legal and cultural systems are commonly encountered (see Chapter 25). These situations can affect access to markets (see Chapter 26) or to resources (see Chapter 24).

Economic issues most commonly interface with industrial ecology and sustainable engineering in green accounting (defining the costs and benefits of IE/SE), cost–benefit analysis of IE/SE actions (in which the discount rate comes into play), and corporate decision making (see Chapter 26). As will be seen, this generates strong incentives to practice cost-free or low-cost IE/SE, actions that are often surprisingly easy to enable.

The Netherlands Approach to Sustainability

The most sophisticated, comprehensive approach to an integrated sustainability policy at this point is probably that of The Netherlands. In 1989, the National Environmental Policy Plan (1989) was released, followed by the National Environmental Policy Plan Plus (1990), the National Environmental Policy Plan 2 (1994), and additional implementation documents released in support of the Plans. The Plans are all explicitly based on the goal of attaining sustainable development in The Netherlands within one generation, with sustainable development being defined as in the Brundtland Report: "Development that meets the needs of the present without compromising the ability of future generations to meet their own needs." The approach is comprehensive: Although the lead is taken by the Ministry of Housing, Physical Planning and Environment, the need to include all other sectors, especially transportation and housing, is explicitly recognized by the relevant ministries and councils in the planning process. Target activities for the Plans include agriculture, traffic and transport, industry and refineries, gas and electricity supply, building trade, consumers and retail trade, environmental trade, research and education, and "societal organizations" (environmental groups, unions, etc.).

At all stages of analysis, the economic impacts of proposed changes are explicitly considered. Moreover, the emphasis is clearly on collaboration with industry and other stakeholders in developing and implementing specific proposals, the goal being scientifically and technically correct decisions. This collaborative approach is borne out by the use of enforceable agreements ("covenants") between industry and the government in order to achieve environmentally desirable ends, rather than the more adversarial legislative process, whenever possible.

The Netherlands, which receives most of its air and water from other countries as a result of regional air and watershed patterns, is not under the illusion that a complex and difficult to define concept such as sustainable development can be achieved by a small country within a few short years. But the selection of sustainable development as the policy goal has enabled a far more sophisticated multidisciplinary approach than would otherwise be possible. Thus, it has encouraged a critical focus on the identification of appropriate metrics to determine progress toward sustainability. It has also resulted in a more sophisticated, comprehensive approach to risk than that usually entailed in traditional environmental risk assessments. As the initial Plan noted, "Making sustainable development measurable . . . is not an easy task, but the results of research on this topic are necessary to enable feedback at the source." In taking this approach, The Netherlands remains a model for industrial ecology policy in action.

Source: The Plans, and other supporting documentation, are available from the Ministry of Housing, Physical Planning and Environment, Department for Information and International Relations, P.O. Box 20951, 2500 EZ The Hague, The Netherlands. Most of the important material is available in English translation.

FURTHER READING

Allenby, B.R., *Industrial Ecology: Policy Framework and Implementation*, Upper Saddle River, NJ: Prentice-Hall, 1999.

Costanza, R., Ed., *Ecological Economics: The Science and Management of Sustainability*, New York: Columbia University Press, 1991.

Dietz, T., E. Ostrom, and P.C. Stern, The struggle to govern the commons, *Science, 302*, 1907–1912, 2003.

Duchin, F., *Structural Economics: Measuring Change in Technology, Lifestyles, and the Environment*, Washington, DC: Island Press, 1998.

Fischer-Kowalski, M., and H. Haberl, Eds., *Socioecological Transitions and Global Change*, Cheltenham, UK: Edward Elgar, 2007.

Grünbühel, C.M., H. Haberl, H. Schandl, and V. Winiwarter, Socioeconomic metabolism and colonization of natural processes in SangSaeng Village: Material and energy flows, land use, and cultural change in Northeast Thailand, *Human Ecology, 31*, 53–86, 2003.

Hall, C., et al., The need to reintegrate the natural sciences with economics, *Bioscience, 51*, 663–673, 2001.

Harvey, D. *Justice, Nature and the Geography of Difference*, Cambridge, MA: Blackwell Publishers, 1996.

Hoffman, A.J., Linking social systems analysis to the industrial ecology framework, *Organization & Environment, 16*, 66–86, 2003.

Liu, J., G.C. Daly, P.R. Ehrlich, and G.W. Luck, Effects of household dynamics on resource consumption and sustainability, *Nature, 421*, 530–533, 2003.

McNeill, J.R. *Something New Under the Sun*, New York: W. W. Norton & Co., 2000.

Princen, T., Consumption and environment: Some conceptual issues, *Ecological Economics, 31*, 347–363, 1999.

Singh, S.J., and C.M. Grünbühel, Environmental relations and biophysical transition: The case of Trinket Island, *Geografiska Annaler, 85B*, 191–208, 2003.

Weiss, E.B., *In Fairness to Future Generations: International Law, Common Patrimony, and Intergenerational Equity*, Tokyo, Japan: The United Nations University, 1989.

Williams, E., et al., Environmental, social, and economic implications of global reuse and recycling of personal computers, *Environmental Science & Technology, 42*, 6446–6454, 2008.

EXERCISES

7.1 Many countries rely on a form of "notice and comment" rule-making in their administration of environmental laws. Such rule-making can take months to years to conduct, and, once rules are in place, they are very hard to change. Present arguments in favor of or against such rule-making procedures in periods of rapid technological change.

7.2 Increasing urbanization is a powerful trend in today's world: Within the next several decades, UN projections show that over half of the world's population will be located in urban centers. This will require the construction of the equivalent of eight cities of ten million inhabitants every year for the foreseeable future. Discuss what this trend implies for flows of products and residues: food, water, sewage, energy, and so on. Overall, is urbanization environmentally advantageous or disadvantageous? Will it continue in any event regardless of environmental impacts?

7.3 In general, democratic nations have better environmental records than communist nations. Why do you think this is? In your opinion, does this reflect differences in developmental stage, ideology, or governance, and why?

7.4 As an industrial ecologist, you have been assigned by your company, a lumber and forest products producer, to begin a dialog with environmentalists who are concerned about your forest management practices. You believe that the available scientific evidence supports your practices, and are concerned that the environmentalists are more interested in headlines than forestry. What kind of presentation would you prepare for your first meeting? What role do you think the governments of the areas within which you operate should play?

7.5 You are the environmental officer of a chemical firm producing commodity polymers. The government of the country in which most of your production facilities are located has just proposed a broad energy tax, substantially higher than in any other developed country.

    (a) List and evaluate the positions your firm can take in response to this public policy initiative.

    (b) Which would you choose and why?

    (c) What data on your firm's operations would be useful in helping you develop your positions?

    (d) What organizational elements of the firm should be involved in helping you develop and implement your positions?

PART III   Implementation

C H A P T E R   8

Sustainable Engineering

8.1   ENGINEERING AND THE INDUSTRIAL SEQUENCE

The *concept* of sustainability was defined in Chapter 2 as "the possibility that human and other forms of life will flourish on the planet forever." With this goal in mind, we begin in this chapter to describe the *practice* of sustainability from the perspective of the engineer. Engineering, of course, is the science by which the properties of matter and the sources of energy in nature are made useful to humans. Sustainable engineering, in the broadest sense, makes resources useful in ways that provide for future generations as well as our own.

The transformation of resources into useful commodities is a sequential process, with different fields of engineering holding sway at different points in the sequence, as shown in Figure 8.1. It is important to follow two paths here: one for organic materials and the other for inorganic. The organic chain begins with an organic feedstock, usually oil or natural gas, but perhaps with biomass of one kind or another. Petrochemical engineers or their counterparts extract the feedstock from its natural reservoir, and chemical engineers then transform it into what might be termed chemical "building blocks" or "semi-products" (box 1). Although there are a very large number of building blocks, seven of them constitute perhaps 90 percent of the present starting materials for organically based products: methane, ethene, propene, butene, benzene, toluene, and xylene.

The building blocks are fashioned into chemical products in box 2, usually by chemists or chemical engineers. The products could be as relatively simple as polymers, or as complex as elaborate pharmaceuticals. There are often several steps involved in the transformations.

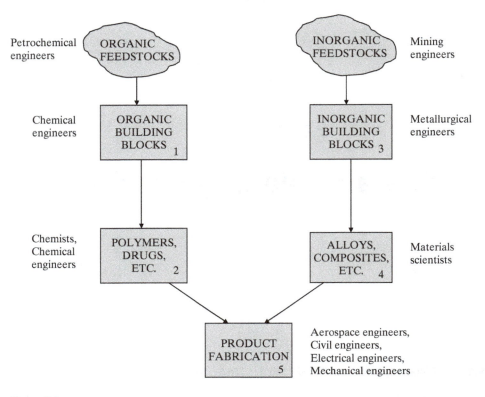

Figure 8.1

The industrial sequence and the engineering specialties most commonly employed at each stage. "Product" is interpreted broadly, to include buildings, infrastructures, and other manifestations of engineering.

A parallel process occurs with inorganic materials, extracted through the services of mining engineers. The materials are processed by metallurgical engineers or other technologists to provide inorganic building blocks, chiefly metals and industrial minerals (box 3). Materials scientists then transform these building blocks into inorganic semi-products such as alloys and composites (box 4). Just as with organics there is a wide variety of inorganic semi-products, but various types of steel are by far the largest on a mass basis.

Organic and inorganic semi-products join forces in box 5 in the manufacture of the products of industry. This is the locus of many common engineering fields: aerospace, civil, electrical, and mechanical. It is worth noting that the products of these engineering specialties have quite different typical lifetimes: 2–5 years for electrical engineers (e.g., cellular telephones), 5–15 years for mechanical engineers (e.g., automobiles), 10–30 years for aerospace engineers (e.g., helicopters), 15–50 years for civil engineers (e.g., bridges and buildings). These time bands are generalities, of course, and are often violated, but they suggest that in her or his career an electrical engineer gets many more opportunities to create fresh designs than does a civil engineer, has more opportunities to employ new materials and new concepts, and has more chances to learn from her or his mistakes. Engineers whose products have long anticipated lifetimes should be particularly sensitive to the goal of "getting it right the first time."

Figure 8.1 points to another difference in the engineering specialties—that some serve the "final customer" who employs the products of box 5, while the specialists in boxes 1–4 (petrochemical, mining, metallurgical, and chemical engineers, and chemists and materials scientists) serve other professionals. (The exceptions here are the chemists and chemical engineers manufacturing drugs in box 2.) This dichotomy means that the box 5 specialists have a much wider range of choices available to them: aluminum or stainless steel, metals or polymers, polymers or composites, and so on. As we will see, the result is that the approaches to sustainable engineering must be different as well.

8.2   GREEN CHEMISTRY

Green chemistry (GC) is the design of chemical products and processes in ways that reduce or eliminate the use and generation of hazardous substances. GC resides primarily in box 2 of the Figure 8.1 sequence. Initially developed in the late 1990s, the specialty has been guided by the 12 principles shown in Table 8.1.

Most of the GC principles relate to facets of chemical synthesis, in which design strategies are developed to manufacture a desired "target" molecule from a chosen set of starting materials. If the target molecule is a complicated one, as is often the case in drug manufacture, for example, the synthesis strategy may involve many steps and many different chemicals. The principles of Table 8.1 guide this process by encouraging efficiency, discouraging the formation of by-products and waste, promoting an informed choice of starting materials and solvents, and minimizing the use of energy. The avoidance or minimization of risk (Chapter 5) is the focus of principle 3.

A traditional chemical process, made up of several reaction steps, has at each step its own temperature and pressure conditions, solvent, and catalyst. Because reactions seldom approach complete conversion of starting materials to products, each step is generally

---

TABLE 8.1   The Twelve Principles of Green Chemistry

---

1. It is better to prevent waste formation than to treat it after it is formed.
2. Design synthetic methods to maximize incorporation of all material used in the process into the final product.
3. Synthetic methods should, where practicable, use or generate materials of low human toxicity and environmental impact.
4. Chemical product design should aim to preserve efficacy whilst reducing toxicity.
5. Auxiliary materials (solvents, extractants etc.) should be avoided if possible, or otherwise made innocuous.
6. Energy requirements should be minimized: Syntheses should be conducted at ambient temperature and pressure.
7. A raw material should, where practicable, be renewable.
8. Unnecessary derivatization (such as protection/deprotection) should be avoided where possible.
9. Selectively catalyzed processes are superior to stoichiometric processes.
10. Chemical products should be designed to be degradable to innocuous products when disposed of and not environmentally persistent.
11. Process monitoring should be used to avoid excursions leading to the formation of hazardous materials.
12. Materials used in a chemical process should be chosen to minimize hazard and risk.

Source: P. T. Anastas, and J.C. Warner, *Green Chemistry: Theory and Practice*, New York: Oxford University Press, 1998.

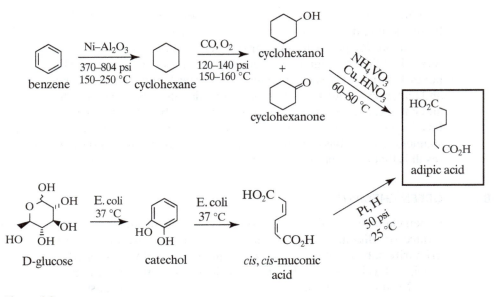

Figure 8.2

The synthesis of adipic acid by the traditional route (top) and a green chemistry route (bottom). Not all by-products are shown. (Adapted from K.M. Draths, and J.W. Frost, Improving the environment through process changes and product substitutions, in *Green Chemistry: Frontiers in Benign Chemical Synthesis and Processes*, P.T. Anastas and T.C. Williamson, Eds., Oxford, UK: Oxford University Press, 1998.)

followed by a separation step to isolate the desired product, followed by a purification step. Solvents are typically needed in these latter steps; often they are organic solvents, and often the solvents are hazardous. Figure 8.2 (top) shows the conventional synthesis of adipic acid, a widely used intermediate chemical. A "green" chemist would object to this approach on at least two grounds: the starting material, benzene, is a biohazard, and the sequence requires high pressures and high temperatures, and thereby lots of energy, violating at least principles 3 and 6, and probably 5. The alternative is shown at the bottom of Figure 8.2. It begins with glucose, a nontoxic material, and the reaction steps are mostly enabled by selected bacteria. Temperature and pressure are moderate, and the reaction yield is about 90 percent.

Green chemists are advocates of *atom economy*—the intention of getting the largest possible fraction of reactants to end up in reaction products. The approach to this goal often involves addition reactions rather than substitution reactions, thus minimizing or eliminating by-products. Alternative approaches are also possible, however. Figure 8.3 (top) shows the traditional synthesis of phenol, starting with sodium benzene sulfonate. Based on the molecular weights of the reactants and principal product, the atom economy is 116/260 = 44.6 percent. The alternative, shown at the bottom of Figure 8.3, begins the key step with cumene hydroperoxide. The addition of a very small amount of acid to aid the decomposition produces phenol and acetone; both these products are useful, and the atom economy is thus 100 percent.

Within the chemical and pharmaceutical industries, these approaches speak to the heart of green chemistry. For those interested in the chemical details, we list several

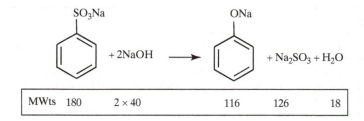

MWts	180	2 × 40		116	126	18

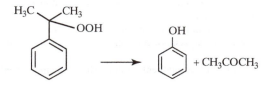

Figure 8.3

The synthesis of phenol by the traditional route (top) and a green chemistry route (bottom). Molecular weights are shown for the traditional route. (Adapted from M. Lancaster, Principles for sustainable and green chemistry, in *Handbook of Green Chemistry and Technology*, J. Clark and D. Macquarrie, Eds., Oxford, UK: Blackwell Science Ltd., pp. 10–27, 2002.)

references at the end of this chapter. The concepts apply much more broadly, however, and any industrial operation that involves a chemical transformation (plating, anodizing, etching, etc.) needs to pay attention to green chemistry approaches.

8.3   GREEN ENGINEERING

Green engineering (GE) is the design and implementation of engineering solutions that take environmental issues into account throughout the life cycle of the design. Although product life cycle considerations are largely the purview of box 5 of the Figure 8.1 sequence, the general philosophy of GE can be employed wherever engineers are transforming materials and using energy, that is, in every box of Figure 8.1.

As with green chemistry, green engineering has its set of 12 principles, given in Table 8.2. Because of the breadth of the engineering profession, the principles are broad as well. The emphasis is not on the molecule, as is the case in green chemistry, but on the process and the final product, be it a portable radio or a giant locomotive. Another important difference is the attention paid in GE to the entire life cycle, from extraction of materials to product obsolescence and recycling.

It is instructive to compare the principles of green engineering with those of green chemistry. When this is done, we see that four of them are essentially the same: GC#1~GE#2, GC#3~GE#1, GC#6~GE#3, and GC#8~GE#12. The common topics relate to the choice of materials (avoid bioactives; use renewables), the minimization of

---

TABLE 8.2   The Twelve Principles of Green Engineering

---

1. Designers need to strive to ensure that all material and energy inputs and outputs are as inherently nonhazardous as possible.
2. It is better to prevent waste formation than to treat it after it is formed.
3. Separation and purification operations should be designed to minimize energy consumption and materials use.
4. Products, processes, and systems should be designed to maximize mass, energy, space, and time efficiency.
5. Products, processes, and systems should be "output pulled" rather than "input pushed" through the use of energy and materials.
6. Embedded entropy and complexity must be viewed as an investment when making design choices on recycle, reuse, or beneficial disposition.
7. Targeted durability, not immortality, should be a design goal.
8. Design for unnecessary capacity or capability (e.g., "one size fits all") solutions should be considered a design flaw.
9. Material diversity in multicomponent products should be minimized so as to promote disassembly and value retention.
10. Design of products, processes, and systems must include integration and interconnectivity with available energy and materials flows.
11. Products, processes, and systems should be designed for performance in a commercial "afterlife."
12. Material and energy inputs should be renewable rather than depleting.

Source: P. T. Anastas, and J. B. Zimmerman, The twelve principles of green engineering, *Environmental Science & Technology, 38*, 94A–101A, 2003.

energy use, and the avoidance of waste. The remainder deal primarily with approaches to chemical synthesis (GC) or the design of products (GE).

Engineering is a diverse profession, and one that operates at many different "trophic levels," as seen in Figure 8.1. As a result, the principles of GE do not apply equally across the engineering specialties. In the case of petrochemical and mining engineers, the choice of the basic material is established by the choice of profession. Thereafter, the materials choices relate to processes. For mining, for example, the practice of cyanide leaching of ore, however efficient, is one that would be avoided by a truly green engineer. And, because the energy requirements of mining and processing ore are so great, many opportunities for energy minimization are present.

Chemical and metallurgical engineers, working at the next higher trophic level, are largely wedded to the starting materials provided by the trophic level below. Some flexibility may exist, however—biomaterials can substitute for petro-chemicals in some instances, for example. At this trophic level, however, the emphasis is clearly on processes and their by-products, the selection of process chemicals, and energy use.

Materials scientist and chemical engineers at trophic level three choose materials for their physical and chemical properties in order to craft a polymer with the desired rigidity or an alloy with outstanding high-temperature performance. Unlike their predecessors in the industrial sequence, however, these technologists should view subsequent stages as more than markets for their products; they are also stages whose performance can be influenced by third-stage choices—is a certain polymer recyclable, is there a second use for a quaternary alloy upon recycling, and so forth.

8.4   **THE PROCESS DESIGN CHALLENGE**

The product manufacture stage is, more than any other, the target for the twelve principles of green engineering. The industrial processes needed to accomplish manufacture constitute a sequence of operations designed to achieve a specific technological result, such as the manufacture of a telephone from metallic and polymeric starting materials. Typical goals for an industrial process designer have traditionally included the following:

- Accomplish the desired technological result.
- Achieve high precision by manufacturing products that consistently fall within desired tolerance limits.
- Achieve high efficiency by manufacturing products in a minimum amount of time.
- Design a process for high reliability over a long period of time.
- Make the process safe for the workers who will use it.
- Design the process to be modular and upgradable.
- Design for minimum first cost (equipment purchase and installation).
- Design for minimum operating cost.

Sustainable engineering imposes on the process designer several additional goals:

- Prevent pollution.
- Reduce risk to the environment.
- Perform process design from a life cycle perspective.

It is seldom that each of these goals can be optimized independently. Rather, the aim is to achieve the optimum balance among the goals.

8.5   **POLLUTION PREVENTION**

One of the central approaches to sustainable industrial process design and operation is termed "pollution prevention" (often referred to as P^2; sometimes termed "cleaner production"). The objective of this activity is to reduce impacts or risk of impacts to employees, local communities, and the environment at large by preventing pollution where it is first generated. The sequence is to identify a problem or potential problem, to locate its source within the manufacturing process, and to change the source so as to reduce or eliminate the problem.

Process evaluation in P^2 deals with sequence flow, the consumption of materials, energy, water, and other resources; the manufacture of desired products; and the identification and quantification of residues. It rests on the construction and analysis of a detailed process flow chart, such as that shown in Figure 8.4. No attempt is made at this stage to quantify any of the flows. The more complete this diagram, the easier the subsequent assessment will be.

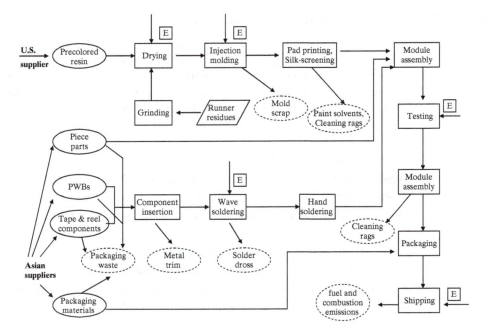

Figure 8.4

A process characterization flow chart for the manufacture of a desktop telephone. (Courtesy of Lucent Technologies.)

P^2 techniques for dealing with issues identified by process characterization include:

1. *Process modification*—changing a process to minimize or eliminate waste generation
2. *Technology modification*—changing manufacturing technology to minimize or eliminate waste generation
3. *Good housekeeping*—changing routine maintenance or operation routines to minimize or eliminate waste generation
4. *Input substitution*—changing process materials to minimize quantity or potential risk of generated waste
5. *On-site reuse*—recycling residues within the facility
6. *Off-site reuse*—recycling residues away from the original facility

The health and environmental risks from different flow streams and processes are, of course, far from equivalent. For example, a detailed study of the intermediate petro-chemicals sector some years ago showed that only a small number of compounds were responsible for most of the potential risk. Although good practice should be applied to all constituents, it is especially important for highly toxic process constituents.

David Allen, now of the University of Texas, has pointed out that P^2 can be addressed at different spatial levels. At the microscale, or molecular level, chemical synthesis pathways and other material fabrication procedures can be redesigned to reduce waste and lower process toxicity. At the process line level, or mesoscale, design considerations include adjustments in temperature, pressure, processing time, and the like, with energy and water use, by-product generation, and inherent process losses as foci. Activities at the mesoscale level are those most commonly termed P^2. Finally, the macroscale, at the sector or intersectoral level, can be addressed through industrial ecology systems perspectives in which by-products find uses outside the facility in which they are generated. Microscale P^2 activities are generally possible only where chemicals are being synthesized; mesoscale and macroscale P^2 can be undertaken anywhere chemicals or other materials are being used.

With the target identified and its flows determined, green chemistry and green engineering approaches can be employed in a straightforward manner:

- Redesign the process to substitute low toxicity materials for those that are highly toxic, or to generate high toxicity materials on-site as needed.
- Minimize process residues.
- Reuse process residues.
- Redesign the process so that unwanted residue streams become streams of useful by-products.

In the case of a process already existing or newly begun, a *waste audit* is generally useful. The approach is to study *all* waste flows from the facility to determine which can be decreased, and how. Industrial solvents, cleaning solutions, and etchants are often good places to begin. An approach that has proven beneficial in a number of cases is the regeneration of chemical solutions, which can often be accomplished by filtration, changes to relax purity requirements, the addition of stabilizers, redesign of process equipment, and so on. An example of the improvement that can be achieved is shown in Figure 8.5 for a peroxide bath, initially used once and discarded. This system was gradually redesigned with improved filtration, recirculation, and testing over a period of years until the solution was eventually replaced only every week. The reduction in cost and decrease in liquid residues that resulted were very large.

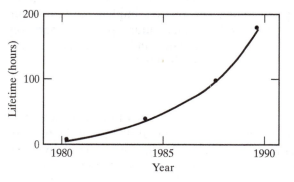

Figure 8.5

The extension in lifetime of a peroxide solution used in the manufacturing of electronic components. (Courtesy of E. Eckroth, AT&T Microelectronics.)

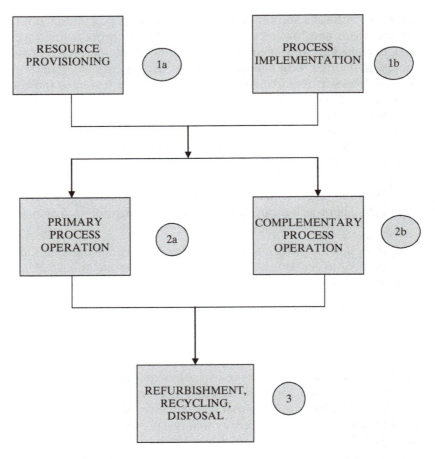

Figure 8.6

The life cycle stages of a manufacturing process.

8.6   THE PROCESS LIFE CYCLE

Industrial processes have life cycles, and green engineering approaches should treat all stage of those cycles, not just the operational stage. The process cycle is comprised of three epochs (Figure 8.6): resource provisioning and process implementation occur simultaneously; primary process operation and complementary process operation occur simultaneously as well; and refurbishment, recycling, and disposal is the end-of-life stage. The characteristics of these stages are described below.

8.6.1   Resource Provisioning

The first stage in the life cycle of any process is the provisioning of the materials used to produce the consumable resources that are used throughout the life of the process being assessed. One consideration is the source of the materials, which in many cases will be extracted from their natural reservoirs. Alternatively, recycled materials may be used. Doing so may be preferable to using virgin materials because recycled materials avoid the environmental disruption that virgin material extraction involves, and often require less energy to recover and recycle than would be expended in virgin material extraction. In addition, the recycling of materials often produces less solid, liquid, or gaseous residues than do virgin materials extractions. Trade-offs, especially in energy use, are always present, however, and the choice must be considered on a case-by-case basis. Another consideration is the methods used to prepare the materials for use in the process. Regardless of the source of a metal sheet to be formed into a component, for example, the forming and cleaning of the sheet and the packaging of the component should be done in an environmentally responsible manner. Supplier operations are thus a topic for evaluation as the process is being developed and, later, as it is being used.

8.6.2   Process Implementation

Coincident with resource provisioning is process implementation, which looks at the environmental impacts that result from the activities necessary to make the process happen. These principally involve the manufacture and installation of the process equipment and installing other resources that are required such as piping, conveyer belts, exhaust ducts, and the like.

8.6.3   Primary Process Operation

A process should be designed to be environmentally responsible in operation. Such a process would ideally limit the use of hazardous materials; minimize the consumption of energy; avoid or minimize the generation of solid, liquid, or gaseous residues; and ensure that any residues that are produced can be used elsewhere in the economy. Efforts should be directed to designing processes whose secondary products are salable to others or usable in other processes within the same facility. In particular, the generation of residues whose hazardous properties render their recycling or disposal difficult should be avoided. Because successful processes can become widespread throughout a manufacturing sector, they should be designed to perform well under a variety of conditions.

An unrealizable goal—but a useful target—is that every molecule that enters a manufacturing process should leave that process as part of a salable product. One's intuitive perception of this goal as unrealistic is not necessarily accurate: Certain of today's manufacturing processes, such as molecular beam epitaxy, come close, and more will do so in the future.

8.6.4   Complementary Process Operation

It is often the case that several manufacturing processes form a symbiotic relationship, each assuming and depending on the existence of others. Thus, a comprehensive process evaluation needs to consider not only the environmental attributes of the primary process itself, but also those of the complementary processes that precede and follow. For example, a welding process generally requires a preceding metal cleaning step, which traditionally required the use of ozone-depleting chlorofluorocarbons. Similarly, a soldering process generally requires a post-cleaning to remove the corrosive solder flux. This step also traditionally required the use of chlorofluorocarbons. Changes in any element of this system—flux, solder, or solvent—usually require changes to the others as well if the entire system is to continue to perform satisfactorily. The responsible primary process designer will consider to what extent his or her process imposes environmentally difficult requirements for complementary processes, both in their implementation and their operation.

8.6.5 Refurbishment, Recycling, Disposal

All process equipment will eventually become obsolete. It must therefore be designed so as to optimize disassembly and reuse, either of modules (the preferable option) or materials. In this sense, process equipment is subject to the same considerations and recommended activities that apply to any product: use of quick disconnect hardware, identification marking of plastics, and so on. Many of these design decisions are made by the corporation actually manufacturing the process equipment, which may well not be the user, but the process designer can control or frustrate many environmentally responsible equipment recycling actions by his or her choice of features on the original process design.

A classic example of the consequences of failing to design for recycling is the Brent Spar oil platform. This North Sea installation was designed and built in the 1960s, well before development of a design for recycling consciousness, as a temporary storage reservoir for crude oil awaiting transfer to refineries. In 1995, Royal Dutch Shell, after a careful review of alternatives, decided to dispose of the then-obsolete platform by scuttling it in the Atlantic Ocean off the west coast of Scotland. This action was seen by Greenpeace, other environmental organizations, and most of Western Europe's citizens as personifying the wasteful society and the disregard of industry for the environment. For several weeks, as options were debated and Greenpeace boats circled the platform, Shell gasoline sales dropped sharply. Finally giving in to the pressure, Shell agreed to bring the platform to shore and pay large sums to have it cut up, the petroleum residues properly treated, and the metal recycled. It made little difference that the overall environmentally preferable choice (in which energy use, exposure to hazardous material, and use of recycled material were considered) may well have been the scuttling originally proposed. The fact was that Shell had paid an enormous price, both monetarily and from a public-relations point of view, for its failure to consider the final life stage in the design of the platform.

8.7 GREEN TECHNOLOGY AND SUSTAINABILITY

A famous edict of William McDonough, architect and environmentalist, is that technology's goals should not be about "being a little less bad." Systems do not go from bad to good overnight, however, and it is difficult to fault the principles of GC and GE: decreased toxicity, lower energy use, design with recycling in mind, and so forth. Nonetheless, GC and GE are *goals of perfection*, never reachable as a consequence of thermodynamics. This is certainly not reason to discard them, but to regard the perspectives of GC and GE as incomplete; that is, they address only part of the broader sustainability challenge.

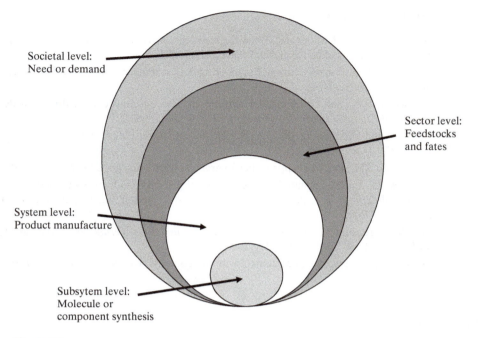

Societal level:
Need or demand

Sector level:
Feedstocks
and fates

System level:
Product manufacture

Subsytem level:
Molecule or
component synthesis

Figure 8.7

Sustainable engineering in its systems context.

Industrial technology and sustainable engineering thus need to be regarded as part of a hierarchy. Figure 8.7 presents the framework (a product-oriented version of Figure 7.1). A molecule or a component may itself be environmentally superior, but it is contained within a higher-level system that we term a product. Higher yet is the environment from which it receives its feedstocks and into which it releases its detritus. Ultimately, the human social system is the driving force that inspired the technology that generates the molecule or component.

Systems thinking is what differentiates the overarching concept of sustainability from GC and GE. And, as William McDonough goes on to say, sustainability (and technology, its instrument) has as its ultimate goal not to be a little less bad, but to "change the story." In the case of technology, Sweden's Natural Step organization suggests three overarching requirements:

1. Substances extracted from the lithosphere must not systematically accumulate in the ecosphere.
2. Society-produced substances must not systematically accumulate in the ecosphere.
3. The physical conditions for production and diversity within the ecosphere must not become systematically deteriorated.

In the chapters to come, we will address the interaction of technology with the natural world, with the aim of considering how these or similar systems principles might usefully be put into practice by the practitioners of modern technology.

FURTHER READING

Allen, D.T., and D.R. Shonnard, *Green Engineering: Environmentally Conscious Design of Chemical Processes*, Upper Saddle River, NJ: Prentice Hall, 2002.

Allen, D.T., Pollution prevention: Engineering design at macro-, meso-, and microscales, *Advances in Chemical Engineering, 113,* 21–323, 1994.

Cano-Ruiz, J.A., and G.J. McRae, Environmentally conscious chemical process design, *Annual Review of Energy and the Environment, 23,* 499–536, 1998.

Clark, J., and D. Macquacrie, *Green Chemistry and Technology*, Oxford, UK: Blackwell Science Ltd., 2002.

DeSimone, J.M., Practical approaches to green solvents, *Science, 297,* 799–803, 2002.

Diwekar, U., and M.J. Small, Process analysis approach to industrial ecology, in *A Handbook of Industrial Ecology,* R.U. Ayres and L.W. Ayres, Eds., Cheltenham, UK: Edward Elgar Publishers, pp. 114–137, 2002.

Mihelcic, J.R., and J.B. Zimmerman, *Environmental Engineering: Fundamentals, Sustainability, and Design*, New York: John Wiley, 2008.

Trost, B.M., On inventing reactions for atom economy, *Accounts of Chemical Research, 35,* 695–805, 2002.

EXERCISES

8.1 Calculate the atom economy for a chemical reaction of your choice.

8.2 Which of the twelve principles of green chemistry seem to you to be the most important? Why?

8.3 Which of the twelve principles of green engineering seem to you to be the most important? Why?

8.4 Choose a manufacturing process, historic or modern, about which you can locate considerable detail concerning its implementation at a specific industrial facility. To the extent applicable and possible, evaluate the process, pointing out its strengths and weaknesses from an industrial ecology standpoint.

8.5 You are the industrial ecologist for a manufacturing company whose leading product is cables for personal computers. The principal components of the cable are copper wire, flexible plastic wire coating, and rigid plastic connectors. What by-product or residue streams do you anticipate? About which should you be most concerned?

8.6 Assume that the peroxide bath whose lifetime is shown in Figure 8.5 is used to make 50,000 silicon wafers per year. The cost of the chemicals for the bath is U.S.$12/liter and the bath is 5 liters in volume. Ten silicon wafers per hour can be processed. How much depleted peroxide bath was generated in 1980, 1983, 1988, and 1990? At an on-site processing cost of U.S.40¢/liter, how much was the cost in each of those years? Was the expenditure of U.S.$3500 for a filtration system in 1988 and U.S.$13,400 for a replenishment system in 1990 justified?

CHAPTER 9

Technological Product Development

9.1 THE PRODUCT DEVELOPMENT CHALLENGE

The principles of sustainable engineering meet reality when new products or new processes are desired. This is because one of the most daunting challenges in modern technology is that faced by the designer of a new product. The products themselves can be enormously complex, some having many more than a million individual parts (see Figure 9.1). Designing a product that does what it is supposed to do is only the beginning for designers, however. Their tasks also include the following:

- *Surveying customers to receive ideas for product characteristics.* A business aphorism is that "products should not merely satisfy the customer, they should *delight* the customer." This is a high standard to live up to, and requires that potential customers and their desires be understood very well.
- *Addressing competitive products.* New designs must meet or exceed those of the competition, or they will be unsuccessful.
- *Complying with regulations.* Product safety, labeling requirements, recyclability, and a host of other legally binding constraints must be considered.
- *Protecting the environment.* Design considerations that have environmental implications are increasingly a topic of interest to customers, regulators, and managers.
- *Producing designs that are attractive, easy to manufacture, delivered on time, and competitively priced.* In today's business world, immediate customer acceptance, efficient manufacturing, and timeliness are crucial.

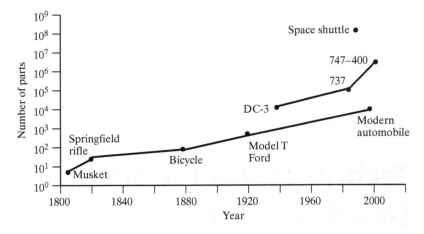

Figure 9.1

The increase in complexity of industrial products over time. (Adapted from a sketch provided by Paul Sheng, University of California, Berkeley.)

Although there are substantial commonalities among design approaches, there are great diversities as well. An important distinction among the industrial sectors is the lifetimes of their products, as indicated in Table 9.1. Some are made with the anticipation that they will function for a decade or more. Others have lives measured in months or weeks. Still others are used only once. A designer must obviously adopt different approaches to these different types of products so far as features such as durability, materials choice, remanufacturing, and recyclability are concerned.

TABLE 9.1 Manufacturing Sectors and Their Products

Manufacturing	Product examples	Product lifetime
Electronics	Computers, cordless telephones, video cameras, television sets, portable sound systems	Long
Vehicles	Automobiles, aircraft, earth movers, snow blowers	Long
Consumer durable goods	Refrigerators, washing machines, furniture, furnaces, water heaters, air conditioners, carpets	Long
Industrial durable goods	Machine tools, motors, fans, conveyer belts, packaging equipment	Long
Durable medical products	Hospital beds, MRI equipment, wheelchairs, washable garments	Long
Consumer nondurable goods	Pencils, batteries, costume jewelry, plastic storage containers, toys	Moderate
Clothing	Shoes, belts, blouses, pants	Moderate
Disposable medical products	Thermometers, blood donor products, medicines, nonwashable garments	Single use
Disposable consumer products	Antifreeze, paper products, lubricants, plastic bags	Single use
Food products	Frozen dinners, canned fruit, dry cereal, soft drinks	Single use

To accomplish their assigned tasks, designers have available to them a large toolbox, developed over many years. Many of these tools were developed before environmental considerations became as important as they are now. Accordingly, in this chapter and several following, we present and discuss several popular design tools and approaches that are environmentally focused or that can incorporate environmental considerations within them. We concentrate not on tools primarily used in product realization (e.g., computer-aided design [CAD]), but on those whose focus is on concepts (such as product definition, product decisions, and product development).

9.2   CONCEPTUAL TOOLS FOR PRODUCT DESIGNERS

9.2.1   The Pugh Selection Matrix

The Pugh Selection Matrix is designed to display the characteristics of a potential product design and to assess how well those characteristics are met by current products of the corporation, by competing products, and by alternative new designs. An example of the rather intricate diagram used in this process is given in Figure 9.2, a room air conditioner being used as the example product.

The left side of the diagram lists existing or desired characteristics of room air conditioners. This list can be as extensive as the design team chooses to make it; we provide

	Product characteristics	Customer importance	Design targets	Alternatives A	B	C	D	Customer survey (∇ Our company, □ Competitor A, ▱ Competitor B, ● Target; scale 1–5)
Performance	Affordable	9	<$253.00	+	+	S	+	∇ (1), □ (3), ● (5)
	Easy to handle	3	<40 lbs	−	+	S	+	□ (2), ∇ (3), ▱● (3.5)
	Quiet	9	Nc30 max	+	+	S	−	□ (2), ▱● (5)
	Not rumbly	9	RC30 max	S	S	S	+	□ (1), ● (4.5)
	Easy to service	1	20 in. lb. torque	−	−	−	+	□ (2), ▱ (3), ∇● (5)
	Air flow	3	200 to 3000 CFM	S	S	S	S	● (5)
	Sound	9	RC30 max	−	−	+	S	∇ (1), □ (2), ▱ (3), ● (4)
	Damper leakage	3	100 cfm at 1" SP	+	+	−	S	∇ (1), □ (3), ● (3.5), ▱ (5)
	Electric heat	1	1 kw per 200 CFM	−	+	S	−	∇ (2), □ (3), ▱● (3.5)
DfE	Energy efficient							
	Recycled materials							
	CFC-free							
	Easy to upgrade							
Cost & Schedule	Fabrication cost	9	< $180.00	+	+	+	+	∇ (1), □ (2), ▱● (4)
	Lead time	9	2 weeks	+	+	−	+	∇ (2), □ (2.5), ▱ (4), ● (4.5)
	Production date	3	10-12-94	−	−	+	+	∇ (1), ▱ (3), □ (4.5), ● (5)
Controls	UL 1096	3		S	S	S	S	● (5)

Figure 9.2

The Pugh Selection Matrix for a room air conditioner. The display's attributes are discussed in the text.

only enough detail here to illustrate how the tool works. The diagram in Figure 9.2 includes a group of characteristics that pertain to environmental and sustainability-related aspects of the design; this is a feature not traditionally included within the Pugh Matrix.

Once the list of characteristics is completed, the next step is to develop design targets for each of the characteristics that the design team wishes to address, and to generate several alternative designs. A customer focus group is asked to rate the characteristics for relative importance (on a high, medium, low, or 9, 3, 1 scale), and to evaluate the *current* product design and those of the principal competitors on a 1–5 scale for its performance relative to the characteristics selected.

When these steps have been taken, the design team is in a position to evaluate each of the alternatives for a *redesigned* product. At each point in the matrix, the design in question is rated + (exceeds the target), *S* (about the same as the target), or − (does not meet the target). With this information, the team can proceed to designate one alternative as the choice for the design concept and proceed to the detailed design process.

9.2.2   The House of Quality

An analytical tool that is an alternative or supplement to the Pugh Selection Matrix is the House of Quality. This tool, shown in Figure 9.3, shares the customer focus of the Pugh Matrix, but goes deeper into the design process. To begin, one enters the "customer room" and lists the product characteristics desired by the customer together with their relative importance. Next, the "environmental foundation" is added, in which desirable environmental attributes are listed. Potential engineering characteristics are added in the "design room."

With the foundation and room characteristics established, the importance of each of the characteristics is rated on a high (H), medium (M), or low (L) scale. The potential features of the product are then related to the alternative design approach being evaluated in the "relationships room." Here the degree to which engineering approaches respond to the desirable characteristics identified by customers are evaluated on a strong (S), moderate (M), or weak (W) scale in order to guide design decisions toward effective responses. Matrix elements with no entries indicate engineering characteristics that do not influence customer or environmental attributes.

The House of Quality is completed by constructing the "roof matrix." The entries in this matrix indicate where a particular design goal reinforces others (X), or where design goals are in conflict (O). If the design room has been constructed in sufficient detail, the roof matrix can often provide insights into design approaches that will satisfy multiple customer preferences and the topics called out by the environmental foundation.

9.3   DESIGN FOR *X*

William Wulf, former President of the U.S. National Academy of Engineering, refers to the focus of the entire engineering profession as performing "design under constraint." In this context, product definition—what the product will be used for, how it will function, what its properties will be, the range of probable cost, and (if appropriate) its aesthetic attributes—provides designers with a substantial range of things to attempt to optimize simultaneously. Modern designers have a list much longer still, however, because they need to consider related product attributes that may, in the end, determine the product's

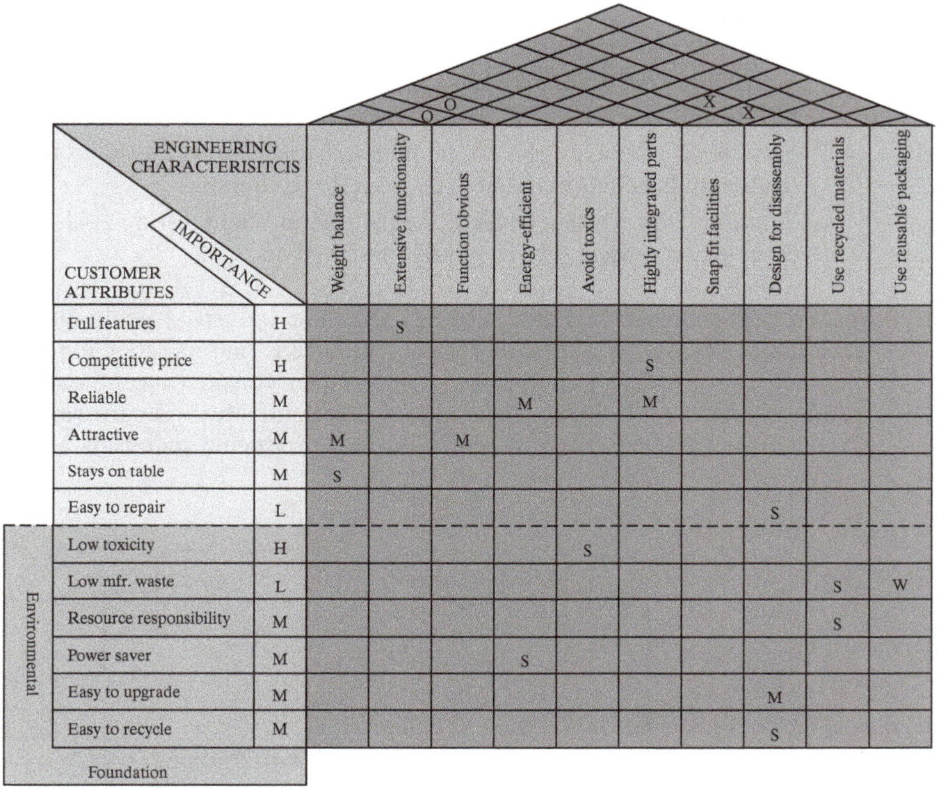

Figure 9.3

The House of Quality for a desktop telephone. The display's attributes are discussed in the text.

success or failure. The paradigm for these latter considerations is termed "Design for *X*" (DfX), where *X* may be any of a number of design attributes, such as:

Assembly (A): Consideration of assemblability, including ease of assembly, error-free assembly, common part assembly, and so on

Compliance (C): Consideration of the regulatory compliance required for manufacturing and field use, including such topics as electromagnetic compatibility

Disassembly (D): Consideration of end-of-life efficiency of product disassembly for refurbishment, recycling, or disposal

Environment and Sustainability (ES): This component of DfX, and the philosophy on which it is based, is a principal subject of this book

Manufacturability (M): Consideration of how well a design can be integrated into factory processes such as fabrication and assembly

Material Logistics and Component Applicability (MC): The topic focuses on factory and field material movement and management considerations, and the reliable supply of components and materials

Reliability (R): Consideration of such topics as electrostatic discharge, corrosion resistance, and operation under variable ambient conditions

Safety and Liability Prevention (SL): Adherence to safety standards and design to forestall misuse

Servicability (S): Design to facilitate initial installation, as well as repair and modification of products in the field or at service centers

Testability (T): Design to facilitate factory and field testing at all levels of system complexity: devices, circuit boards, and systems

The major characteristic that distinguishes the industrial ecology approach from traditional environmental regulatory compliance is that its scope extends far beyond the factory walls. This perspective is captured by considering the entire life cycle of products, as shown in Figure 9.4. Stage 1, premanufacture, is performed by suppliers, drawing on (generally) virgin resources and producing materials and components.

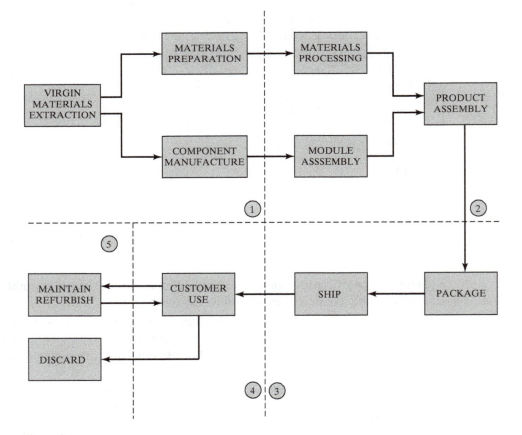

Figure 9.4

Activities in the five life cycle stages (circled numbers) of a product manufactured for customer use. The "maintain and refurbish" activity is sometimes treated as a sixth life cycle stage. In an environmentally responsible product, the environmental impacts at each stage are minimized, not just those in Stage 2.

Stage 2 is the manufacturing operation directly under the control of the corporation making the product under consideration. Stage 3, product delivery, will generally be under the control of the manufacturer, although complex products containing many components and subassemblies may involve a global web of shippers, dealers, and installers. Stage 4, the customer use stage, is influenced by how products are designed and by the degree of continuing manufacturer interaction. In Stage 5, a product no longer satisfactory because of obsolescence, component degradation, or changed business or personal decisions is refurbished, recycled, or discarded.

DfX practices that address all aspects of the life cycle are already being implemented by leading manufacturing firms. Accordingly, the least difficult way to ensure that environmental principles are internalized into manufacturing activities in the short term is to develop and deploy Design for Environment and Sustainability (DfES, see Chapter 10) as a module of existing DfX systems. Moreover, the fact that DfES is intended to be part of an existing design process acts as a salutary constraint, requiring that DfES methods and resulting recommendations be implementable in the real world.

An increasingly important aspect of the design process is the degree to which it is linked to the computer. Most modern industrial design teams utilize computer-aided design/computer-aided manufacturing (CAD/CAM) software, which can incorporate standard component modules into a design, check a design for spatial clearances, produce lists of materials, and so forth. To the degree that DfES can be integrated into these design tools, it will become automatically a part of the physical design process. DfES incorporation into CAD/CAM is far from universal, and diligent effort will be needed to bring it to a high level of development.

9.4   PRODUCT DESIGN TEAMS

The focus on the customer, on cost, and on all the topics raised by the DfX process requires that product and process design can no longer be performed by an isolated group of engineers. Rather, modern industrial practice is to form product design teams made up of individuals from a wide range of specialties. An example of team structure is given in Figure 9.5. In addition to the appropriate mix of engineering specialists, such teams often include environmental experts, packaging engineers (i.e., those who can determine how safely and inexpensively products can be delivered to customers), manufacturing engineers (i.e., those that understand the intricacies of product fabrication), marketing specialists, business planners, and perhaps financial and purchasing experts. In a reflection of today's increasingly complex economy, many design teams include members from strategic partners such as critical suppliers or customers.

The practice of using diversified teams for product design considerably complicates the process in its initial stages. The benefits, however, arise from the early consideration of the variety of attributes that will ultimately determine the success or failure of the product. Most features of a product design are effectively frozen-in very early in the design process. The presence of an industrial ecology specialist on every design team is currently the single most effective way to improve the environmental responsibility of the products of our modern technological society. In the longer term, one would expect

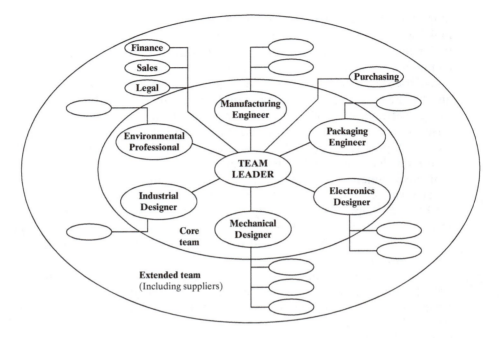

Figure 9.5

The structure of a typical design team for an electromechanical product. (Adapted from K.T. Ulrich, and S.D. Eppinger, *Product Design and Development*, New York: McGraw-Hill, 1995.)

all team members to be versed on such topics as life cycle assessment, green engineering, and similar concepts.

9.5 THE PRODUCT REALIZATION PROCESS

Modern managers wish to stimulate their design and development staffs to generate numerous ideas for new products, in the hope that a few really successful products will result. Carrying every product idea through from concept to manufacture is too expensive to be feasible, however, and so a structured process, the "product realization process (PRP)," has been developed to guide business decisions at each step along the way.

There are a number of versions of PRP, some labeled IDS (integrated development system), some IPD (integrated product development). They vary in level of detail and in the number of sequence steps, and many corporations have developed handbooks to guide their design teams. All share the general approach, if not each specific step, shown in Figure 9.6. Eight steps in the product realization process, from idea to obsolescence, are indicated in the figure. The transition from one step to the next passes through "gates": opportunities for the managers to decide whether to permit the product development to proceed. In the formal structure of the product realization process, a review is held when a product idea under development reaches each gate in the sequence. The

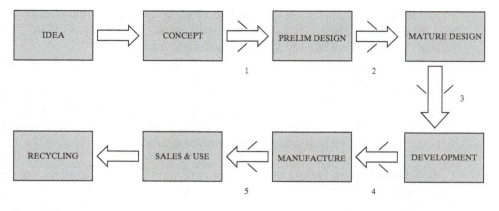

Figure 9.6

The product realization process. The successive points at which "go-no go" decisions are reached (termed "gates") are numbered from 1 to 5.

review team typically includes representatives from design, manufacturing, purchasing, marketing, and other appropriate organizations within the corporation. The outcome of the review determines whether or not gate passage will be allowed.

The items considered at each gate review include marketability (do we still think our customers want this product?), manufacturability (can we make the product as envisioned?), economics (can we make a profit on this item?), strategy (are we ahead of our competitors?), and a variety of other factors. Cost is a major influence on decision making, because the financial investment required to move to the next step of the product development increases as one moves from early gates to later gates. By Gate 4 or 5, if the product is then judged unpromising, a substantial unrecoverable investment will have been made. The goal of the oversight process is to let promising products move quickly to manufacture, but to close gates early on projects that will consume investment dollars without the probability of substantial financial return.

Information of all kinds becomes more detailed as a product progresses from early to later stages of development: concepts are transformed to designs; materials are specified; sizes and features are determined; costs can be more and more accurately calculated; and customer response can be better estimated.

Gate 1: From Concept to Preliminary Design. The first gate controls the transition from concept to preliminary design. The business questions at this gate are very basic: Does this concept appear to meet a customer need? Is it consistent with the corporate product line? Does it have the potential to compete effectively?

Gate 2: From Preliminary Design to Mature Design. The initial or concept stage of product development typically involves only a handful of people, and the only expense is their time. At the next stage, preliminary product design, the size of the group expands but its activities are still limited to computers, communications, and pencil sketches, so the embedded development expense is still modest. At Gate 2 the major design decisions have been made, but few details are available.

The typical business questions at Gate 2 are formulated from the perspective of the preliminary design: Do the estimated performance specifications meet the product goals? Is the design visually attractive? Is the product likely to be profitable? Have any hazardous or scarce materials been specified? This is often the most critical decision point in the product development process, because the corporate investment in a product that passes the second gate begins to increase rapidly.

Gate 3: From Mature Design to Development.   At Gate 3, the design team presents a fully worked product design together with moderately detailed information on manufacturing processes. The product itself can then receive a reasonably detailed environmental review. In the case of processes, if the product is relatively similar in type and materials to other products of the corporation, there may be little need for a new review of the environmental implications of manufacturing. For a new process or set of processes, however, the manufacturing review will be extensive.

The Gate 3 product review is in all cases quite detailed. From a business standpoint the questions become more focused then at earlier stages: Are there technical impediments to development? Do the manufacturing processes satisfy green engineering guidelines? Are the electrical and mechanical goals for the product fully realized? Will the product have customer appeal?

Gate 4: From Development to Manufacture.   By the time the Gate 4 review committee meets, the design is finalized, the manufacturing process set, all materials and components chosen, all suppliers at least tentatively identified, and all costs established. The decision at this gate is whether to proceed with manufacture, the most costly of all the stages.

The business decisions at Gate 4 are obvious and important: Have the cost estimates been met? Is product manufacturability satisfactory? Has a reliable set of suppliers been identified, and are the suppliers capable of supporting the environment and sustainability goals of the product? Does the product as it will emerge from the manufacturing process retain the desirable characteristics identified at Gate 3?

Gate 5: From Manufacture to Sales and Use.   The Gate 5 review is often ceremonial, especially if decisions at previous gates have been sufficiently thoughtful and comprehensive. Provided that no unexpected and unwelcome information has arisen, the product is released for sale and use. The business questions involve a review of the degree to which manufacturing meets expectations and the ways in which the marketing campaign should move forward.

In a departure from past practice, today's PRP procedures and today's managers explicitly include environment and sustainability issues at every gate, to greater or lesser degrees. This requires extensive knowledge of supply chains, product life cycles, and materials choices. Industrial ecology and sustainable engineering, once bystanders during the PRP, are now fully integrated into the PRP activities of more advanced corporations.

FURTHER READING

Boothroyd, G., P. Dewhurst, and W. Knight, *Product Design for Manufacture and Assembly*, New York: Marcel Dekker, 1994.

Hauser, J., and D. Clausing, The house of quality, *Harvard Business Review,* 63–73, May–June, 1988.

Lee, G.H., and A. Kusiak, The house of quality for design rule priority, *International Journal of Advanced Manufacturing Technology, 17,* 288–296, 2001.

Pugh, S., *Total Design*, Reading, MA: Addison-Wesley, 1990.

Ulrich, K.T., and S.D. Eppinger, *Product Design and Development*, New York: McGraw-Hill, 1995.

EXERCISES

9.1 Choose a common but sophisticated household appliance such as a refrigerator, a television set, or a washing machine. Describe the life stages of the appliance, including identifying who is primarily responsible for the environmental concerns at each life stage.

9.2 Develop a House of Quality diagram for a toaster, as on Figure 9.3, and evaluate the matrix elements to the degree possible.

9.3 Choose one of the following products: a bar of soap, a bicycle, a car wash, an ocean cargo ship. For the product selected, construct a materials and energy flow diagram of the type of Figure 8.4 (but with minimal detail in the manufacturing stage and enhanced detail in the product use stage). What do you see as the major green chemistry/green engineering issues that will need to be addressed?

9.4 You are the leader of a project to design and manufacture a super-lightweight, high-performance bicycle. It will be made largely from honeycomb aluminum and carbon fiber and will be marketed in various sizes to fit children from age six up to adults. What are the principal challenges you anticipate at each step of the product realization process?

CHAPTER 10

Design for Environment and Sustainability: Customer Products

10.1 INTRODUCTION

The approaches to product design discussed in Chapter 9 provide the design team with a sense of its customer, of various constraints on the design as it emerges, and on the process for transforming a design into a product for a customer or an organization. What they do not do is to provide much in the way of technological or aesthetic advice. Exactly what is an "attractive" telephone? What are "ecomaterials"? When has "easy to upgrade" been achieved? What, exactly, does "recycled materials" mean? There is, of course, no definitive answer to the question of attractiveness, and it turns out that definitive answers are hard to come by for other questions as well. Nonetheless, doing better at approaching the goal of responsible Design for Environment and Sustainability (DfES) is a crucial component of a more responsible society, and this requires detailed consideration of the entire life cycle of each product. In this chapter, we provide some initial guidance on the many topics related to this task, directed at products manufactured in multiple identical or similar copies. More detail is available in the Further Reading list.

10.2 CHOOSING MATERIALS

Materials influence the functioning of a product, its ruggedness, its appearance, and numerous other characteristics. In many cases any of a number of different materials could be chosen for a particular application. The initial considerations of the product designer so far as materials choice is concerned are obvious and important:

- Does the material have the desired physical properties (strength, conductivity, index of refraction, etc.)?
- Does the material have the desired chemical properties (solubility, photosensitivity, reactivity, etc.)?
- Is the cost reasonable?

In the modern, increasingly complex world, the designer must also pay attention to a number of additional materials considerations. Those of particular relevance to sustainable engineering are discussed below.

Materials choices are often limited by toxicity concerns. Other things being comparable for a specific application, a designer's objective should be to select materials that have the least significant potential risk. Government environmental agencies generally define those materials that merit concern from a toxicity standpoint, and their lists are a good starting point for the physical designer. In Table 10.1, we reproduce the 17 chemicals or chemical groups targeted for reduction in the U.S. EPA's Industrial Toxics Project, a voluntary effort by industrial corporations to reduce emissions of targeted chemicals. Most of these materials have also been restricted by the European Union and other governmental bodies. The list includes both product chemicals and process chemicals. Cadmium, chromium, lead, nickel, and their compounds are sometimes used industrially such that the metals end up as part of the products that are produced, often as platings or coatings. Most of the remaining materials are process chemicals which may be used as either solvents or cleaners. Chlorinated solvents and monoaromatic species comprise most of the listed items. Cyanide solutions generally employed in metal plating also appear. More expansive lists are produced from time to time as old materials are phased out and some newer ones are shown to be of possible concern. The physical design team thus needs to consider two facets of materials choices involving toxic materials: the potential for materials substitution in products and the potential for process changes.

TABLE 10.1   Chemicals Identified in EPA's Industrial Toxics Project

Benzene	Cadmium and compounds
Carbon tetrachloride	Chloroform
Chromium and compounds	Cyanides
Dichloromethane	Lead and compounds
Mercury and compounds	Methyl ethyl ketone
Methyl isobutyl ketone	Nickel and compounds
Tetrachloroethylene	Toluene
Trichloroethane	Trichloroethylene
Xylenes	

The most widespread and important legislative approach to challenges of toxicity is the European Union's REACH (Registration, Evaluation, and Authorisation of Chemicals) program. This action, which came into force on June 1, 2007, requires that chemicals manufactured in quantities greater than 1 metric ton must be registered, those manufactured in quantities greater than 100 metric tons be evaluated for toxicity, and those of high concern (such as carcinogens or mutagens) must be specially authorized for use. Downstream users of the chemicals are involved as well as those who make the chemicals. REACH has made it much more difficult and expensive to utilize previously unauthorized materials and provides a substantial incentive for attempts to meet design criteria without utilizing problematic new materials or new material combinations.

A second issue of which designers need to be aware, especially when designing products such as buildings or infrastructure that have very long lifetimes, is the prospect of resource scarcity. That topic is discussed in detail in Chapter 24. We merely mention here that a number of materials are available only from a few repositories in nature, are often acquired only as by-products of parent materials, and lack efficient recycling networks. Designers should avoid using any of these "materials of concern" unless absolutely required by performance specifications for the product; if they employ them, they should make provision for a suitable recycling system as well.

A third consideration, often overlooked, is that the design team should endeavor to specify that materials come from post-consumer recycled stock if at all possible. The use of such material avoids the mining and processing of virgin resources, which not only depletes long-term supply but requires more energy, water, and other resources than using recycled stock and generates substantial amounts of "hidden flows" (of waste rock, for example). The subject of hidden flows is discussed in more detail in Chapter 18.

A final issue is that of dematerialization. Because industrial sectors tend to operate independently, they are unlikely to appreciate concerns that primarily fall on other sectors. However, an integrated assessment shows that many products require a hundred times their weight in virgin extraction under today's technologies; an example is shown in Figure 10.1. This jet engine analysis reveals a "buy-to-sell" ratio of 10:1, as was

100,000 Pounds Metal
(*forgings and bar stock*)

100   **10**   **1**

(avoidable by using   (largely avoidable by
recycled material)   near net-shape casting)

Figure 10.1

The weight of the mining and metal residues that result from the manufacture of a single jet engine are roughly one hundred times the weight of the engine itself, but recycling and advanced casting can minimize this hidden flow of materials. (Modified from a diagram furnished by R. Tierney, Pratt & Whitney, Inc.)

the case also for the upstream processor of the metal ingots. Reducing the buy-to-sell ratio by careful attention to product design and manufacturing involving each material in a product not only saves money but inevitably has salutary implications for the environment and sustainability as well. One reason is that extraction of materials is extremely energy intensive and tends to be destructive of local ecological habitats. A second reason is that many light-weight products, such as automobiles or aircraft, require less energy when in use than do their normal-weight components.

10.3 COMBINING MATERIALS

It is not uncommon for a designer who thoughtfully selects individual materials to then meet with complications of one kind or another when considering how those materials are to be combined. In this regard, a designer's goal should be to enable reuse and recycling. However, modern technology often makes it difficult to achieve this intention. Today's electronics industry, for example, employs roughly half the periodic table of the elements in the manufacture of an integrated circuit, up from about 11 elements in the 1980s (Figure 10.2). The capabilities of the different chemical elements in performing specialized tasks have greatly improved the performance of everything from computers to cellular telephones. Nonetheless, with only milligrams of most of the elements on each circuit, the opportunities for recovery and reuse of any but the most valuable are remote.

Even in the best situations, separation is seldom perfect, and in some cases impurities can seriously degrade the properties of recycled materials. Castro and coworkers have developed this concept from a thermodynamic perspective (Table 10.2). They

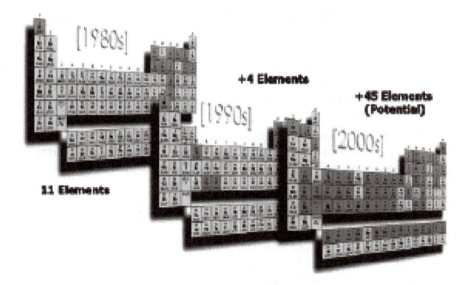

Figure 10.2

The increase over time in the number of elements incorporated into a typical integrated circuit. (Courtesy of T. McManus, Intel Corporation.)

TABLE 10.2   The Matrix for the Thermodynamic Evaluation of the Recycling Compatibility of Industrial Materials (Right Side), with Considerations Related to the Average Passenger Vehicle at the Left.

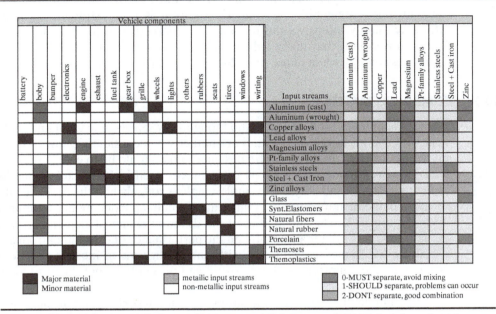

Reprinted with permission from M.J.B. Castro, J.A.M. Remerswaal, M.A. Reuter, and U.J.M. Boin, A thermodynamic approach to the compatibility of materials combinations for recycling, *Resources, Conservation, and Recycling, 43*, 1–19, 2004.

identify three possible decisions related to the recycling of the individual components of mixed metals:

- *Must separate.* This decision relates to a material combination where valuable resources are lost or seriously degraded if separation is not carried out.
- *Should separate.* The value of the separated resources may or may not justify special attention to recycling.
- *Don't separate.* The resources are under no long-term supply constraints; the value is low at typical concentrations in scrap; they do no harm if retained; and separation will consume large amounts of energy and/or other important resources.

With respect to the automobile example, Table 10.2 indicates that iron *must* be separated from aluminum; precious metals *should not* be separated from copper; and *copper* must be separated from steel.

Of course, materials are generally combined in order to provide improved physical properties of one sort or another. In the case of metals, the half-dozen common alloy groupings are shown in Figure 10.3. In a number of cases (e.g., manganese as a component of steel), almost the entire use of a material is in a combination of this sort. Once alloyed or otherwise combined, it is usually prohibitive from a financial or technological

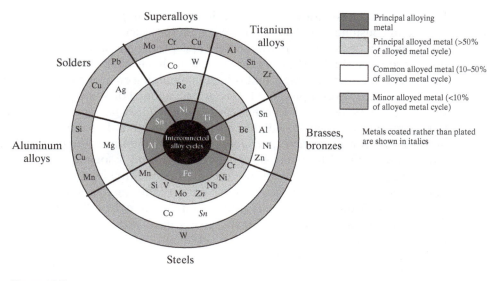

Figure 10.3

The wheel of metal linkages in modern alloys.

standpoint to separate the original components. Rather, the alloy or other combination must be reused *as such* if it is to be reused at all. New alloys or composites for which recycling networks do not exist are likely to result in discard after a single use, and should be specified with caution, particularly if one or more of the constituents might have come under supply constraints (see Chapter 24).

Fastening technology is another aspect of design worth careful attention. Ideally, fasteners should be quickly and easily reversible—snap fits are preferable to screws, screws preferable to rivets. Research on the products of automobile shredding have shown aluminum with rubber inserts, aluminum still bolted to steel, and so forth—products of design decisions to combine materials without corresponding consideration of separation technology.

A final item to note is that used parts are often quite desirable in newly assembled products. One advantage of utilizing evaluated used parts is financial—Xerox saves very large amounts of money by reusing selected copy machine assemblies, for example, and the reuse of decorative building features is common. A second is that in some cases the average "experienced" or "well-treated" part is more reliable than the average new part. This is particularly true in the electronics industry, where there is clear evidence that reused parts are often more reliable than new parts that cannot be fully "burned in" before being given to the customer.

10.4   PRODUCT DELIVERY

Where detailed assessments have been made, some 30 percent of all municipal solid waste has been found to be packaging material. Indeed, it has been estimated that about one-third of all plastics production is for short-term disposable use in packaging.

For many products—convenience food items, for example—packaging is the primary residue of consumer use. The use of toxic materials such as heavy metals in packaging inks may be a first-order environmental impact for some products. Improved packaging of products of all sorts, from large-volume chemicals to small consumer personal care items, thus plays an important role in maintaining environmental sustainability.

There are several possible levels of packaging. For some products, no packaging at all may be required. In other cases, only *primary packaging*, that is, packaging that is in physical contact with the product, is needed. Less certain is the need in many cases for *secondary packaging* (a supplementary shipping container) or *tertiary packaging* (the outer shipping container and associated material). Product packaging should always aim to use the minimal number of stages. However, different applications impose different packaging requirements. Some food packaging, for example, is quite complex: A potato chip bag may be a "sandwich" of seven or eight different components and many layers, each with a separate function. To the degree that any packaging stage can be eliminated or simplified, the residue stream will be reduced and shipping and storage expenses for the producer and consumer will be minimized.

A suggested order of precedence for approaches to packaging, in decreasing order of preference, is

- No packaging
- Minimal packaging
- Consumable, returnable, or refillable/reusable packaging
- Recyclable packaging

This list is only a guide, because innovative packaging solutions for specific products may outweigh the precedence order, but it is a good starting point for the packaging engineer.

A list of causes for overpackaging has been presented by the Institute of Packaging Professionals in Herndon, VA:

- An overly cautious approach to the protection of the packaged contents
- Increasing the package size to deter shoplifting
- Overly conservative environmental test specifications
- Requirements of packaging machinery
- Decorative or representational packaging
- Increasing packaging size to provide space for regulatory information, customer information, or bar coding

Many of these causes are capable of being surmounted by thoughtful packaging design. For example, the use of electronic theft protection systems is able to mitigate the need for some of the overpackaging that has traditionally been employed.

Once the amount of packaging is minimized, the next thing to consider is whether the packaging can be reused. Reusable packaging is not limited to foam "peanuts"; it

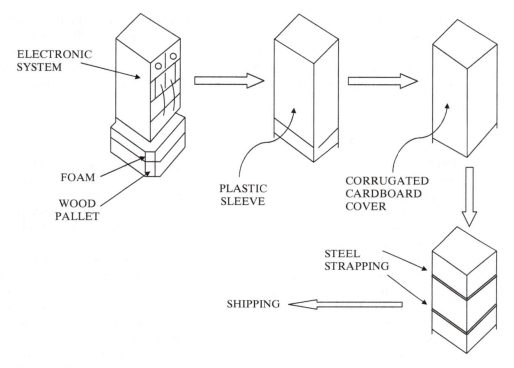

Figure 10.4

The sequence of packaging circa 1995 for a large electronic equipment module.

includes innovations such as Ametek Corporation's "couch pouch," a composite of polypropylene sheet foam and polypropylene fabric that can be repeatedly reused for shipping furniture.

A high degree of materials diversity in packaging is not uncommon and occurs because of the different physical and chemical properties of packaging materials, in combination with a lack of concern about packaging recyclability. An example is sketched in Figure 10.4 for a large electronic equipment module. First, the module is placed on a wooden pallet that is constructed with polymer cushioning to minimize shock damage during shipment. Next, a plastic sleeve is placed over the equipment for dust protection. Styrofoam blocks are added to cushion the top of the module and restrict its motion. Over the styrofoam and plastic is placed a corrugated cardboard cover to protect the module during shipment. The cover is held in place by steel strapping. This packaging system is very effective at protecting the product during shipment. At installation, however, the customer must deal with wood, cardboard, steel, two types of foam, and plastic sheeting. The chances of most of this material being recycled are small. It is likely that a packaging engineer working with the product design team and striving for both safe delivery and environmental superiority in the packaging operation could have reduced the materials diversity substantially.

Does each product require its own package? Not necessarily, say packaging engineers who can often minimize the total number of packages needed. A common example is the sale of window cleaner and liquid soap refills in a size inappropriate for normal use but suitable for replenishing containers in service. The use of refills also eliminates the need to manufacture, use, and discard spray heads for each bottle of product.

A final rule in packaging is to put package recycling information on (or in) each container and to help provide an infrastructure for the return of packaging for reuse or recycling.

10.5   THE PRODUCT USE PHASE

DfES also considers the environmental impact produced by products when they are used. Unlike the extraction or manufacturing environments, which are under the direct control of corporations, the use and maintenance of a product after it passes to the customer is largely constrained only by the product design. This circumstance places special responsibilities on the designer to envision aspects of design that minimize impacts during the entire useful life of the product.

Ideally, a product's use will not involve the purchase and eventual disposal of consumable supplies, such as cartridges for laser printers or lubricants for gears. If these seem necessary, the devices should be designed to encourage reuse or efficient recycling. An example of the latter is the approach used by a number of corporations to recycle the laser printer cartridges. To the degree that solid residues from consumables can be eliminated by innovative design, the environmental benefit is obvious. In any case, consumables should have little or no inherently toxic materials within them.

Products whose use involves such processes as the venting of compressed gas or the combustion of fossil fuels require the industrial ecologist to explore design modifications to minimize or eliminate these emissions. The automobile's internal combustion engine is perhaps the most common example of such a product and one whose cumulative emissions are very substantial, but anything that emits an odor during use is generating gaseous residues; examples include vapors from carpet adhesives, polymer stabilizers from plastics, and vaporized fluids from dry cleaners. Replacements for the volatile chemical constituents will often be available if the designer looks for them.

A number of products use energy when in operation. The energy may be electrical, as with refrigerators or hair dryers, or furnished by fossil fuel, as with power lawn mowers. Recent redesigns have produced lower energy consumption during use for many products, and legislation in an increasing number of countries will enhance such efforts. Energy-efficient designs may sometimes involve new approaches; the result provides not only lower cost operation but also improved product positioning (from a sales standpoint), particularly in areas of the world that are energy-poor.

Longer product lifetimes, and thus the consumption of resources, can sometimes be enhanced by remanufacturing. This involves the reuse of nonfunctional products by retaining serviceable parts, refurbishing usable parts, and introducing replacement components (either identical or upgraded). Such a process is frequently cost-effective and almost always environmentally responsible. It requires close relationships between customer and supplier, frequently on a lease basis; these relationships are often competitive advantages in any case. Remanufacture requires thoughtful design, because the process is often made possible or impossible by the degree to which products can be readily disassembled and readily modified.

Dissipative Products: Avoid if Possible

Many products are designed to be dissipative in use, that is, to be eventually lost in some form to the environment with little or no hope of recovery. Examples include surface coatings such as paints or chromate treatments, lubricants, personal-care products, and cleaning compounds. Attempts are being made to minimize both the packaging volume and the product volume in some of these situations, as in the recent introduction of superconcentrated detergents (though this does not change their basic dissipative nature). Alternatively, some liquid products that are dissipated when used can be designed to degrade in environmentally benign ways. Over the past several years this approach has successfully been adopted for a number of pesticides and herbicides. A recent demonstration of design for biodegradability is the development of a biodegradable synthetic engine oil, designed specifically for inefficient two-cycle engines that in operation emit approximately 25 percent of the gasoline-oil mixture unburned to the environment. An even better design approach, if possible, is to avoid the possibility of dissipation (making color an integral part of a plastic component rather than using paint, for example).

Another common example of a potentially dissipative product is fertilizer for crops, where any excess spread on fields is dissipated to local and regional ground and surface waters. The amount used by farmers has traditionally shown great variation—for the same crop yield, fertilizer application has ranged over some two orders of magnitude, or, to look at the data in another way, the same level of fertilization can produce manifold variations in yield. Much of this variation is due, of course, to the qualities of the different soils and differences in climate, but it is generally agreed by experts that farmers in the developed countries tend to use more fertilizer than can be justified by the resulting crop yields, and farmers in the developing world tend to use too little (often because of the expense). To the extent that excess fertilizer is used, it has a threefold negative impact: excessive extraction of raw materials, the accrual of financial penalties to the farmers, and dissipative release and negative impacts of excess fertilizer on proximate water supplies.

10.6   DESIGN FOR REUSE AND RECYCLING

10.6.1   The Comet Diagram

The concept of industrial ecology is one in which the components or constituents of products that have reached the end of their useful life re-enter the industrial flow stream and become incorporated into new products. The efficiency with which this cyclization occurs is highly dependent on the design of products and processes; it thus follows that Designing for Recycling (DfR) is one of the most important aspects of DfES.

Alternative end-of-life strategies are shown in the "comet diagram" of Figure 10.5. Users are at the perihelion point in the product's orbital path. During the approach to perihelion (the portion of the orbit at the top of the diagram), materials are formed,

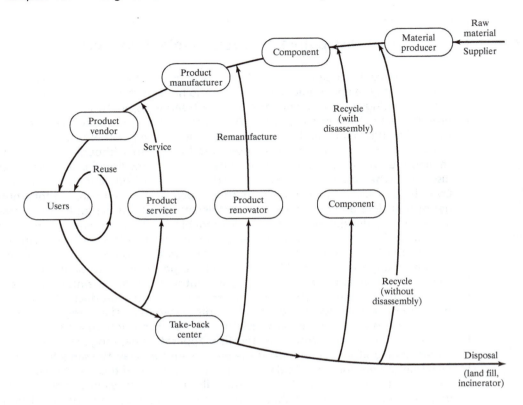

Figure 10.5

The "comet diagram," showing end-of-life reuse, refurbishment, and recycling strategies at different stages in the product life cycle. (Adapted from C.M. Rose, *Design for Environment: A Method for Formulating Product End-of-Life Strategies,* PhD dissertation, Stanford University, 2000.)

converted into components, then to products, and then marketed. During the retreat from perihelion, the products or their constituents are either reused or discarded.

A closed loop is obviously preferable from an environmental viewpoint, and the shorter the loop the better, because short loops retain the materials and energy embodied in products during their manufacture. Most of the loops require service agents who enable resources to move from the outgoing segment of the orbit to the incoming segment.

The original designer of a product defines the loop options available to the user and potential recycler. The ideal design permits renovation and enhancement to be accomplished by changing a small number of subassemblies and recycling those that are replaced. Next best is a design that requires replacement of the product but permits many or most of the subassemblies to be recovered and recycled into new products. Usually the least desirable of the alternatives is complete disassembly followed by recovery of the separate materials in a product (or perhaps some of the embodied energy, if the product is best incinerated) and the injection of the materials or energy back into the industrial flow stream.

10.6.2 Approaches to Design for Recycling

A directly practical reason for all industries to practice DfR is the trend for governments and other consumers to require or give preference to products incorporating the DfR philosophy. In 1991, for example, U.S. Government Executive Order 10780 was promulgated. It requires all agencies of the government (when combined, the agencies are the country's largest consumer) to buy products made from recycled materials and to encourage suppliers to participate in residue recovery programs. In the same year, the State of New York issued a Request for Proposal for personal computers for its offices in which it stated that recyclability would be a factor in choosing the successful bidder. The German "Blue Angel" environmental seal routinely includes such requirements in its assessments. Such actions provide a graphic and easily communicated rationale for DfR.

A central consideration in DfR is to minimize the number of different materials and the number of individual components used in the design. (This design strategy is independently known as design for simplicity.) To get a sense of the importance of this recommendation, picture yourself responsible for recycling hundreds of television sets or photocopy machines or refrigerators every week. If you need to locate, sort, clean, and provide efficient recycling for 2 or 3 metals and 2 or 3 plastics, you are far more likely to be successful than if you must deal with 5 metals, 12 alloys, 20 plastics, and miscellaneous items such as glass or fabric. The functional and aesthetic demands of design sometimes make it difficult to limit materials diversity or complexity too greatly, but minimization should be a central focus for every designer.

A second general goal is to consider the fate at the end of product life of any hazardous materials. This topic has been discussed earlier with respect to the extraction or manufacture of materials, and their dissemination during industrial processes. The goal is equally important in the product recycling arena, where the presence of such materials is a deterrent to detailed disassembly, eventual reuse, or, if necessary, safe incineration and energy recovery. Where hazardous materials must be utilized in a design they should be easily identifiable, and the components that contain them readily separable.

Another general recommendation for the designer is not to join dissimilar materials in ways that make separation difficult. A simple example of a product not designed for recycling is the glass container for liquids whose top twists off while leaving a metal ring affixed; small cutting pliers are required for the conscientious housekeeper to properly sort the materials if the local recycling facility is unable to quickly and cheaply do so. More complex variations on this theme are metal coatings applied to plastic films, plastic molded over metal or over a dissimilar plastic, and the "up-scale" automobile dashboard, which is a complex mixture of metal, wood, and plastic. Any time a designer uses dissimilar materials together, she or he should picture whether and how they can eventually be easily separated, an important concept because labor costs tend to be a significant barrier to recycling.

Research by C.M. Rose and colleagues (2002) has shown that the factors most important in end-of-life strategies are (1) product wear-out life, (2) technology cycle (length of time that a product will be on the leading edge of technology), (3) number of parts, and (4) design cycle (how often do companies design new products). A particularly useful metric turns out to be the ratio between the wear-out life λ and the technology cycle τ. Three groups of products can be identified by this ratio, and their probable end-of-life treatment specified:

$\lambda/\tau \geq 4$ These products are long-lasting, but the technology embodied within them changes rapidly. Television sets are examples. This product group should generally be recycled at end of life.

$1 < \lambda/\tau < 4$ These products have a range of lifetimes, but all the lifetimes are longer than the technology cycle. Digital copiers and washing machines are examples. For $\lambda < 10$ years, recycling is appropriate. For $\lambda > 10$ years, individual evaluation should be used to decide if remanufacturing or recycling is appropriate.

$\lambda/\tau \leq 1$ These products are short-lived, but the technology cycle is long. Examples include vacuum cleaners and shipping containers. They should generally be remanufactured at end of life.

If recycling is the choice, two complementary approaches should be considered: *closed-loop* and *open-loop*. As seen in Figure 10.6, closed-loop recycling involves reuse of the materials to make the same product over again (sometimes called "horizontal recycling"), while open-loop recycling reuses materials to produce different products (sometimes called "cascade recycling"). (Typical examples could be aluminum cans to aluminum cans in the first instance, office paper to brown paper bags in the second.) The mode of recycling will depend on the materials and products involved, but closed-loop should generally be preferred.

10.6.3 Recycling Complexities

Ever since the beginnings of human cultural society, resources that are scarce for one reason or another have been more valued than those that are abundant. It turns out, in fact, that there is approximately an inverse correlation between price and the concentration of resources in the natural environment. Among the consequences of this relationship is that some materials in discarded products are more likely to be recovered and recycled than others. Another consequence is that materials present in discarded products at concentrations lower than those currently being mined are unlikely to be recycled. Figure 10.7 shows that many materials from automobiles do indeed undergo recycling (those near or above the "natural resource" line). (Mercury is recovered not because of its abundance in automobiles, but because regulations require it.) Silver, tin, and tungsten, present at low concentrations, are typically not recovered.

Another design feature that has a major import on recycling probability is the structural intricacy of the product design. A useful analytical tool to assess this feature is the Reverse Fishbone Diagram. The standard fishbone diagram, in common use in industrial engineering, is a graphical illustration of the sequence in which the components of a product are assembled from materials or lower-level components, and the sequence in which the final product is assembled from the components. The reverse fishbone diagram is a picture of the ideal *dis*assembly process, showing the order of removal and separation of the parts. An example is shown in Figure 10.8. By constructing such a diagram,

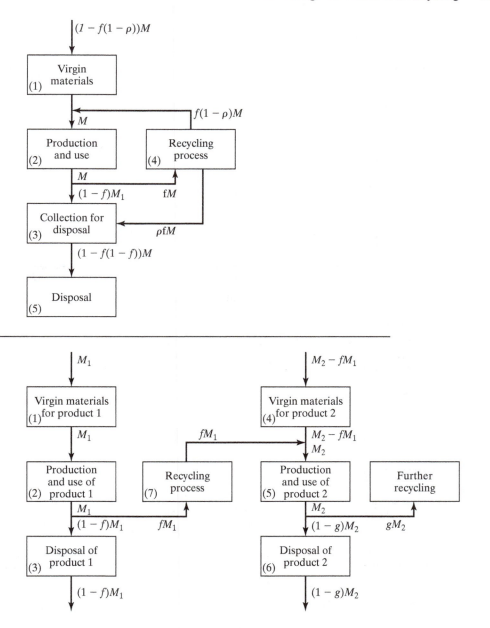

Figure 10.6

Closed-loop (top) and open-loop (bottom) recycling of materials. In the diagrams, M refers to mass flows, f and g to the fractions of the flows delivered to the recycling process, and ρ to the fraction of those flows rejected as unsuitable for recycling. (Adapted from B.W. Vigon, D.A. Tolle, B.W. Cornaby, H.C. Latham, C.L. Harrision, T.L. Boguski, R.G. Hunt, and J.D. Sellers, *Life-Cycle Assessment: Inventory Guidelines and Principles*, EPA/600/R-92/036, US Environmental Protection Agency, Cincinnati, OH, 1992.)

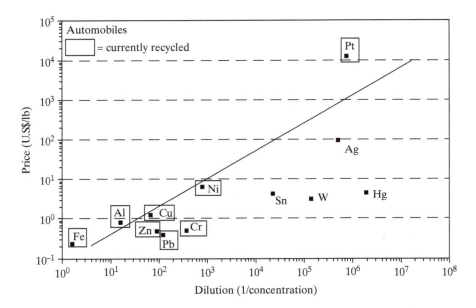

Figure 10.7

Metal concentrations in automobiles in relation to price. The straight line is derived from the price–dilution relationship for minimum profitable ore grades of metals mined in 2004. (Reproduced with permission from J. Johnson et al., Dining at the periodic table: Metals concentrations as they relate to recycling, *Environmental Science & Technology, 41*, 1759–1765, 2007.)

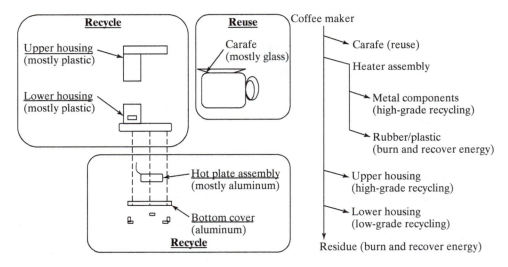

Figure 10.8

The reverse fishbone diagram for a coffeemaker. (Courtesy of Kosuke Ishii, Stanford University.)

the engineer can often discover opportunities for increasing product recyclability while retaining other desirable product characteristics.

A quantitative evaluation of disassembly efficiency has been devised by Dahmus and Gutowski (2007). They treat binary separation steps in disassembly by analogy with information theory, defining a material mixing parameter H as

$$H = \sum_{i=1}^{M} c_i \ln c_i \qquad (10.1)$$

where M is the number of materials and c_i is the concentration of material i. They then derive H values for a number of different products, and plot H against recycled material value (Figure 10.9). The results suggest the existence of an "apparent recycling boundary," below which recycling is unlikely to occur.

A final challenge to recycling has to do with the particular combinations of elements in technological products. Figure 10.3 demonstrated that metals are frequently used in alloy form in which perhaps 5–40 percent of one or more elements are added

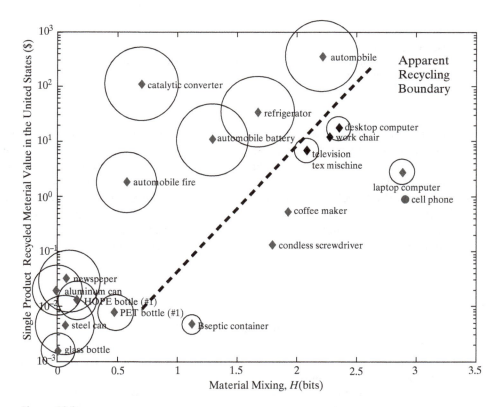

Figure 10.9

Single product recycled material values, material mixing, and recycling rates, the last as indicated by the area of the circles, for 20 products in the United States. (Reproduced with permission from J.B. Dahmus, and T.G. Gutowski, What gets recycled: An information theory based model for product recycling, *Environmental Science & Technology, 41,* 7543–7550, 2007.)

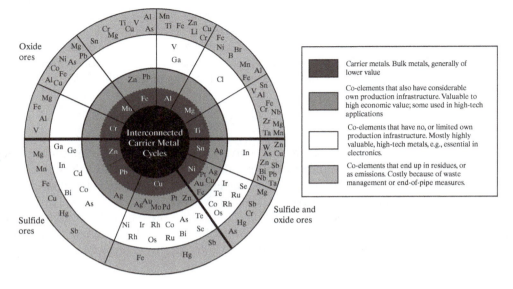

Figure 10.10

The wheel of metal linkages in natural resource processing. (Reproduced with permission from E.V. Verhoef, G.P.J. Dijkema, and M.A. Reuter, Process knowledge, system dynamics, and metal ecology, *Journal of Industrial Ecology, 8* (1–2), 23–43, 2004.)

to a predominant element. The result generally is a material possessing enhanced physical or chemical characteristics. These alloys are creations of humans, however, not of nature; nature forms its material deposits in much different combinations, as shown in Figure 10.10. Humans have mined and processed metal ores for millennia, and separation technologies for dealing with the material combinations given in Figure 10.10 are well developed. The same cannot be said for some of the newer alloy combinations, and therefore alloys are generally recycled only if they are kept separate from contamination and reprocessed and reused as such.

"Just-in-Case" Designs

Among the more unusual topics that enter into DfES is that concerning products designed in the hope and expectation that their use may seldom or never be required. Spare parts are an obvious example, but entire products such as air packs or medical supplies, sprinkler systems for fire suppression, and backup safety devices for elevators can fit the definition as well. The range of technology in such equipment extends to the most sophisticated, as in intercontinental ballistic missiles with computer-guided control systems.

(*continued*)

Just as society needs "green products" in its everyday activities, so too it needs "green spares," "green rarely used items," and (an intriguing oxymoron) "green weapons." In other words, just because an item is designed with a reasonable expectation that it will sit around for a decade or two without being used and then be discarded, but must work the first time if needed, the designer of that item is not excused from the responsibilities of DfES. Such features as materials selection, modularity, Design for Disassembly, and especially Design for Maintenance need to be given special attention for "just-in-case" products. This topic has received little attention to date, but the enormous inventory of materials and products involved suggests that the time for focused activity is well overdue.

10.7 GUIDELINES FOR DfES

Many authors have devised sets of rules to aid the product designer in the development of environmentally superior products. To some extent, these rules differ according to the industrial sector being addressed. A generic list is useful, however, and we provide one in Table 10.3. Designers who carefully follow these rules will produce much "greener" products than would otherwise be the case, and will contribute to the sustainability of their firm and their planet.

TABLE 10.3   A Generic Set of Design for Environment and Sustainability Guidelines

1. Choose readily available materials, e.g., those that avoid the "region of danger" on the supply-risk importance of use diagram (Chapter 24).
2. Minimize or avoid hazardous substances, and utilize closed loops for necessary but hazardous constituents.
3. Minimize materials diversity.
4. Minimize energy and resource consumption in the production phase (Chapters 13 and 14).
5. Design the manufacturing process to minimize the buy-to-sell ratio.
6. Use structural features and high-quality materials to minimize product weight.
7. Employ recycled materials and recycled parts whenever possible.
8. Package products in an environmentally sound manner.
9. Minimize energy requirements and resource consumption and dissipation in the use phase.
10. Choose materials, surface treatments, and structural arrangements to inhibit dirt, wear, and corrosion, thereby enhancing product life.
11. Encourage upgrading, repair, remanufacturing, and recycling through accessibility, labeling, modules, breakage points, and manuals.
12. Design for disassembly by using as few joining elements as possible and emphasizing screws, snap fits, and geometric locking.
13. Avoid material combinations that force time consuming and otherwise unnecessary separation at the recycling stage.
14. Follow guidelines 1–13 down the supply chain so that all parts of your product meet the DfES standard.

Source: Developed from information earlier in this chapter and from C. Luttropp, and J. Lagerstedt, EcoDesign and the ten golden rules: Generic advice for merging environmental aspects into product development, *Journal of Cleaner Production, 14*, 1396–1408, 2006.

FURTHER READING

Design for Environment—general

Graedel, T.E., and B.R. Allenby, *Design for Environment*, Upper Saddle River, NJ: Prentice Hall, 1996.

Kutz, M., Ed., *Environmentally Conscious Mechanical Design*, Hoboken, NJ: John Wiley, 2007.

Materials choice

Ashby, M.F., *Materials Selection in Mechanical Design*, 3rd ed., Amsterdam: Elsevier, 2005.

Castro, M.J.B., J.A.M. Remerswaal, M.A. Reuter, and U.J.M. Boin, A thermodynamic approach to the compatibility of materials combinations for recycling, *Resources, Conservation, and Recycling, 43*, 1–19, 2004.

Product packaging

Hellström, D., and M. Saghir, Packaging and logistics interactions in retail supply chains, *Packaging Technology and Science, 20*, 197–216, 2007.

Stilwell, E.J., R.C. Canty, P.W. Kopf, and A.M. Montrone, *Packaging for the Environment: A Partnership for Progress*, New York: American Management Association, 1991.

Remanufacturing

Ijomah, W.L., and S.J. Childe, A model of the operations concerned in remanufacture, *International Journal of Production Research, 45*, 5857–5880, 2007.

Klausner, M., W.H. Grimm, and C. Hendrickson, Reuse of electric motors in consumer products, *Journal of Industrial Ecology, 2* (2), 89–102, 1998.

Truttmann, N., and H. Rechberger, Contribution to resource conservation by reuse of electrical and electronic household appliances, *Resources, Conservation, and Recycling, 48*, 249–262, 2006.

Recycling

Craig, P.P., Energy limits on recycling, *Ecological Economics, 36*, 373–384, 2001.

Dahmus, J.B., and T.G. Gutowski, What gets recycled: An information theory based model for product recycling, *Environmental Science & Technology, 41*, 7543–7550, 2007.

Reuter, M.A., A. van Schaik, O. Ignatenko, and G.J. de Haan, Fundamental limits for the recycling of end-of-life vehicles, *Minerals Engineering, 19*, 433–449, 2006.

Rose, C.M., A. Stevels, and K. Ishii, Method for formulating product end-of-life strategies for electronics industry, *Journal of Electronics Manufacturing, 11*, 185–196, 2002.

EXERCISES

10.1 Select a wrench or other simple hand tool and, to the degree possible, evaluate it qualitatively for each of its "Design for X" features.

10.2 Repeat Exercise 10.1 for a kitchen electrical appliance of your choice.

10.3 Form a design team with four other students. Choose one of the following products: 10-cup coffee maker, overhead projector, bicycle, power lawn mower. Each member of your team will play one of the following roles: mechanical designer, manufacturing engineer, environmental specialist, marketing specialist, corporate lawyer. Develop the House of

Quality for the product of your choice, and describe the diagram and the resulting product design concept in a 4–5-page report.

10.4 Discuss options for packaging of the following consumer products: motor oil, grapefruit juice, toothpaste, magazines, shirts, decongestant tablets.

10.5 You are the designer of a table to be used for sorting fruit in a field near a cannery. The table is to have a steel surface and wooden legs. The surface is to be covered with a soft foam top to reduce fruit damage. It is expected that the foam top and the legs will need to be replaced periodically and the cannery owner, who expects to purchase several hundred tables, wants component replacement to be quick and efficient. With the help of your local hardware store (if needed), design the table for optimum disassembly.

CHAPTER 11

Design for Environment and Sustainability: Buildings and Infrastructure

11.1   THE (INFRA)STRUCTURES OF SOCIETY

Buildings and infrastructure are in many ways not only the products of modern technological societies, but also the enablers of those societies. Unlike the individual products discussed in Chapter 10—computers, automobiles, and the like—buildings and infrastructure have different attributes that call them out to the sustainable engineer as worthy of special attention. The first is their long lives—Roman water infrastructures span many centuries, and buildings constructed two and three hundred years ago are not too uncommon in many places. If these entities were poorly designed at the start, the consequences of the designs resonate with such characteristics as profligate energy use or poor water quality over very long periods of time.

A second feature of buildings and infrastructure is the sheer mass of the resources they require. Figure 11.1 for the United States is typical in that some 80 percent of all resources (by mass) are employed in the construction, modification, and refurbishment of buildings and infrastructure.

A third attribute of buildings and infrastructure is that they are built in place, and are generally (but not always) one-time designs. Each project is thus an opportunity for a unique contribution to responsible technology and sustainability.

Notwithstanding these distinctions, many features of building and infrastructure design are essentially identical to those of smaller, shorter-lived products. Informed material choice is certainly one commonality, with historic mistakes like mercury switches in automobiles and asbestos insulation in buildings illustrating the issue. Another is manufacture, in which the environmental disruptions and emissions from

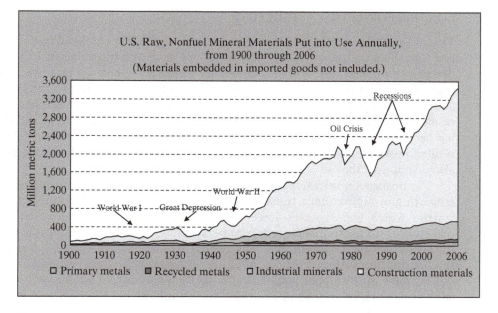

Figure 11.1

The use of materials in the United States, 1900–2006. (Courtesy of S. Sibley, U.S. Geological Survey.)

building and road construction are widely recognized challenges. The environmental implications of product use, often significant for products such as air conditioners or automobiles, are absolutely crucial for very long-lived buildings and infrastructure, as are designs that promote maintainability.

Infrastructure differs from other products in that the services it provides are common-pool resources that are developed when a geographic concentration of people expresses a need for them and creates institutions to build and administer them. Four types of infrastructure are commonly identified: electric power provisioning, water supply and treatment, communications, and transport (roads, railroads, airports). Water and transport facilities are generally owned by government entities, and energy and communications by private entities, though arrangements differ from place to place. The key factor is an institution of some kind that constructs, maintains, and administers the infrastructure system.

Partly because of their long lives, and partly because of technological progress, few buildings and few elements of infrastructure live out their lives without change. As a consequence, while the initial design and construction have a strong sustainability component, the evolution of buildings and infrastructure over time and their eventual demise are at least equally important, probably more so. Thus, design for recycling is vital, particularly in view of the large volumes of discarded construction and demolition debris that are an inevitable attribute of a rapidly developing society. The products of the architect and the civil and environmental engineers, and the degree of flexibility and adaptability in their designs, are thus major focal points in any discussion of design for environment.

11.2   ELECTRIC POWER INFRASTRUCTURE

Virtually nothing in today's world is possible without reliable, high-quality electric power, and furnishing that power is a central feature of modern infrastructures. Electric power designers consider three system components: generation (in which power is produced), transmission and distribution (in which power is transferred from generators to users), and use (where power is employed). The system is sketched in Figure 11.2. Generation can result from a variety of technologies—fossil fuels, hydropower, solar, wind, and so on, and has been well covered elsewhere (see Further Reading). It will not be discussed here, as it is often provided independently rather than as a component of public infrastructure. We also do not treat the use of electrical power here, but in Chapter 19.

Transmission infrastructure is generally divided into the *transmission system*, consisting of high-voltage transmission lines and transformers, and the *distribution system*, which encompasses lower-voltage lines and transformers that form the connections to users. From a material perspective, aluminum is the usual conductor in the transmission system, whereas copper is used in transformers, bus bars, and the distribution system. Portions of either system may be underground. Such installations have much higher installation costs than above-ground systems, but are not subject to outages in storms. In urban areas, underground distribution is generally the only feasible approach given space restrictions.

The classic means of energy provisioning is to provide large centralized facilities such as coal, nuclear, or hydropower facilities. These facilities provide (in principle, at least) oversight that encourages good economic, environmental, and safety performance.

National and international transmission grids provide substantial power reserves and enable power to be shifted rapidly from one user or one region to another. Because user needs vary continuously, the rate of generation must also vary continuously, and the degree of interconnection generally renders the supply of electrical power stable and reliable. The potential penalty paid for this interconnectivity is that failures in one part of the system have the potential to quickly cause failures elsewhere.

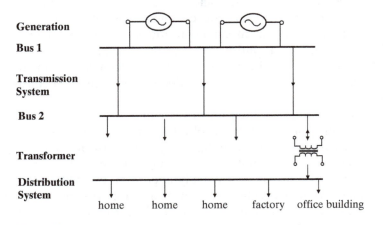

Figure 11.2

The electrical power infrastructure system. The transformation and distribution components are indicated by examples only. (Buses are conducting bars that connect electrical current to several different circuits.)

A modern alternative to large centralized facilities is *distributed generation*, in which electricity is generated from many small energy sources. These sources often use clean fuels such as sunlight and wind to generate electricity, or they may adopt combined-cycle natural gas to generate heating or cooling as well. Added benefits are that energy loss in transmission (and construction of extensive transmission facilities) can be minimized in distributed facilities, and the insertion of many small energy sources in a neighborhood grid can often increase reliability and power quality (i.e., avoidance of "brownouts" and power dips and surges).

With the advent of electric cars, planners are now envisioning a grid that charges automobile batteries at night, when electrical demand is low, and draws excess power from those batteries during the day, when demand is higher. So long as the batteries are fully charged when the vehicle is needed (a data and management challenge), such an approach would allow energy to be stored and utilized much more efficiently than is now possible.

11.3   WATER INFRASTRUCTURE

Supplying and treating water is a second major element of infrastructure. Unlike buildings, energy, communications, and highways, the amount of water use is not directly related to urban areas: some 70 percent of water is used in agriculture.

The water must be transported, either by gravity or by the use of energy, and is generally used without treatment. For the other 30 percent and especially where human consumption of water is involved, various degrees and types of purification are employed. A typical system is sketched in Figure 11.3. Unlike electrical power, which can travel at elevation above ground, water moves from one place to another at or below ground, and systems of canals and pipes constitute major infrastructure projects.

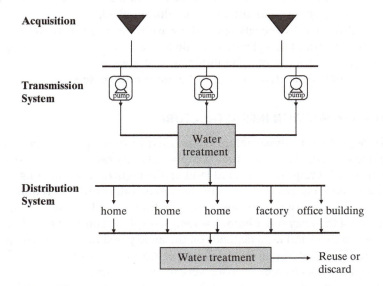

Figure 11.3

The water infrastructure system. Pumps are employed very widely to acquire and move water; only a few are shown here. Factories often perform water treatment in addition to that shown.

The complex water infrastructure is comprised of a number of components: acquisition from freshwater, groundwater, or saline sources; transmission; treatment; distribution; use; wastewater treatment; and possible reuse. Where acquisition is from freshwater (rivers, lakes, snowmelt, etc.), the infrastructure generally involves storage behind dams and transport in aqueducts. Groundwater acquisition requires pumping from depth. Saltwater must be acquired through pumping and piping.

Water treatment is designed to remove pathogens and other undesirable trace constituents. If saline water is being used, desalination (the removal of excess salts) is required as well. All of these steps, especially the latter, employ large amounts of energy. The technologies for water purification and desalination are varied and are discussed in Further Reading; we do not go into detail here because the installations are best regarded as products whose designers should consider the guidelines given in Chapter 10.

Water infrastructures use large amounts of energy and large quantities of materials. Such attributes provide strong incentives for water reuse and for dedicated maintenance in order to limit the amount of water needed and the amount of water lost. The industrial ecology of water is discussed in more detail in Chapter 20.

Water quality is an increasingly important topic at both the initial use stage and the discharge stage. Modern water systems utilize tertiary (three-stage) treatment, an approach not contemplated when many systems were initially built. The now common detection of low levels of bioactive drug products in water suggests that even this level of treatment will be inadequate in the future, a prospect that appears to make flexibility of design in any new construction a priority.

Following treatment, the distribution system transports water to the final customers—residential, commercial, and industrial. The construction and maintenance of the distribution system is a major component of infrastructure. Dedicated maintenance is important, because water lost in distribution must be made up by additional supply, with all the energy and infrastructure thus implied.

After use by the customer, the wastewater must be treated again to render it safe for reuse or discharge. Depending on the reuse (toilet flushing, irrigation of vegetation, etc.), the treatment may be minimal. If the wastewater is to be restored to drinking water quality, the treatment must be much more extensive.

11.4   TRANSPORTATION INFRASTRUCTURE

Transportation infrastructure is designed to move people and goods efficiently from place to place. There are four components: roadways, railways, ports and harbors, and airports. All require dedicated land, initial construction, and ongoing maintenance and renovation. Extensive provisions for public safety are also required.

The construction of streets and highways involves the movement and use of more material than any other human activity except the construction of buildings. The material that is moved and used depends on the anticipated degree and intensity of the use of the road. (The total distances of different categories of roads in the United States are shown in Figure 11.4.) Roads receiving very modest use will normally have a gravel surface, the maintenance of which requires the periodic hauling of substantial amounts of stone and gravel. Roads with more traffic, such as local and collector roads, are generally paved with asphalt. Arterial roads and expressways often have concrete surfaces.

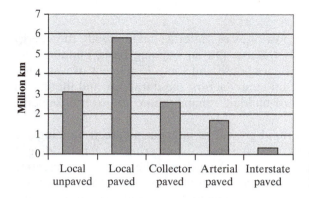

Figure 11.4

The 2005 U.S. road network, by type of road. Source: From *National Transportation Statistics*, 2005, Bureau of Transportation Statistics, Washington, DC.

There is great diversity and magnitude in the materials contained in a modern roadway. In addition to the anticipated items, large quantities of lumber are used as framing for concrete; culverts and pipes are incorporated routinely; and fuel and lubricant use for construction equipment is substantial.

Railroads are major users of steel in rails and in electrical distribution systems. Tunnels may be required, either in mountainous country or for urban transit systems. The necessary rolling stock is designed on Chapter 10 guidelines.

Ports, harbors, and airports require extensive infrastructure to handle freight and people. These facilities are often in environmentally sensitive areas, and their development must be approached with care. The siting and development considerations discussed below for green buildings are often applicable here.

11.5   TELECOMMUNICATIONS INFRASTRUCTURE

Telecommunications networks consist of links and nodes arranged so that messages may be passed from one part of the network to another. Some networks are private, as with in-house computer networks, but the infrastructure example is the public switched telephone network. This network has historically consisted entirely of landlines, but has now evolved into a combination of landline and radio (cellular telephone) facilities. (In many developing countries, only the latter is employed.)

There are three overlaid parts to telecommunication networks:

- The control plane, which carries signaling information
- The user plane, which carries messages
- The management plane, which carries network management information

As with electrical power, telecommunications infrastructure can be on overhead or underground facilities. (In the case of cellular telephones, once a message reaches a base station, it then travels just as a landline message does.) It often shares poles and underground passages and access entryways with the electrical power infrastructure. While vital to modern society, the material and energy requirements for telecommunication infrastructures are far less than those for energy, water, or transportation.

11.6   GREEN BUILDINGS

As with all products, buildings are repositories of materials and embody the energy consumed to prepare those materials for use. On an ongoing basis, buildings are responsible for some 30–40 percent of energy use and some 15 percent of water use worldwide; these rates are dependent on many factors: the durability and lifetime of building components, rates and types of building rehabilitation, and overall building lifetimes.

Recent years have witnessed a rapid growth of interest in "green buildings," a development well within the purview of sustainable engineering. Green building is defined by the U.S. Environmental Protection Agency as "the practice of creating structures and using processes that are environmentally responsible and resource-efficient throughout a building's life cycle, from sitting to design, construction, operation, maintenance, renovation, and deconstruction." The emphasis is on the careful choice and efficient use of materials, energy, and water, and the minimization of environmental effects of all kinds.

New buildings are the instinctive focus of green building thinking, but such a view is far too narrow. In many of the more developed countries, renovation and refurbishment activities are at least as important as new building construction. It is not uncommon for residences to become small business offices, for industrial buildings to become apartments, or for university classroom buildings to become offices. Often these alternative uses can be anticipated when the buildings are originally designed. As a consequence, architects and engineers should design buildings that are easy to alter and renovate, and choice of materials should be made from the perspective of this flexibility.

Green building design avoids the traditional architecture approach of establishing strict barriers with the outside world. Rather, the goal is to take advantage of local resources—using natural light instead of artificial light where possible, outside air circulation rather than heating or air conditioning where feasible, solar and geothermal power where practicable, and so forth. Consequently, energy use is minimized; water is recycled and reused in various ways; and interior materials are selected to minimize or avoid health problems.

It is not unusual for green building construction costs to be a few percent more than those of a traditional building. This added cost tends to be rapidly recovered in reduced energy and water costs, and in increased productivity of building occupants.

Ideally, buildings should be designed with end of life in mind. Although some structures last for centuries, most do not reach their potential service life. This is because economics, not structural decay, usually define the realized lifetime, which reaches an end when the market value of the property plus any demolition costs are exceeded by the value of the cleared site for a new use. Building design for disassembly recognizes this fact in making it easier to recover useful elements of the building fabric and interior.

All of these facets of green building design are captured, at least in principle, by rating systems such as Australia's Green Star system or the United State's Leadership in Energy and Environmental Design (LEED) program. These scoring systems focus on measurable design attributes that can be certified by independent auditors.

The project checklist for the LEED system, which is typical of those in many other countries, is shown in the attached display. It has six principal components. The sustainable sites component promotes reuse of previously used land, use of public transportation, and operational management of storm water and other local environmental influences. The water efficiency and energy and atmosphere components promote minimalism in the use of water and energy. The fourth component is materials and those acquired nearby. The fifth component, indoor environmental quality, aims to make the environment for the occupants of the building as healthy as possible. Finally, innovations in green building design not otherwise addressed are captured as the sixth component.

The LEED system has versions of its checklist for different types of building activities: homes, existing building, schools, and so forth. The maximum score is 69 points, and four levels of certification are awarded:

Certified	26–32 points
Silver	33–38 points
Gold	39–51 points
Platinum	52–69 points

The transparent and widely publicized green building systems around the world have transformed architecture in recent years. Innovations such as energy-efficient elevators, vegetative roofs for storm water management, and the use of low-emission interior materials are increasingly common. Because traditional buildings tend to be profligate uses of materials, energy, and water, green building design and operation is a vital component of long-term sustainability.

11.7   INFRASTRUCTURE AND BUILDING MATERIALS RECYCLING

Infrastructures and buildings share a great commonality in their use of materials:

Concrete: Widely used in buildings, arterial highways and expressways, bridge decks, water culverts, and pipes. Usually reinforced with galvanized steel bars (i.e., steel bars plated with zinc to inhibit corrosion). In sensitive uses such as bridge decks, stainless steel reinforcing bars are sometimes used—more expensive, but less likely to corrode.

Aggregate: The principal constituent of asphalt, used widely on roadways, and the principal constituent of concrete.

Cement: The binding agent in concrete.

Asphalt: The typical surface treatment on local and collector roads.

Steel: Widely used throughout all facets of infrastructure and buildings, generally in galvanized form.

Zinc: A relatively inexpensive and largely effective anticorrosion coating applied to most steel used in infrastructures and buildings.

Copper: The principal electrical conductor for electrical power transformers and distribution systems and for much of the telecommunications networks (although

L E E D™
LEADERSHIP IN ENERGY & ENVIRONMENTAL DESIGN

Version 2.1 Register ed Project Checklist

Project Name
City, State

Yes	?	No			
			Sustainable Sites		**14** Points
Y			Prereq 1	**Erosion & Sedimentation Control**	Required
			Credit 1	**Site Selection**	1
			Credit 2	**Urban Redevelopment**	1
			Credit 3	**Brownfield Redevelopment**	1
			Credit 4.1	**Alternative Transportation**, Public Transportation Access	1
			Credit 4.2	**Alternative Transportation**, Bicycle Storage & Changing Rooms	1
			Credit 4.3	**Alternative Transportation**, Alternative Fuel Vehicles	1
			Credit 4.4	**Alternative Transportation**, Parking Capacity and Carpooling	1
			Credit 5.1	**Reduced Site Disturbance**, Protect or Restore Open Space	1
			Credit 5.2	**Reduced Site Disturbance**, Development Footprint	1
			Credit 6.1	**Stormwater Management**, Rate and Quantity	1
			Credit 6.2	**Stormwater Management**, Treatment	1
			Credit 7.1	**Landscape & Exterior Design to Reduce Heat Islands** , Non-Roof	1
			Credit 7.2	**Landscape & Exterior Design to Reduce Heat Islands**, Roof	1
			Credit 8	**Light Pollution Reduction**	1

Yes	?	No			
			Water Efficiency		**5** Points
			Credit 1.1	**Water Efficient Landscaping**, Reduce by 50%	1
			Credit 1.2	**Water Efficient Landscaping**, No Potable Use or No Irrigation	1
			Credit 2	**Innovative Wastewater Technologies**	1
			Credit 3.1	**Water Use Reduction**, 20% Reduction	1
			Credit 3.2	**Water Use Reduction**, 30% Reduction	1

Yes	?	No			
			Energy & Atmosphere		**17** Points
Y			Prereq 1	**Fundamental Building Systems Commissioning**	Required
Y			Prereq 2	**Minimum Energy Performance**	Required
Y			Prereq 3	**CFC Reduction in HVAC&R Equipment**	Required
			Credit 1	**Optimize Energy Performance**	1 to 10
			Credit 2.1	**Renewable Energy**, 5%	1
			Credit 2.2	**Renewable Energy**, 10%	1
			Credit 2.3	**Renewable Energy**, 20%	1
			Credit 3	**Additional Commissioning**	1
			Credit 4	**Ozone Depletion**	1
			Credit 5	**Measurement & Verification**	1
			Credit 6	**Green Power**	1

			Materials & Resources	13 Points
Y			Prereq 1 **Storage & Collection of Recyclables**	Required
			Credit 1.1 **Building Reuse**, Maintain 75% of Existing Shell	1
			Credit 1.2 **Building Reuse**, Maintain 100% of Shell	1
			Credit 1.3 **Building Reuse**, Maintain 100% Shell & 50% Non-Shell	1
			Credit 2.1 **Construction Waste Management**, Divert 50%	1
			Credit 2.2 **Construction Waste Management**, Divert 75%	1
			Credit 3.1 **Resource Reuse**, Specify 5%	1
			Credit 3.2 **Resource Reuse**, Specify 10%	1
			Credit 4.1 **Recycled Content**, Specify 5% (post-consumer + ½ post-industrial)	1
			Credit 4.2 **Recycled Content** , Specify 10% (post-consumer + ½ post-industrial)	1
			Credit 5.1 **Local/Regional Materials**, 20% Manufactured Locally	1
			Credit 5.2 **Local/Regional Materials**, of 20% Above, 50% Harvested Locally	1
			Credit 6 **Rapidly Renewable Materials**	1
			Credit 7 **Certified Wood**	1

Yes ? No

			Indoor Environmental Quality	15 Points
Y			Prereq 1 **Minimum IAQ Performance**	Required
Y			Prereq 2 **Environmental Tobacco Smoke** (ETS) **Control**	Required
			Credit 1 **Carbon Dioxide** (CO_2) **Monitoring**	1
			Credit 2 **Ventilation Effectiveness**	1
			Credit 3.1 **Construction IAQ Management Plan**, During Construction	1
			Credit 3.2 **Construction IAQ Management Plan**, Before Occupancy	1
			Credit 4.1 **Low-Emitting Materials**, Adhesives & Sealants	1
			Credit 4.2 **Low-Emitting Materials**, Paints	1
			Credit 4.3 **Low-Emitting Materials**, Carpet	1
			Credit 4.4 **Low-Emitting Materials**, Composite Wood & Agrifiber	1
			Credit 5 **Indoor Chemical & Pollutant Source Control**	1
			Credit 6.1 **Controllability of Systems**, Perimeter	1
			Credit 6.2 **Controllability of Systems**, Non-Perimeter	1
			Credit 7.1 **Thermal Comfort**, Comply with ASHRAE 55-1992	1
			Credit 7.2 **Thermal Comfort**, Permanent Monitoring System	1
			Credit 8.1 **Daylight & Views**, Daylight 75% of Spaces	1
			Credit 8.2 **Daylight & Views**, Views for 90% of Spaces	1

Yes ? No

			Innovation & Design Process	5 Points
			Credit 1.1 **Innovation in Design**: Provide Specific Title	1
			Credit 1.2 **Innovation in Design**: Provide Specific Title	1
			Credit 1.3 **Innovation in Design**: Provide Specific Title	1
			Credit 1.4 **Innovation in Design**: Provide Specific Title	1
			Credit 2 **LEED™ Accredited Professional**	1

Yes ? No

			Project Totals (pre-certification estimates)	69 Points

Certified 26-32 points   **Silver** 33-38 points   **Gold** 39-51 points   **Platinum** 52-69 points

gradually being supplanted by fiber optics in the latter use). Commonly used for plumbing throughout infrastructure and buildings, (although gradually being supplanted by plastic).

Aluminum: Used extensively for building exterior components such as window frames and siding. Aluminum is the usual conductor in electrical power transmission systems (Figure 11.5).

Infrastructure components can be recycled as can any large industrial product. Among current practices in a number of locations are the reuse of aluminum signs following stripping and cleaning (at one-third to one-fifth the cost of new signs), the reconstruction of metal-beam guardrails, and the reuse at the same or different locations of hardware such as manhole covers and frames, lighting standards, and chain-link fencing. For hardware components, especially, design for environment approaches can be helpful in making recycling easier and more profitable.

In general, the approach to green infrastructure design is no different than for any other product: all infrastructures should be designed ab initio to be modular, easy to disassemble, and capable of being reused at the end of life. Horizontal recycling is preferable to vertical recycling, but the economics for this may be difficult given the low cost of much of the virgin material used for roads and highways. As with the

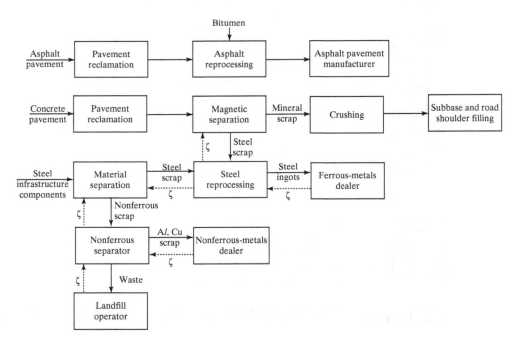

Figure 11.5

The infrastructure recycling sequence. Solid lines indicate flows of materials. Dashed lines and the generic symbol ζ indicate flows of money.

automobile, the recycling of infrastructure occurs in several stages, each stage with its own actors. The sequence is shown in Figure 11.5.

The infrastructure material recycled in the largest amounts is asphalt pavement. This process involves pavement fragmentation, followed by rotor milling to reduce the pavement chunks to very nearly the size of the original aggregate. Depending on the availability of equipment and the ease of transportation, the milling may be done at the construction site or the paving fragments may be taken to a rotor mill at a fixed site. The resulting milled material is then reheated, reformulated with the addition as necessary of new bitumen, and reapplied to the base surface material. It is often quicker to recycle pavement on-site than to import and use virgin materials, a characteristic that permits minimum disruption of traffic flow during repaving activities.

The potential for recycling of asphalt pavement can be high if the proper materials were used in the pavement in the first place. The key ingredient is the aggregate, which must be nonporous and have good integrity so that it will retain a rough surface and contribute to a road surface with good traction. Such aggregates are said to resist "polish." Where attention has been given to incorporating high-quality aggregate, and where the rather expensive rotor-milling equipment is available, recycling rates of 90 percent or higher are not unusual. This is particularly true in urban areas. Overall, the asphalt recycling rate is probably in the neighborhood of 50 percent and is growing.

Concrete undergoes end-of-life processing by crushing, followed by magnetic separation to separate the steel reinforcing bar and rod fragments from the mineral matrix. These metal fragments are then recycled as with any relatively impure industrial steel. The mineral–cement matrix has several potential uses. The best, if energy costs do not make it prohibitive, is sending lightly crushed concrete to an aggregate supplier for final crushing. The resulting material is then used just as is granular material from natural sources. Alternatively, the crushed matrix material can be used as filler for highway construction or modification.

11.8   GREEN DESIGN GUIDELINES

By analogy with the principles of green chemistry and green engineering presented in Chapter 8, we give in Tables 11.1 and 11.2 some principles for green infrastructure and green buildings. All four sets of principles share as common goals the choice of materials, the minimization of energy use, and the avoidance of waste. For buildings and infrastructure, however, the guidelines emphasize efficient operation, the flexibility to adapt, enlarge, or renovate with high efficiency and low environmental penalty, and the realization that long lifetimes impose special challenges. Because buildings and infrastructure will likely be in place for decades (or perhaps for centuries), the designers and operators of the future must be given energy opportunity to take advantage of new technologies and new lifestyles while discarding as little as possible of the material and energy investment made when the products were constructed. This is an ongoing design challenge, only weakly addressed by today's practitioners.

TABLE 11.1   The Principles of Green Infrastructure Design and Operation

1. Sites and rights-of-way should be chosen to minimize ecosystem disruption.
2. Material and energy inputs should be renewable rather than depleting.
3. To the degree possible, recycled material should be used.
4. Material and energy inputs and outputs should be as inherently nonhazardous as possible.
5. Processes and systems should be designed to maximize mass, energy, and space efficiency.
6. It is better to prevent waste formation than to treat it after it is formed.
7. Maintenance and refurbishment should be facilitated.
8. Universal design (e.g., "one size fits all") solutions should be avoided.
9. Targeted durability, not immortality, should be a design goal.
10. Designs should be as flexible as possible to facilitate future renovation and expansion.
11. Systems should be designed to enable and encourage recycling at end of life.

TABLE 11.2   The Principles of Green Building Design and Operation

1. Sites and rights-of-way should be chosen to minimize ecosystem disruption.
2. Material and energy inputs should be renewable rather than depleting.
3. To the degree possible, recycled material should be used.
4. Material and energy inputs and outputs during construction and operation are to be as inherently nonhazardous as possible.
5. Buildings should be designed to maximize energy use efficiency.
6. Buildings should be designed to maximize water use efficiency.
7. It is better to prevent waste formation during construction and operation than to treat it after it is formed.
8. Universal design (e.g., "one size fits all") solutions should be avoided.
9. Targeted durability, not immortality, should be a design goal.
10. Designs should be as flexible as possible to facilitate future renovation and expansion.
11. Buildings and building systems should be designed for recycling during remodeling or at end of life.

FURTHER READING

Green buildings:

Ali, M.M., Energy efficient architecture and building systems to address global warming, *Leadership and Management in Engineering, 8* (3), 113–123, 2008.

Brand, S., *How Buildings Learn*, New York: Viking Penguin, 1995.

Kibert, C.J., *Sustainable Construction: Green Building Design and Delivery*, Hoboken, NJ: John Wiley, 2005.

Kohler, N., and U. Hassler, The building stock as a research object, *Building Research and Information, 30*, 226–236, 2002.

Orr, D., *The Nature of Design: Ecology, Culture, and Human Intention*, Oxford, UK: Oxford University Press, 2002.

U.S. Green Buildings Council, *LEED Rating Systems*, www.usgbc.org/DisplayPage.aspx?CMS PageID=222, accessed November 30, 2008.

Green infrastructure, General:

Hendrickson, C.T., and A. Horvath, Resource use and environmental emissions of U.S. construction sectors, *Journal of Construction Engineering and Management, 126*, 38–44, 2000.

Horvath, A., Construction materials and the environment, *Annual Review of Environment and Resources, 29*, 181–204, 2004.

Energy:

Energy Information Administration, *International Energy Outlook 2007*, U.S. Department of Energy, Washington, DC, 2007.

Shaw, R.W., Jr., Microgeneration technology: Shaping energy markets, *The Bridge, 33* (2), 29–35, 2003.

Water:

Angement, L.T., K. Karim, M.H. Al-Dahlan, B.A. Wrenn, and R. Dominguez-Espinosa, Production of bioenergy and biochemicals from industrial and agricultural wastewater, *Trends in Biotechnology, 22*, 477–485, 2004.

Levine, A.D., and T. Asano, Recovering sustainable water from wastewater, *Environmental Science & Technology, 38*, 201A–208A, 2004.

Otterpohl, R., U. Braun, and M. Oldenburg, Innovative technologies for decentralized water, wastewater, and biowaste management in urban and peri-urban areas, *Water Science and Technology, 48*, 23–32, 2003.

Viessman, W., Jr., and M.J. Hammer, *Water Supply and Pollution Control*, 7th ed., Upper Saddle River, NJ: Prentice Hall, 2005.

Transportation:

Schipper, L., Sustainable urban transport in the 21st century: A new agenda, in *Transportation, Energy, and Environmental Policy*, D. Sperling and K. Kurani, Eds., pp. 42–62, Washington, DC: The National Academies, ISBN 0-309-08571-3, 2003.

Sperling, D., and D. Gordon, Advanced passenger transport technologies, *Annual Review of Environment and Resources, 33*, 63–84, 2008.

Telecommunications:

Schaefer, C., C. Weber, and A. Voss, Energy usage of mobile telephone services in Germany, *Energy, 28*, 411–420, 2003.

EXERCISES

11.1 Figure 11.1 shows many interesting material-use patterns over time. Comment on those that seem significant and suggest explanations for them.

11.2 Given the patterns of Figure 11.1, how do you expect use of nonfuel mineral materials to evolve in the United States over the next few years? Why?

11.3 You are the industrial ecologist for a large telecommunications firm and have been asked by your CEO to create a chart of the various sustainability impacts arising from the company's operations, from those attributable to your manufacture of cell phones, to those attributable to network operation, and to those attributable to your services. How do the data needed at different scales differ? How certain and objective are your findings at different levels? If the public buys your products and services and surveys show that they are happy with them, what does this imply for your sustainability analysis?

11.4 Identify and review a local infrastructure project. Evaluate it using the DfES principles of Table 11.1.

11.5 Review the LEED Project Checklist. Which five of the items seem most important to you? Defend your choices.

11.6 You are the engineer in charge of two office building designs—one in the American Southwest, where there is very little rain and hot temperatures for much of the year, and one in British Columbia, where it is frequently rainy and foggy, and winter, the longest season, is quite cold. How good is the guidance provided by the LEED green building standards in each case? How could you make this better?

11.7 Identify and review a local building project. Evaluate it from the DfES principles of Table 11.2.

11.8 Your civil engineering construction firm, which prides itself on designing and building "sustainable infrastructure," is given two projects by a local city. One involves replacing a mile of old water pipe in the center of a downtown business and residential area. The second involves building a water system for a new community that, in an effort to be more sustainable, involves construction of a town center with closely spaced town homes around it.

   (a) What elements of materials selection and design constitute "sustainable" choices for each project, and how do they differ?

   (b) As lead design engineer, you note that you are seriously constrained in your water pipe replacement project by the existing design and operation of the municipal water system. How might this lead you to rethink your new design?

   (c) You design a very energy efficient water infrastructure for the new build community, only to be told by the city that your initial costs are too high. How do you respond, and what data would you use to support your response?

   (d) From reading engineering journals, you are increasingly aware of the trend to build "intelligent" infrastructure, such as the "smart grid" for electricity. You therefore propose to the city that it construct an "intelligent water infrastructure" for the new community. What sorts of information systems and functions might such an intelligent water infrastructure include?

CHAPTER 12

An Introduction to Life Cycle Assessment

12.1 THE CONCEPT OF THE LIFE CYCLE

The environmentally related activities of the 1970s and 1980s focused on the manufacturing facility and its emissions. This was a necessary concentration of effort, but it ignored some other obvious effects of its operations—the use of resources mined and processed elsewhere, the creation of products that may have environmental impacts when used, and so forth. In the 1990s, the scope of interest was enlarged to consider the entire life cycle of products and their associated flows and impacts, as sketched in Figure 12.1.

The components of a product life cycle can be defined in various ways depending on the goals and level of detail desired, but the four numbered stages in Figure 12.1 are typical: (1) acquisition and processing of the necessary resources, (2) manufacture, (3) use, and (4) reuse/recycling/disposal. The generation of reusable discards in manufacturing stimulates a "prompt scrap" subcycle as well (upper right on the diagram). Resources, either from primary ("virgin") or secondary (recycled) sources, are required to a greater or lesser degree at a number of points in the cycle, and emissions occur at a number of points as well.

The goal of life cycle assessment (LCA) is to quantify or otherwise characterize all of these material flows, to specify their potential environmental impacts, and to consider alternative approaches that can change those impacts for the better. The product cycle itself and the LCA that studies it are complex, but LCAs have become widely practiced, and many gains have been made as a result.

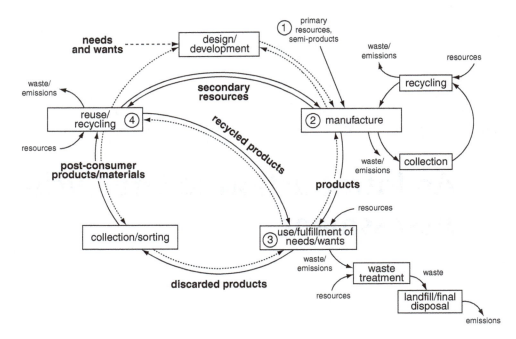

Figure 12.1

A representation of the generic life cycle of a product. Solid arrows represent energy and material flows, dashed arrows flows of information. (Adapted from G. Rebitzer, et al., Life cycle assessment Part 1: Framework, goal and scope definition, inventory analysis, and applications, *Environment International, 30*, 701–720, 2004.)

12.2   THE LCA FRAMEWORK

The formal structure of LCA has been delineated by the International Standards Organization; in its basic form it contains three stages: *goal and scope definition*, *inventory analysis*, and *impact analysis*. The concept is pictured in Figure 12.2. First, the goal and scope of the LCA are defined. An inventory analysis and an impact analysis

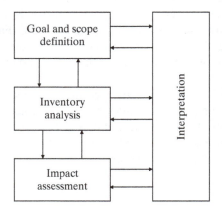

Figure 12.2

Phases in the life cycle assessment of a technological activity. The arrows indicate the basic flow of information. At each stage, results are interpreted, thus providing the possibility of revising the environmental attributes of the activity being assessed. (Adapted from International Standards Organization, *Environmental Management-Life-Cycle Assessment, Principles and Framework*, Geneva, 1997.)

are then performed. The *interpretation of results* that follows these three steps guides an analysis of potential improvements (which may feed back to influence any of the stages, so that the entire process is iterative).

There is perhaps no more critical step in beginning an LCA evaluation than to define as precisely as possible the questions to be answered (the goal) followed by choosing the evaluation's scope: what materials, processes, or products are to be considered, and how broadly will alternatives be defined? Consider, for example, the question of releases of chlorinated solvents during a typical dry-cleaning process. The purpose of the analysis is to reduce environmental impacts. The scope of the analysis, however, must be defined clearly. If it is limited, the scope might encompass only good housekeeping techniques, end-of-pipe controls, administrative procedures, and process changes. Alternative materials—in this case, solvents—might be considered as well. If, however, the scope is defined broadly, it could include alternative service options: Some data indicate that a substantial number of items are sent to dry-cleaning establishments not for cleaning per se but simply for pressing. Accordingly, offering an independent pressing service might reduce emissions considerably. One could also take a systems view of the problem: Given what we know about polymers and fibers, why are clothing materials and designs that require the use of chlorinated solvents for cleaning still being provided? Among the considerations that would influence the choice of scope in cases such as the above are: (a) who is sponsoring and who is performing the analysis, and how much control they can exercise over the implementation of options; (b) what resources are available to conduct the study; and (c) what is the most limited scope of analysis that still provides for adequate consideration of the systems aspects of the problem.

The resources that can be applied to the analysis should also be assessed. Most traditional LCA methodologies provide the potential for essentially open-ended data collection and, therefore, virtually unlimited expenditure of effort. As a general rule, the depth of analysis should be keyed to the degrees of freedom available to make meaningful choices among options, and to the importance of the environmental or technological issues leading to the evaluation. For example, an analysis of using different plastics in the body of a currently marketed portable disk player would probably not require a complex analysis, because the constraints imposed by the existing design and its market niche make the options available to a designer quite limited. On the other hand, a government regulatory organization contemplating limitations on a material used in large amounts in numerous and diverse manufacturing applications would want to conduct a fairly comprehensive analysis, because the degrees of freedom involved in finding substitutes could be quite numerous and the environmental impacts of substitutes could be significant.

The second component of LCA, inventory analysis (sometimes termed "LCI"), is by far the best developed. It uses quantitative data to establish the levels and types of energy and materials used throughout the lifetime of a product, process, or system, and the environmental releases that result. The approach is based on the idea of a family of materials budgets, in which the analyst measures the inputs and outputs of energy and resources. The assessment is done over the entire life cycle. The products of this activity are a comprehensive flow diagram of the manufacturing process (often involving suppliers and sometimes industrial customers), and a list (by mass) of resources used and of emissions to air, water, and soil, all detailed by mass flow and chemical speciation.

The third stage in LCA, the impact analysis, involves relating the outputs of the system to environmental impacts, or, at least to the stresses being placed on the environment by the outputs. Aspects of this difficult and potentially contentious topic are discussed in the next chapter.

The interpretation of results phase is where the findings from one or more of the three stages are used to draw conclusions and develop recommendations. The output from this activity is often the explication of needs and opportunities for reducing environmental impacts as a result of industrial activities being performed or contemplated. It follows ideally from the completion of stages one through three, and occurs in two forms: Design for Environment and Sustainability (the proactive activities discussed in Chapters 10 and 11) and Pollution Prevention (the "best current practice" activities discussed in Chapter 8).

12.3   GOAL SETTING AND SCOPE DETERMINATION

A common LCA goal is to derive information on how to improve environmental performance. If the exercise is conducted early in the design phase, the goal may be to compare two or three alternative designs. If the design is finalized, or the product is in manufacture, or the process is in operation, the goal can probably be no more than to achieve modest changes in environmental attributes at minimal cost and minimal disruption to existing practice.

It is possible, though not nearly so common, for an LCA target to be much more ambitious than the evaluation of a single product or process. This usually occurs with the evaluation of an organization of some sort: the operation of an entire facility or corporation, for example, or of an entire governmental entity. In such a case, it is likely that alternative operational approaches can be studied, but not alternative organizations. In addition, an organization that makes a logical entity from an LCA viewpoint may involve more than one implementer (an entire supply chain, for example), so collaborative goal setting may be required. If a goal can be quantified, such as "achieve a 20 percent decrease in overall environmental impact," it is likely to be more useful and the result more easily evaluated than with qualitative goals. Quantification of the goal requires quantification of each assessment step, however, and quantitative goals should be adopted only when one is certain that adequate data and assessment tools are available.

The scope of the assessment is perhaps best established by asking a number of questions: "Why is the study being conducted?" "How will the results be used, and who will use them?" "Do specific environmental issues need to be addressed?" "What level of detail will be needed?" It is useful to recognize that LCA is an iterative process, and that the scope may need to be revisited as the LCA proceeds.

12.4   DEFINING BOUNDARIES

The potential complexity of comprehensive LCAs is nowhere better illustrated than by the problem of defining the boundaries of the study. There are many potential issues for discussion in this regard, and no consensus on the best ways of approach. The discussion that follows explores a number of these issues and concludes with some general recommendations concerning choices of boundaries in LCA.

12.4.1 Level of Detail Boundaries

How much detail should be included in an LCA? An analyst frequently needs to decide whether effort should be expended to characterize the environmental impacts of trace constituents such as minor additives in a plastic formulation or small brass components in a large steel assembly. With some modern technological products containing hundreds of materials and thousands of parts, this is far from a trivial decision. One way it is sometimes approached is by the *5 percent rule*: If a material or component comprises less than 5 percent by weight of the product, it is neglected in the LCA. A common amendment to this rule is to include any component with particularly severe environmental impacts. For example, the lead-acid battery in an automobile weighs less than 5 percent of the vehicle, but the toxicity of lead makes the battery's inclusion reasonable. Potential items for inclusion in this way could be ozone-depleting fire suppressants or radioactive materials.

12.4.2 The Natural Ecosystem Boundary

A natural ecosystem issue that arises when choosing LCA boundaries is that of biological degradation. When industrial materials are discarded, as into a landfill, biodegradation produces such outflows as methane from paper, chlorofluorocarbons from blown foam packaging, and mobilized copper, iron, and zinc from bulk metals. LCA approaches to these complications have included incorporating these flows in the inventory, excluding landfill outflows completely, or including those flows for a specific time period only. Flows from landfills are generally difficult to estimate, so one is faced with a trade-off between comprehensiveness and tractability.

A second example of the natural/industrial boundary issue is the process of making paper from wood biomass, as shown in Figure 12.3. Here the assessor has several possible levels of inventory detail to choose from. The basic analysis is essentially a restriction of the inventory to life stage 2. The energy envelope incorporates some of the external flows related to the production of energy. The extended envelope includes all life cycle stages and flows directly connected with the industrial system. The comprehensive envelope adds the natural processes of biomass formation and the degradation of materials in a landfill. None of these options is inherently correct or incorrect, but the choice that is made could determine the amount of effort required for the LCA, as well as the results that emerge.

12.4.3 Boundaries in Space and Time

A characteristic of environmental impacts is that their effects can occur over a very wide range of spatial and temporal scales. The emission of large soot particles affects a local area, those of oxides of nitrogen generate acid rain over hundreds of kilometers, and those of carbon dioxide influence the planetary energy budget. Similarly, emissions causing photochemical smog have a temporal influence of only a day or two, the disruption of an ecosystem several decades, and the stimulation of global climate change several centuries. LCA boundaries may be placed at short times and small distances, long times and planetary distances, or somewhere in between. The choice of any of these boundary options in space and time may be appropriate depending on the scope of the LCA.

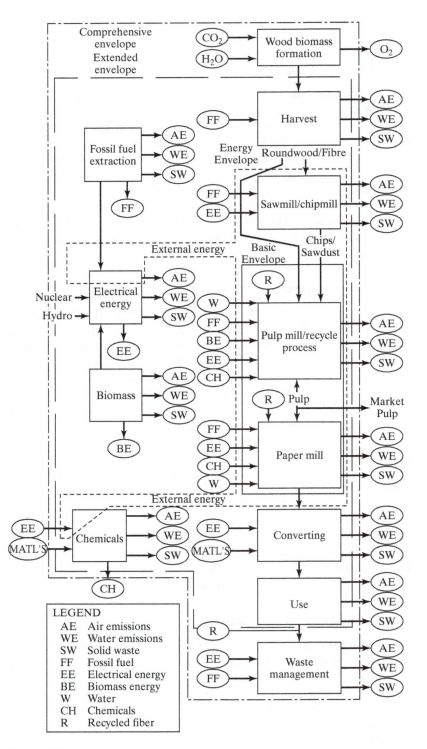

Figure 12.3

A simplified quantitative inventory flow diagram for the manufacture of paper. Four levels of possible detail are shown. (Adapted from a diagram provided by Martin Hocking, University of British Columbia.)

12.4.4 Choosing Boundaries

It should be apparent that the choice of LCA boundaries can have enormous influence on the timescale, cost, results, meaningfulness, and tractability of the LCA. The best guidance that can be given is that the boundaries should be consistent with the goals of the exercise. An LCA for a portable radio would be unlikely to have goals that encompass impacts related to energy extraction, for example, both because the product is not large and because its energy impacts will doubtless be very modest. A national study focusing on flows of a particular raw material might have a much more comprehensive goal, however, and boundaries would be drawn more broadly. The goals of the LCA thus define much of the LCA scope, as well as the depth of the inventory and impact analyses.

12.5 APPROACHES TO DATA ACQUISITION

Once the scope of the LCA has been established, the analyst proceeds to the acquisition of the necessary data. Data acquisition for a product is begun by constructing, in cooperation with the design and manufacturing team, a detailed manufacturing flow diagram. The aim is to list, at least qualitatively but preferably quantitatively, all inputs and outputs of materials and energy throughout all life stages. Figure 8.4 showed an example of such a diagram for the manufacture of a desktop telephone in which the housing is molded in the plant from precolored resin; the electronics boards are constructed from components furnished by suppliers; and those parts and others (microphone, electronic jacks, batteries, etc.) are assembled into the final product. The diagram indicates a number of material and energy by-products (the latter being mostly unused heat). Once the inventory flow diagram is constructed, in as much detail as possible, the actual inventory analysis can begin.

Some of the information needed for an inventory analysis is straightforward, such as the amounts of specific materials needed for a given design or the amount of cooling water needed by a particular manufacturing process. Quantitative data obviously have advantages: They are widely utilized in high-technology cultures; they offer powerful means of manipulating and ordering data; and they simplify choosing among options. However, the state of information in the environmental sciences may not permit the sound quantification of environmental and social impacts because of fundamental data and methodological deficiencies. The result of inappropriate quantification might be that those concerns that cannot be quantified would simply be ignored—thereby undercutting the systemic approach inherent in the LCA concept.

Case Study 1: The Upscale Automobile

The first stage of conducting a life cycle inventory on a product being manufactured is to assess the product itself. What is it made of? How much of each material does it contain? If the product is assembled from components supplied by others, it may be necessary to deal with suppliers to get a complete picture. Especially where potentially hazardous materials are involved, the overall effort can become quite detailed.

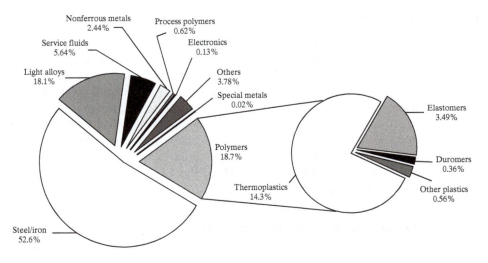

Figure 12.4

The material composition of the Mercedes-Benz S-class sedan. (M. Finkbeiner, et al., Application of life cycle assessment for the environmental certificate of the Mercedes-Benz S-class, *International Journal of Life Cycle Assessment, 11*, 240–246, 2006.)

An example of a material assessment is shown in Figure 12.4 for the Mercedes-Benz S-class sedan. About half the weight is seen to be iron and steel, and another 16 percent in a variety of light metal alloys. Nonferrous metals are largely zinc anticorrosion coatings, amounting to 2–3 percent of the weight. Polymers constitute some 19 percent; of this quantity, more than two-thirds is made up of thermoplastics, which have a high recycling potential.

This straightforward diagram by itself provides significant input to the product design team. If similar information is available with each new design, it is possible to track transitions to new mixes of metals and plastics, and the level of diversity of materials within the products of a corporation.

Case Study 2: The 1.7 Kilogram Microchip

The microchips that are at the heart of modern electronics are small, as are their power requirements. This would suggest that their environmental impacts are small as well, but such is not necessarily the case when the full life cycle is considered. Microchips are formed on silicon wafers, and the water used in processing must be very pure. The subsequent fabrication of the transistors and other components on the chip require a large suite of chemicals and frequent deposition, etching, and washing stages, all involving substantial material and energy flows (Figure 12.5a). Once manufactured, the chip consumes energy when it is used, but even over a four-year life its energy requirements are only half those used in its manufacture (Figure 12.5b).

Microchips have been regarded as praiseworthy examples of dematerialization, in which a function (computing, in this case) is performed by a product comprising reduced

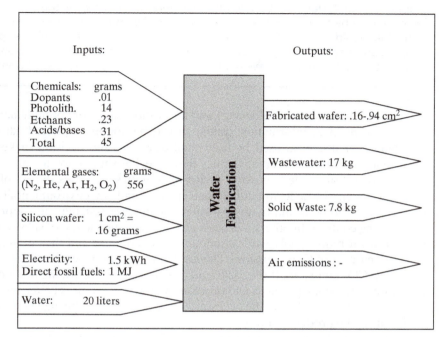

(a)

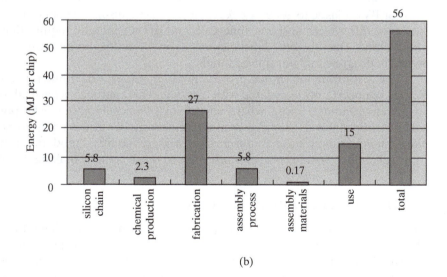

(b)

Figure 12.5

Life cycle inventory information for a 32 MB DRAM memory chip. (a) Input and output material flows for the silicon wafer manufacturing process; (b) Energy consumption in production and use. (Reproduced with permission from E.D. Williams, et al., The 1.7 kg microchip: Energy and material use in the production of semiconductor devices, *Environmental Science & Technology, 36*, 5504–5410, 2002.)

amounts of material. The LCA results of Figure 12.5a indicate that this impression is incomplete at best and that, if all the materials used in its manufacture were added up, the microchip would "weigh" 1.7 kg! As the researchers state: "increasingly complex products require additional secondary materials and energy to realize their lower entropy form." This is an insight that would not have made itself known without a life cycle inventory analysis.

In order to maximize efficiency and innovation and avoid prejudgment of normative issues, an LCA information system should be nonprescriptive. It should provide information that can be used by individual designers and decision makers given the particular constraints and opportunities they face, but should not, at early stages of the analysis, arbitrarily exclude possible design options. In some cases, the use of highly toxic materials might be a legitimate design choice—and an environmentally preferable choice from among the alternatives—where the process designer can adopt appropriate engineering controls. In others, a process choice involving the use of substantial amounts of lead might require only modest amounts of energy use and thus be responsible for modest amounts of CO_2 emissions. The alternative might be less lead, more energy use, and more CO_2 emissions. If the toxic lead can be well contained, the first option may be preferable. Designing products and processes inherently requires balancing such considerations and constraints, and the necessary trade-offs can only be made on a case-by-case basis during the product realization process.

In the ideal case, LCA data at different hierarchical levels should be mathematically additive; for example, LCA information for copper wire could be combined with that for PVC plastic to get an LCA result for plastic-insulated copper wire. In practice, however, differences in scope, timescale, and so on generally require that every LCA stand alone. This obvious deficiency in the methodology emphasizes that LCAs are works in progress and not finished tools.

LCA information should provide not only relevant data but, if possible, also the degree of uncertainty associated with that data. This approach is particularly important in the environmental area, where uncertainty, especially about risks, potential costs, and potential natural system responses to emissions of various types, is endemic. Often the relatively simple ordinal indicators—"high reliability," "moderate reliability," and "low reliability"—will be of substantial use to those actually making design decisions.

12.6 THE LIFE CYCLE OF INDUSTRIAL PRODUCTS

The life-stage outline assumes that a corporation is manufacturing a final product for shipment and sale directly to a customer. Often, however, a corporation's products are intermediates—process chemicals, steel screws, brake systems—made for sale to and incorporation in the products of another firm. How does that concept apply in these circumstances?

Picture the detailed process of manufacture as shown in Figure 12.6. Three different types of manufacture are illustrated: (A) the production of intermediate materials from raw materials (e.g., plastic pellets from petroleum feedstock or rolls of paper from bales of recycled mixed paper), (B) the production of components from intermediate materials (e.g., snap fasteners from steel stock or colored fabric from cotton), and (C) the processing

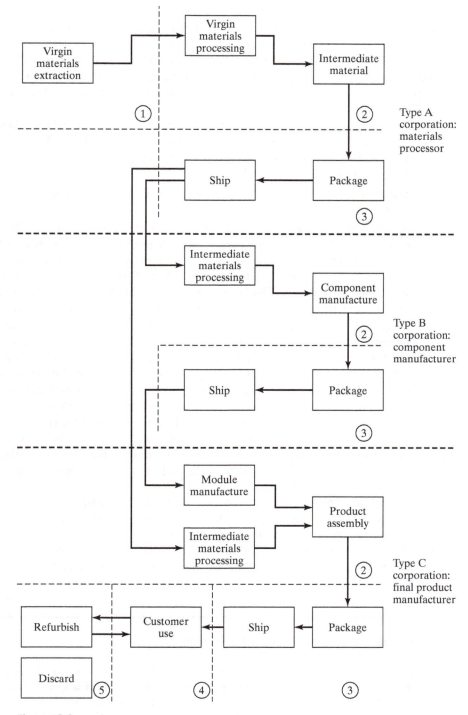

Figure 12.6

The interrelationships of product life stages for corporations of Type A (materials processors), Type B (component manufacturers), and Type C (final product manufacturers).

of intermediate materials (e.g., cotton fabric) or the assembly of processed materials (e.g., plastic housings) into final products (e.g., shirts or tape recorders). An operation of Type C is one in which the design and manufacturing team virtually has total control over all product life stages except Stage 1: Premanufacture. For a corporation whose activities are of Type A or B, the perspective changes for some life stages, but not for others:

> *Stage 1, Premanufacture.* Unless a Type A corporation is the actual materials extractor, the concept of this life stage is identical for corporations of Types A, B, and C.
>
> *Stage 2, Manufacture.* The concept of this life stage is identical for corporations of Types A, B, and C.
>
> *Stage 3, Product Delivery.* The concept of this life stage is identical for corporations of Types A, B, and C.
>
> *Stage 4, Product Use.* For Type A corporations, product use is essentially controlled by the Type B or C receiving corporation, though product properties such as intermediate materials purity or composition can influence such factors as by-product manufacture and residue generation. For Type B corporations, their products can sometimes have direct influence on the in-use stage of the Type C corporation final product, as with energy use by cooling fans or lubricant requirements for bearings.
>
> *Stage 5, Refurbishment, Recycling, or Disposal.* The properties of intermediate materials manufactured by Type A corporations can often determine the potential for recyclability of the final product. For example, a number of plastics are now formulated with the goal of optimizing recyclability. For Type B corporations, the approach to the fifth life stage depends on the complexity of the component being manufactured. If it can be termed a component, such as a capacitor, the quantity and diversity of its materials and its structural complexity deserve review. If it can be termed a module (such as an electronic circuit board made up of many components), the concerns are the same as those for a manufacturer of a final product ease of disassembly, potential for refurbishment, and the like.

Thus, Type A and B corporations can and should deal with LCAs of their products much as should Type C corporations. The considerations of the first three life stages are, in principle, completely under their control. For the last two life stages, the products of Type A and B corporations are influenced by the Type C corporation, with which they deal and, in turn, their products influence the life stages 4 and 5 characteristics of Type C products.

Case Study 3: Energy Use in Buildings

The use of LCA for buildings and infrastructure is somewhat more challenging than for smaller products. Each construction product can be "one of a kind," the lifetime is long, the minor constituents (electrical equipment, insulation, etc.) may not be clearly specified, and geographical location is important. Most of the studies accomplished to date relate to energy use. An example result is shown below for a three-story generic office building with wood framing in Vancouver, Canada. Even for a revised design

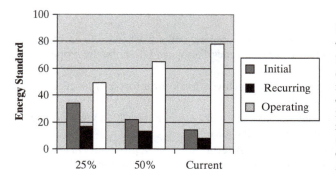

Figure CS-1

The energy use fractions for a three-story office building over 25 years due to initial embedded energy, energy for recurring replacement and repair activities, and operating energy. The calculations are for typical current designs, and for designs that use 25 percent or 50 percent as much energy in operation as current designs. (From R.J. Cole, and P.C. Kernan, Life cycle energy use in office buildings, *Buildings and Environment, 31*, 307–317, 1996.)

that uses only 25 percent of the current standard of energy use in operation, more than half of the overall energy use over the life cycle occurs during the use phase. This result emphasizes the importance of building designs that require little or no operating energy (Figure CS-1).

12.7   THE UTILITY OF LIFE CYCLE INVENTORY ANALYSIS

The greatest benefit of life cycle inventory analysis, in the minds of many product designers, is that it expands the breadth of their thinking. It is not instinctive to contemplate the flows of materials, energy, and water needed to extract and purify the resources used to manufacture a product, nor to consider a product's use of energy, or its fate at end of life. Merely the recognition of the entire life cycle is enough to stimulate many designers to make environmentally beneficial changes in their designs.

A second benefit comes when quantification of flows is performed. This step enables the analyst and/or designer to answer some relevant questions: What are the relative sizes of emissions? At which life stage is use of energy the most? Can substitute materials minimize any of the environmental aspects of the product? While environmental impact has not been fully analyzed, life cycle inventory studies nonetheless have been shown to raise issues of concern and to stimulate productive responses. As a result, they are standard practice in many corporations, and lessened corporate environmental footprints often follow.

FURTHER READING

Curran, M.A., *Environmental Life-Cycle Assessment*, New York: McGraw-Hill, 1996.

Guinée, J., et al., *Handbook on Life Cycle Assessment–Operational Guide to the ISO Standards*, Dordrecht, The Netherlands: Kluwer Academic Publishers, 2002.

Reap, J., F. Roman, S. Duncan, and B. Bras, A survey of unresolved problems in life cycle assessment. Part 1: Goal and scope and inventory analysis, *International Journal of Life Cycle Assessment, 13*, 290–300, 2008.

Science Applications International Corporation, *Life Cycle Assessment: Principles and Practice*, Report EPA/600/R-06/060, Cincinnati, OH: U.S. Environmental Protection Agency, 2006.

EXERCISES

12.1 You are the LCA analyst for a papermaking company and are asked to do an LCA for a new type of paper to be used for printing currency. Define and describe the scope of your assessment.

12.2 Repeat Exercise 12.1 for the situation in which you work for a forest products company that supplies wood fiber for the paper.

12.3 Several alternative LCA boundary choices are indicated in Figure 12.3. What do you see as the advantages and disadvantages of each choice?

12.4 Reap and colleagues (2008) identify what they term "unresolved problems" in goal, scope, and inventory analysis. Which do you think is potentially the most serious, and why?

C H A P T E R   1 3

The LCA Impact and Interpretation Stages

13.1 LCA IMPACT ANALYSIS

The previous chapter discussed the component of LCA termed "inventory analysis." Quantitative information on materials and energy flows is acquired at that stage in some cases, qualitative information in others. The data presentations in the previous chapter made it obvious that some aspects of life cycle analysis had the potential to be more problematical than others, but the approach begged the question of priorities. One could easily foresee a situation where alternative designs for a product or process each had similar materials use rates, but used different materials. How does the analyst make a rational, defensible decision among such alternatives? The answer is that (1) the influences of the activities revealed by the LCA inventory analysis on specific environmental properties must be accurately assessed, and (2) the relative seriousness of changes in the affected environmental properties must be given some sort of priority ranking. Together, these steps constitute LCA's impact assessment.

Assessing environmental influences is a complicated procedure, but it can, in principle at least, be performed by employing relationships between stressors, which are items identified in the inventory analysis that have the potential to produce changes in environmental properties, and the degree of change that is produced (e.g., the generation of carbon dioxide as a result of energy use). The relationships between stressors and the environment are developed by the environmental science community. By combining LCA inventory results with these relationships, a manufacturing process might be found, for example, to have a minimal impact on local water quality, a modest impact on regional smog, and a substantial impact on global climate change. The life cycle impact analysis (LCIA) uses that information to evaluate the relative importance of those impacts.

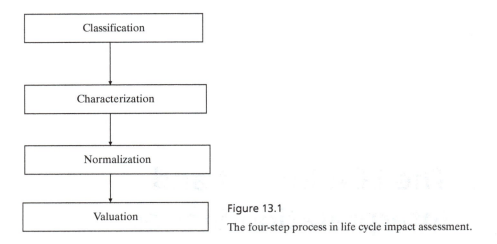

Figure 13.1

The four-step process in life cycle impact assessment.

The LCIA procedure is a four-step process, as shown in Figure 13.1 and discussed below.

Classification. Classification begins with the raw data on flows of materials and energy from the inventory analysis. Given those data, the classification step consists of identifying the environmental concerns ("categories" or "themes") suggested by the inventory analysis flows. For example, emissions from an industrial process using a petroleum feedstock may be known to include methane, butene, and formaldehyde. Classification assigns the first to the global warming category, the second to the smog formation category, the third to the human toxicity category. Table 13.1 lists those impact categories addressed in most LCIAs. Others, such as loss of biodiversity resulting from land development, or waste heat in power plant cooling water, may be added as needed.

TABLE 13.1   A Hypothetical Impact Analysis Including Normalization and Weighting

Impact category	S_j value (kilogram equivalent)	N_j value (year)	Ω_j value	W_j value (year)
Depletion of abiotic resources	3.5 antimony	2.2×10^{-11}	0.01	2.2×10^{-13}
Climate change	3.5 CO_2	2.7×10^{-14}	2.4	1.4×10^{-13}
Human toxicity	3.5 1,4-DCB	1.8×10^{-16}	1.1	1.9×10^{-16}
Freshwater aquatic toxicity	3.5 1,4-DCB	6.7×10^{-15}	0.2	1.3×10^{-15}
Terrestrial ecotoxicity	3.5 1,4-DCB	6.2×10^{-18}	0.4	3.9×10^{-18}
Photooxidant formation	3.5 ethylene	2.6×10^{-15}	0.8	2.1×10^{-15}
Acidification	3.5 SO_2	1.1×10^{-13}	1.3	1.4×10^{-13}
Eutrophication	3.5 phosphate	3.7×10^{-15}	1.0	3.7×10^{-15}

Source: Adapted from J.B. Guineé, Ed., *Handbook on Life Cycle Assessment*, Dordrecht, Netherlands, 2002.

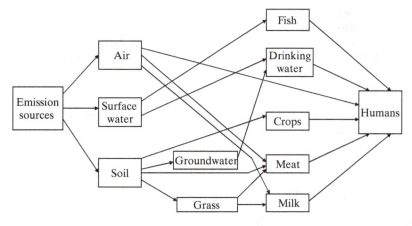

Figure 13.2

A map of the links between emissions and human exposure routes. (Adapted from J.B. Guineé, Ed., *Handbook on Life Cycle Assessment*, Dordrecht, Netherlands, 2002.)

Characterization. Characterization is the process of quantitatively determining the impact resulting from the stress indicated by the inventory values, that is,

$$S_j = \sum_i C_{i,j} \cdot E_i \qquad (13.1)$$

where E_i is the mass flow identified for species i in the inventory assessment, $C_{i,j}$ is the "characterization factor" for species i and category j (i.e., what level of environmental stress of category j is caused by the emission of a unit mass of species i), and S_j is the category stress indicator for category j.

An example of a category could be human toxicity; the summation in Equation 13.1 reflects the fact that there could be several flows from the LCA inventory with impact on that category, as depicted in Figure 13.2.

In the *Handbook of Life Cycle Assessment*, an example is given of the results of a hypothetical inventory analysis. We reproduce portions of this example in column 2 of Table 13.1. Note that the S_j values are quantified in terms of a common unit for each category so that the $E_i \cdot C_{i,j}$ products may be summed. Because the common units are so different, there is no sense from the table as to which impacts are important and which are not.

Case Study 1: Alternative Solders

Even a partial LCIA can provide useful information to designers and policy makers. An example is a study of alternative solder compositions for the electronics industry. Inspired by regulations banning the use of lead in solder because of lead's toxicity, the study compared traditional tin–lead solder with a lead-free solder (95.5 weight percent tin, 3.9 percent silver, 0.7 percent copper). The LCIA was restricted to global warming potential (GWP).

The results of the analysis were twofold: (1) use of the lead-free solder eliminates lead from the solder life cycle—an obvious conclusion, and (2) for an equivalent amount of soldering, the lead-free option has 10 percent higher carbon dioxide emissions. The latter occurs because the lead-free solder has a higher melting point and thus requires increased energy use. The two results thus provide useful information for policy, although they comprise only a small portion of an overall LCIA.

Source: T. Ekvall, and A.S.G. Andrae, Attributional and consequential environmental assessment of the shift to lead-free solders, *International Journal of Life Cycle Assessment, 11*, 344–353, 2006.

Case Study 2: Women's Shoes LCA

The production of leather footwear and its subsequent use and end-of-life stages form the basis of a life cycle assessment designed to show the environmental impacts of various stages of the life cycle. Most of the processes of interest refer to the raising of animals and the acquisition and treatment of the hides, but textiles and paper must also be taken into account (Figure CS-1).

The life cycle stages were defined as (1) cattle raising, (2) slaughterhouse, (3) tanning, (4) footwear manufacture, (5) waste management, and (6) transportation. We will not present the inventory results here; they are available in the reference given below. During impact assessment, however, the input/output list items were classified and their contributions to a small number of impacts characterized. The results, expressed as percentages of the total impacts, are shown in Figure CS-2. Normalization and valuation were not performed as part of this process.

The agricultural phase of the life cycle turned out to be important for ecologically related impacts: global climate change, acidification potential, and eutrophication potential. In the case of water consumption, the tannery stage is most important; the tannery stage is also highly significant for eutrophication potential and the depletion of nonrenewable materials. Footwear manufacture is the largest energy-consuming stage, and its impacts are especially significant for energy-related metrics: air pollution, human toxics potential, and fossil-fuel depletion. Thus, two life stages that were not thought particularly significant in environmental terms, agriculture and footwear manufacturing, were identified by the LCA as deserving enhanced attention.

L. Milà, et al., Application of life cycle assessment to footwear, *International Journal of Life Cycle Assessment, 3*, 203–209, 1999.

Normalization. The goal of this step in LCIA is to relate the S_j values derived at the characterization step to some sort of reference value R_j and thereby to arrive at a normalized indicator N_j:

$$N_j = \frac{S_j}{R_j} \tag{13.2}$$

The purpose is to put the S_j values into a broader perspective. The reference value may be selected in a number of different ways, the one chosen being important to the organization conducting the LCIA. For example, a national government might

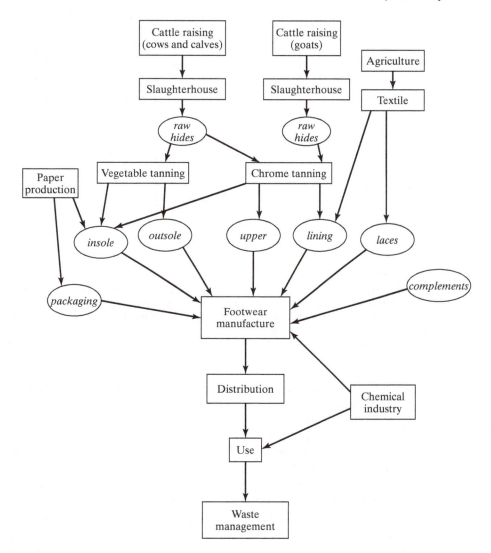

Figure CS-1

A life cycle process diagram for women's footwear. (Reproduced with permission from L. Milà, et al., Application of life cycle assessment to footwear, *International Journal of Life Cycle Assessment, 3,* 203–209, 1999.)

choose the national climate change potential, while a corporation might choose the climate change potential of emissions from individuals, corporations, and governments within the region where it manufactures its products. A typical choice for climate change might be the average global per capita CO_2 emission rate, in kg/yr.

The example of Table 13.1 is continued in column 3 of the table, where R_j values in terms of flow rates (expressed in kg/yr) have been applied. The results

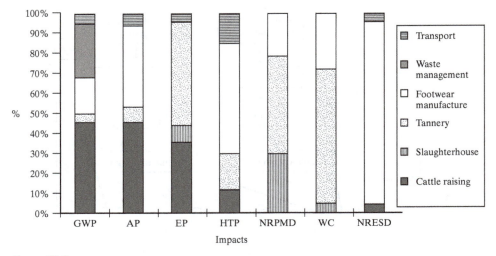

Figure CS-2

Contributions of different life cycle stages in women's footwear to major environmental concerns. GWP = global warming potential; AP = air pollution; EP = eutrophication potential; HTP = human toxicity potential; NRPMD = nonrenewable primary materials depletion; WC = water consumption; NRESD = nonrenewable energy sources depletion. (Reproduced with permission from L. Milà, et al., Application of life cycle assessment to footwear, *International Journal of Life Cycle Assessment, 3,* 203–209, 1999.)

are now in a common unit, and we find depletion of abiotic resources to have the highest value, followed by acidification.

Because the N_j values are dimensionless ratios, they permit the LCIA characterizations to be directly compared with each other. Without this step, one is left with non-normalized S_j values, and it is unclear how they should be interpreted (other than "more important is worse than less important," perhaps). Nonetheless, the choice of the reference values is potentially contentious, and normalization is often omitted from LCIAs.

Valuation. Valuation is the process of assigning weighting factors to the different impact categories based on their perceived relative importance as set by social consensus. For example, an assessor, an international standards organization, or a stakeholder panel might choose to regard climate change impacts as twice as important as acidification, and apply weighting factors to the normalized impacts accordingly. Mathematically, this provides a weighted indicator, given by

$$W_j = \Omega_j N_j \tag{13.3}$$

where the Ω_j values are the weighting factors.

If one wishes, an overall life cycle impact evaluation can then be calculated as

$$I = \sum_j W_j \tag{13.4}$$

The example of Table 13.1 is completed in the last two columns of the table, where weighting factors developed by a panel of participants are listed and applied. Depletion of abiotic resources and acidification continue to be quite important, but climate change is now highlighted as well.

Case Study 3: Palm Oil in Malaysia

An LCA study that employs both normalization and weighting has been performed for the production of crude palm oil in Malaysia. The procedure involves agriculture, transportation, and another industry (for milling of the palm kernels), and involves more than one-third of the country's total cultivated area. The goal of the LCA was to determine the environmental consequences of palm oil production and to serve as an improvement guide for oil palm plantations and palm oil mills.

A very detailed inventory was conducted for the functional unit of 1000 kg of crude palm oil. This inventory was then used as input for LCA software that used generic reference values and weighting factors to compute a final result, which demonstrated that fertilizer production for the palms was the most severe of the impacts, with transportation and boiler emissions also important. The most significant impacts were human toxicity (respiratory inorganics) and depletion of fossil fuels. Climate change, acidification, and eutrophication impacts, which might have been expected to be highly significant, were shown not to be so.

S. Yusoff and S.B. Hansen, Feasibility study of performing an life cycle assessment on crude palm oil production in Malaysia, *International Journal of Life Cycle Assessment, 12*, 50–58, 2007.

The application of weighting factors is controversial, because doing so involves making social, political, and ethical choices. As a result, LCIA evaluations suitable for a particular culture or location or time are unlikely to be useful in other circumstances. Because of this limitation, weighting is often omitted from LCIAs. (Doing so, however, is equivalent to making an implicit weighting with all Ω_j values the same.)

13.2 INTERPRETATION

13.2.1 Identify Significant Issues in the Results

A comprehensive LCA generates a substantial amount of results, only some of which is important. To identify the important issues, the analyst typically addresses the following questions:

- Do particular life stages dominate the results?
- Do particular processes dominate the results?
- Which environmental impacts are identified, and which are likely to be of most concern?
- Are any of the results particularly unusual or surprising?

The result of this review is a short list of issues in the product design or manufacture that deserve special attention.

13.2.2 Evaluate the Data Used in the LCA

With the significant issues identified, the next step in the LCA interpretation phase is to evaluate the completeness and consistency of the data. The goal is to ensure that each identification of significance is backed by adequate, reliable information. This is particularly important if alternative product designs are being evaluated, because the designs need to have comparable bases for comparison.

At the data completeness step, one wishes to confirm that all product life stages have been addressed, as well as all relevant environmental impacts. This information should be verified as meeting the system boundaries established at the beginning of the study, and that the significant raw materials and releases have been incorporated. Next, the uncertainties in the data are reviewed to see if the determination of significant issues is robust. If it is determined that the data are satisfactorily complete, consistent with project goals, and within acceptable uncertainty limits, the analyst can be comfortable in moving on to conclusions and recommendations.

13.2.3 Draw Conclusions and Recommendations

The final step in the LCA improvement stage is to use the information flowing from the LCA inventory and impact stages to develop a set of conclusions relating to the activity under study. An example is the conclusion from the solder study discussed earlier that solder composition substitution decreased lead exposure but increased global warming emissions. If a specific product or alternative products comprise the focus of the LCA, recommendations for improvement are also developed. The intention is to produce environmental benefits or, at least, minimize environmental liabilities. If the LCA stopped at the characterization stage, however (as was the situation with the women's shoes case study discussed earlier), the identification of significant issues is substantially constrained. The results and conclusions can still be useful from the perspective of comparisons against targets, for example, or of identifying issues to be brought to the attention of product designers.

13.3   LCA SOFTWARE

Because of the complexity of the LCA process, a number of research organizations and private consulting companies have developed software to facilitate life cycle assessment. To prepare for employing the software, the user develops a comprehensive description of the product and of the materials involved in its manufacture. If it will require resources when used (gasoline for an automobile, paper for a printer, for example), those resources and their anticipated rates of consumption are identified. In the typical approach, the user enters into a database the identity of the materials used in the product under study, together with the quantities of each. The software, taking advantage of internal databases that relate materials to impacts of various types and at various stages of the life cycle, then computes the stress indicators S_j and, if desired, the normalization indicators Nj, the weighting indicator Ω_j, and the overall impact evaluation I.

LCA software packages are continuously being refined. Many include extensive databases and are quite easy to use. Perhaps their greatest weakness tends to be the need to quantify data of uncertain validity and to compare unlike risks, in the process making assumptions that may gloss over serious value and equity issues. For this reason, some have argued that the uncertainties related either to data or subjective judgment or both in normalization and (especially) in valuation are often so high that it is preferable to make decisions based on the more reliable information at earlier stages of the LCA sequence.

13.4   PRIORITIZING RECOMMENDATIONS

Assume that a set of recommendations has emerged from one or another of the possible approaches to LCA. Those recommendations will be based solely on the perceived importance of the environmental impacts, but a corporation must consider many other factors in determining its LCA-inspired actions. If it is not possible to act upon all the recommendations, or at least not to react to them simultaneously, how might the actions be reviewed and prioritized?

13.4.1   Approaches to Prioritization

Complex products tend to generate long lists of recommendations. For example, here are selections from a list resulting from the LCA of a telecommunications product:

Manufacturing

- Rewrite specifications for equipment frames to encourage or mandate the use of some recycled material in their manufacture.
- Work with suppliers to minimize the diversity of packaging material entering the facility, so that recycling of solid waste may be optimized.
- Use nitrogen inerting on wave-solder machines to reduce solder dross buildup.
- Minimize the diversity of materials in outgoing equipment packaging, and develop labels to indicate appropriate recycling procedures to the customer.
- Develop reusable shipping containers that satisfy physical and electrostatic protective criteria and are ultimately recyclable.

Design

- Eliminate the use of chromate as a metal preservative in favor of removable organic coatings.
- Review specifications and requirements with the goal of using as few different plastics as possible and of using thermoplastics instead of thermosets.
- Mark all plastic parts using ISO standards.

Product Management

- Implement a customer information online service to contain not only the operator's manual but also instructions on recycling of parts, components, and packaging during service life and of the entire unit at end of life.
- Develop and implement a strategy for the recovery of used batteries from the field.

In developing a list of recommendations based on LCA results, it is important for the assessor to be inclusive, and to range widely. Recommendations that subsequently prove to be infeasible for one reason or another will be identified and discarded at the prioritization step, the second activity in improvement analysis. Some items, such as the marking of plastic parts, will not require the procedure of a full LCA to indicate their desirability, but would normally be at least implied by LCA results if not explicitly called out. Both more obvious and less obvious recommendations should be considered.

It is worth noting that some recommendations are very specific (i.e., avoid the use of chromate), while others are much more diffuse (i.e., minimize the diversity of packaging materials). Both types are important to include. The highly specific recommendations are easier to generate, and their accomplishment is more easily measured. The diffuse recommendation may be more difficult to deal with, but may in some cases be very important; their inclusion is crucial to a successful implementation of the LCA improvement stage.

The environmental performance of an assessed product can usually be substantially improved by adopting the bulk of the recommendations made in the assessment report. Complete implementation may not be possible for a variety of reasons, however, and in any case the recommended actions cannot be accomplished simultaneously. Prioritization is thus useful, and in order to prioritize the recommendations one should consider more than just environmentally related characteristics. Some researchers have proposed that the LCA recommendations be prioritized on the basis of how much environmental benefit will result. This procedure does not take into account, however, the fact that industrial decision making incorporates many factors in addition to environmental ones. Thus, actions suggested as a result of an LCA process are properly regarded as a subset of possible actions, both environmental and nonenvironmental.

A broadly tractable prioritization approach is to discard quantification and deal with the "binning" of recommendations, that is, dividing them into a small number of categories on the basis of expert information. For example, one can rank each recommendation on a "+/−" scale ("++" being the most desirable score and "− −" being the least desirable score) across the following product constraints:

- *Technical Feasibility:*   Rates the technical facility of implementing a particular recommendation; "++" means the recommendation presents no technical challenges and is therefore very easy to implement.
- *Environmental Improvement:*   Judges to what extent implementation of a recommendation will respond to an important environmental concern, the situation being evaluated on both a scientific and social basis; "++" means implementation will strongly support desirable environmental initiatives.
- *Economic Benefit:*   Rates the net financial impact for an organization of implementing a particular recommendation; "++" means the product will cost less if the recommendation is incorporated. Here the total life cycle cost to the manufacturer is considered. For example, some parts may cost more due to DfE constraints but will also yield a higher residual value when an item of leased equipment is returned to the manufacturer for recycling.

- *CVA Impact:*   Accounts for the customer-perceived value added by implementing a particular recommendation; "++" means the DfE attribute has a very high perceived value.
- *Production Management:*   Estimates the production schedule impact or other manufacturing management influence resulting from implementing a particular recommendation; "++" means adoption of the recommendation would reduce the amount of time required to develop and/or manufacture the product; +/− means it would have no significance.

An example of prioritization of the recommendations listed above is given in Table 9.1. The individual scores were assigned by the LCA assessor and the recommendations were then sorted in order of decreasing overall value to the manufacturing organization in each of the three categories: manufacturing, design, and management.

13.4.2   The Action-Agent Prioritization Diagram

Although the prioritization table is helpful in developing additional supporting information relative to LCA recommendations, its extensiveness may make the most significant information difficult to extract readily, particularly if the number of recommendations is larger than shown here. An alternate display of the information is with a prioritization diagram, as shown in Figure 13.3. The first step in constructing the diagram is to normalize the assessment sum of Table 13.2 by reducing each sum by 10; the philosophy is that the maximum score is 20, and a score at or below 10 reflects neutral or negative overall

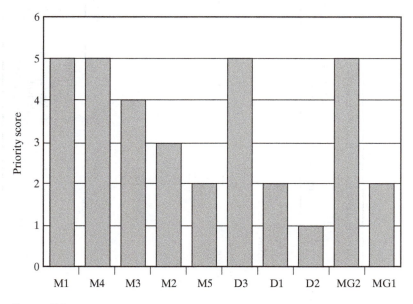

Figure 13.3

An action-agent prioritization diagram of the recommendations from the streamlined life cycle assessment of a telecommunications product. The designations on the x-axis refer to the recommendations (in the order given in Table 13.2) for manufacturing, design, and management. On the *y*-axis, higher numbers indicate greater priority.

TABLE 13.2 A Prioritization Table for DfE Recommendations

Recommendations	Life stage	Technical feasibility	Environmental sensitivity	Economic impact	CVA impact	Production management	Total
Manufacturing:							
Recycled metal specs.	L1.1	++	++	+/–	+	+/–	15
Packaging diversity-inflow	L2.1	++	+	+/–	+/–	+/–	13
Packaging diversity-outflow	L3.1	++	+	+/–	+	+/–	14
Reusable ship containers	L3.2	++	+	+/–	+	+/–	14
Solder bath N_2 inerting	L2.2	++	++	–	+/–	–	12
Design:							
Avoid chromate	L1.2(5)	+	+	+/–	+/–	+/–	12
Less plastic diversity	L5.1	+/–	+	+/–	+	–	11
Mark plastic parts	L5.2	++	++	+/–	+	+/–	15
Management:							
Online information	L4.1	++	+	–	+	–	12
Battery recovery	L4.2	++	++	–	++	+/–	15

Symbol	Value	Points
++	Very good/high	4
+	Good/high	3
+/–	Moderate, average	2
–	Little/bad	1
– –	Very little/bad	0

impacts and thus can be regarded as pertaining to a recommendation that would produce little net benefit. The practical effect of the adjustment is to make it easier to distinguish between and choose among the more highly rated recommendations. The adjusted prioritization sums are plotted in three groups, each group representing recommendations that would need to be carried out by specific "action agents": manufacturing engineers, design engineers, or management personnel.

The highest priority recommendations are quickly distinguished from those of lower priority in Figure 13.3. In the manufacturing area two actions have the highest priority rating: (1) Specify that major metal parts contain recycled content, and (2) use reusable shipping containers for modules and components. Several other actions listed in the table are rated high (though not highest) in priority; accomplishing these would also be well justified. The economic impact for all these actions is small to negligible. In the design area, the recommendation that stands out is to mark the major plastic parts with ISO symbols (as discussed in Chapter 10). For management, one priority action is also identified: the development of a program to efficiently take back discharged batteries from the field.

13.4.3 The Life-Stage Prioritization Diagram

As with the action agent diagram, the basic information is taken from Table 13.2 and normalized. The recommendations are then divided into five groups, one for each life stage: premanufacture, manufacturer, product delivery, product use, and end of life. If a recommendation pertains to more then one life stage, it is included in each life-stage group to which it pertains. The result for the telecommunications product example is shown in Figure 13.4.

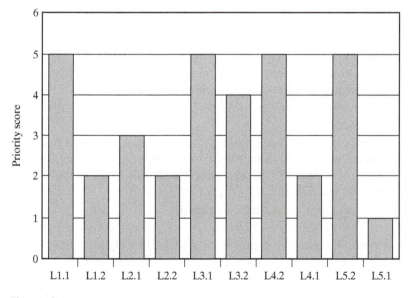

Figure 13.4

A life-stage prioritization diagram of the recommendations from the streamlined life cycle assessment of a telecommunications product. The first digit under each bar refers to the life stage; the second is a recommendation identification number (see Table 13.2).

The life-stage diagram provides a different perspective on the recommendations, one that varies in time and space rather than in the action agents. The environmental aspects of the manufacturing stage, for example, are seen as relatively benign, as the priority scores of the applicable recommendations are low. In contrast, the end-of-life stage has recommendations with higher priority scores. Attention is also indicated for the product use stage. The latter two stages are under the direct control of product designers. The premanufacture stage merits activity that requires the participation of the procurement organization in working with suppliers.

13.5   THE LIMITATIONS OF LIFE CYCLE ASSESSMENT

The great asset of the LCA concept is that it appears inherently to be the ideal way to quantitatively assess the range of environmental impacts attributable to a specific product. As seen in this chapter, however, there prove to be many limitations in practice. The drawing of assessment boundaries is difficult, the specification of the functional unit is not obvious, and it is difficult to recommend a consistent approach to either. Data collection and analysis are also limitations to accuracy and completeness. As a result, inventory analyses by different assessment teams can produce different, though perhaps equally defendable, results.

A partial list of the challenges related to the impact assessment stage includes the following:

- LCIAs do not incorporate locational information (e.g., they assume that emissions of a certain quantity of smog-forming chemicals into the air is just as significant in Oslo as in Los Angeles).
- LCIAs do not incorporate temporal information (e.g., they assume that emissions of a certain quantity of smog-forming chemicals into the air is just as significant at midnight as in midmorning).
- LCIAs routinely omit consideration of environmental impacts for which no agreed-upon characterization factor is available.
- LCA inventory data are often too general (e.g., "VOCs," "metals") to perform an adequate LCIA.
- Linearity of impacts is assumed (e.g., the impact of a 500 g emission of a certain chemical is assumed to be 100 times that of a 5 g emission). This excludes consideration of nonlinear responses and thresholds that are known to exist for some materials (see Chapter 6).
- Recycling loops are difficult to include.

To demonstrate some of the problems with LCIAs at their present state of development, let us elaborate on two examples where difficulties are real and obvious. The first is depletion of abiotic resources such as metals. There is indeed a general supposition that resources are being depleted at excessive rates in some cases. However, there are no satisfactory reference indicators for resource depletion available. Some researchers have multiplied the average concentration of the resource in Earth's crust by the mass of the crust; this is unsatisfactory because mining average crustal rock is unfeasible—

miners mine enriched deposits instead. Other workers use estimates of economic resources as a reference; this is unsatisfactory as well, because increased demand as reflected in price, or improved mining technology, tends to increase reserves and thus means that contemporary reserve numbers are likely to be serious underestimates.

A second example relates to ecotoxicity, in which two emissions of equal amount are assumed to have equal impacts. The ecosystems and organisms that receive those emissions can be very different, however. In some cases, organisms can withstand and ultimately reject small doses of a substance. In others, organisms may ingest a material such as copper that is biologically essential rather than harmful (both of these cases violate the principal of linearity). Finally, ecosystems differ in their ability to sequester materials, so neglecting that difference does not take spatial location into account.

Notwithstanding this daunting list of challenges, the most difficult issues of all doubtless relate to normalization and valuation, in which the absolute assignment of value to different environmental impacts is thwarted by differences in societal structure and preferences. These constraints lead at least a few practitioners to say that LCAs can only study burdens placed on the environment and not environmental impacts (as seen in two of the case studies in this chapter). Because quantifying and prioritizing impacts is the purported reason for doing LCAs, however, a retreat to burdens is, in a sense, a retreat from the desired quantitative approach. Finally, no matter how sophisticated a quantitative analysis may be, if it has a subjective basis or uses subjective data, it gives subjective results.

LCA software programs generate results involving the inventories they are given and the environmental impacts for which they are programmed. They may include normalization and valuation steps, which require that they have incorporated "expert opinion" of some kind. The naïve user of the software is often unaware of these nuances, thereby assuming that the results that are presented are as rigorous as an engineering determination of stress or strain. This is potentially dangerous business, especially if results for two rather different ways of satisfying a customer's need are being compared.

Is, then, the concept of a product-level comprehensive LCA, with its scoping, inventory, impact, and interpretation phases, infeasible, at least as a routine tool? In LCA's present form, the answer is "probably." Nonetheless, those who have performed almost any of the types of the LCAs mentioned above have found benefits both for the product being assessed and the environment being affected, because issues are raised that would otherwise be overlooked. That fact suggests that a less doctrinaire and simpler version of an LCA might have substantial utility, whether or not it meets all the lofty goals of the ultimate LCA. This simpler approach, termed "streamlined LCA" (SLCA), is the subject of Chapter 14.

FURTHER READING

Blengini, G.A., Life cycle of buildings, demolition and recycling potential: A case study in Turin, Italy, *Building and Environment, 44,* 319–330, 2009.

Ehrenfeld, J., The importance of LCAs–Warts and all, *Journal of Industrial Ecology, 1* (2), 41–49, 1997.

Finnveden, G., Valuation methods within LCA–Where are the values? *International Journal of Life Cycle Assessment, 2,* 93–99, 1997.

Finnveden, G., On the limitations of life cycle assessment and environmental systems analysis tools in general, *International Journal of Life Cycle Assessment, 5,* 229–239, 2000.

Guinée, J., et al., *Handbook on Life Cycle Assessment–Operational Guide to the ISO Standards,* Dordrecht, The Netherlands: Kluwer Academic Publishers, 2002.

Lenzen, M., Uncertainty in impact and externality assessments: Implications for decision-making, *International Journal of Life Cycle Assessment, 11,* 189–199, 2006.

Reap, J., F. Roman, S. Duncan, and B. Bras, A survey of unresolved problems in life cycle assessment. Part 2: Impact assessment and interpretation, *International Journal of Life Cycle Assessment, 13,* 374–388, 2008.

EXERCISES

13.1 The LCIA sequence can be stopped at any point in the chain of Figure 13.1. How might you defend a decision to terminate the analysis after the characterization step or the valuation step?

13.2 The recommendations that result from the interpretation stage of the LCA of a telecommunications product (Section 13.4.1) are completely qualitative. Is this realistic or does the lack of precision limit the usefulness of the results?

13.3 Reap and colleagues (2008) identify what they term "unresolved problems" in impact assessment and interpretation. Which do you think is potentially the most serious, and why?

CHAPTER 14

Streamlining the LCA Process

14.1   NEEDS OF THE LCA USER COMMUNITY

It is useful at this point to ask, "Who are those who can most benefit from a life cycle assessment?" One group is the policy makers, who find it useful, for example, to know that generic lead-free solder use trades lead exposure for enhanced global warming emissions. A second group consists of academics who are exploring the interactions between technology and the environment. A third group, larger and more focused than either of the others, consists of product designers, process designers, and their managers. This group is interested in whether the technology for which they are responsible has any notable environmental concerns, or, sometimes, whether product A is "greener" than product B.

We discussed earlier in this book the ways in which environmentally informed design takes place, but it is appropriate to present a brief review at this point. A key feature of the process is that product realization involves what has been called "design under constraint"—the art and science of dealing simultaneously with requirements for size, performance, cost, reliability, appearance, and so forth. The time available for design is tightly limited in almost every case. As a result, decisions that will strongly influence many of a product's characteristics are often made on the basis of past experience and rough design concepts. The consequence is that at the time a decision needs to be made, it is often the case that material choices, energy use, and other factors essential to a full LCA are not available. Decisions that have environmental relevance are indeed part of the design optimization at this stage, but the information available is generally qualitative or, at best, semiquantitative.

Life cycle assessment is useful in the early design stage if it is designed with the realization of the constraints of the industrial design process. This fact inspires streamlined approaches to LCA. In this chapter, we present typical tools of this type and conclude by discussing stages and types of assessment for which the different LCA approaches are most useful.

14.2   THE ASSESSMENT CONTINUUM

If no limitations to time, expense, data availability, and analytical approach existed, a comprehensive LCA as described in Chapters 11 and 12 would provide the ideal advice for assessing and improving environmental performance. In practice, however, these limitations are always present. As a consequence, although very extensive LCAs have been performed, a complete, quantitative LCA has probably never been accomplished, nor is it ever likely to be. There are many compromises of necessity, among which have often been the use of averages rather than specified local values for energy costs, landfill rates, and the like; the omission of analyses of catalysts, additives, and other small (but potentially significant) amounts of material; neglect of capital equipment such as chemical processing hardware; and the failure to include materials flows and impacts related to supplier operations. As a consequence, detailed LCAs cannot be regarded as providing rigorous quantitative results, but rather as providing a framework upon which more efficient and useful methods of assessments can be developed. In addition, detailed LCAs are expensive and time-consuming—characteristics that do not endear them to users, especially corporate users.

The question of data availability as it relates to product design and manufacture deserves added discussion. Experts agree that perhaps 80 percent of the potential environmental impacts of a product are determined at the design stage, and that modifications at later stages of product development will have only very modest effects. The ideal time to conduct an LCA analysis is thus early in the design phase. At that point, however, the characteristics of the product tend to be quite fluid—materials may not have been selected, no manufacturing facility may have been built, no packaging approach may have been determined, and so forth. Hence, there is often no possible way to complete a quantitative LCA *at the precise time when one would be most useful*.

Techniques that purposely adopt some sort of simplifying approach to life cycle assessment are termed "streamlined life cycle assessments" (SLCAs). As shown in Figure 14.1, the family of assessment techniques forms a continuum of effort, the degree of detail and expense generally decreasing as one moves from the left extreme toward the right. The region termed "extensive LCA" is that of detailed, quantitative LCAs such as the women's footwear analysis discussed in Chapter 13. The *scoping* or *ecoprofile* regions are those that are purposely sketchy, done to ensure that no truly disastrous design choices have been made or to determine whether additional assessment is needed. Somewhere within the SLCA region is the ideal point—where the assessment is complete and rigorous enough to be a definite guide to industry and an aid to the environment, yet not so detailed as to be difficult or impossible to perform.

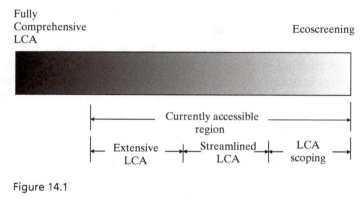

Figure 14.1

The LCA/SLCA continuum.

14.3   PRESERVING PERSPECTIVE WHILE STREAMLINING

If streamlining is to be a universal characteristic of LCAs, how should it be done, and how can one tell if an SLCA has not streamlined away the LCA's legitimacy? A number of options have been suggested:

- *Screen the product with an "inviolates" list.* This approach treats some activities or choices as so obviously incorrect from an environmental standpoint that no design or plan to which they apply should be allowed to go forward. Examples of inviolates are the use of mercury switches in a product or of ozone-depleting chemicals in manufacturing. While an inviolates list is useful as an assessment tool, limiting an assessment to the use of such a list obviously has the potential to overlook many life stages and environmental concerns.

- *Limit or eliminate life cycle stages.* Some studies limit the LCA to practices occurring within an industrial facility. This "gate-to-gate" approach amounts to a version of pollution prevention. While meritorious, it clearly does not satisfy the criterion of treating the entire life cycle. A second common approach is to limit or eliminate only upstream stages (resource extraction, for example). This approach is more defendable than gate-to-gate, especially if the evaluation of upstream stages is limited rather than eliminated.

- *Include only selected environmental impacts.* Some studies limit the LCA to impacts of highest perceived importance or those that can be readily quantified. Such choices tend to be responsive to public pressure rather than to environmental science, and to be anthropocentric rather than balanced.

- *Include only selected inventory parameters.* This is a variation of the approach immediately above, because if only selected impacts are of interest, only the inventory data needed to evaluate those impacts will be gathered.

- *Limit consideration to constituents above threshold weight or volume values.* An assessment may be limited only to major constituents or modules. This limitation overlooks small but potent constituents (it would fail as a tool for an SLCA of medical radioisotope equipment, for example), but may sometimes be justifiable from the standpoint of efficiency and tractability. It obviously applies only to quantitative assessments.

- *Limit or eliminate impact analysis.* Impact analysis is a major component of LCA, and eliminating it clearly abridges the process. The result is that the overall assessment can rely only on a "less is better" philosophy. While pursuing such an approach will probably result in some useful actions, the result provides absolutely no connection between the knowledge base of environmental science and the recommendations made by this abridged LCA.

- *Use surrogate data.* It is sometimes possible to use data on a similar material, module, or process when the specific data desired for an assessment are not available. The use of surrogate data is often contentious and has many of the same limits in usefulness as qualitative data.

- *Eliminate interpretations or recommendations.* In some studies, inventory and impact results are provided in detailed reports, with the recipient left to devise actions that should be taken in response to the report. If an SLCA is to be useful, however, specific recommendations should be provided by the assessment team, and a method for implementing those recommendations developed.

- *Use qualitative rather than quantitative information.* Quantitative data are often difficult to acquire, or may not even exist. However, qualitative data can often be sufficient to reveal the potential for environmental impacts at different life stages. The qualitative approach makes it difficult or impossible, however, to compare one product with another or with a new design.

Arguments can be made for each of these alternatives. However, because we believe that ease of analysis, speed, and cost avoidance have proven to be vital in achieving a functional LCA, we recommend the qualitative option and describe a realization of it below.

14.4   THE SLCA MATRIX

Many SLCA approaches adopt a matrix format in which the several life stages are evaluated for their potential impacts on a number of environmentally related concerns. An ideal assessment system for environmentally responsible products should have the following characteristics: it should lend itself to direct comparisons among rated products, be usable and consistent across different assessment teams, encompass all stages of product life cycles and all relevant environmental concerns, and be simple enough to permit relatively quick and inexpensive assessments to be made. Clearly, it must explicitly treat the five life cycle stages in a typical complex manufactured product.

The assessment system recommended here was developed by the authors in 1993 at AT&T; similar ones have been devised elsewhere since that time. The AT&T system has as its central feature a 5×5 matrix, the Environmentally Responsible Product Assessment Matrix, of which one dimension is the life cycle stage, and the other is environmental concern (Table 14.1). In use, the assessor studies the product design, manufacture, packaging, in-use environment, and likely disposal scenario and assigns to each element of the matrix an appropriate value. There is no a priori reason why the matrix element values must be continuous. Expert systems of various kinds often use data that are quantized: the values

TABLE 14.1 The Environmentally Responsible Product (ERP) Assessment Matrix (The numbers are matrix element designations.)

Life stage	Environmental concern				
	Material choice	Energy use	Solid residues	Liquid residues	Gaseous residues
Resource extraction	1,1	1,2	1,3	1,4	1,5
Product manufacture	2,1	2,2	2,3	2,4	2,5
Product delivery	3,1	3,2	3,3	3,4	3,5
Product use	4,1	4,2	4,3	4,4	4,5
Refurbishment, recycling, disposal	5,1	5,2	5,3	5,4	5,5

may be either binary (as in problem/no problem decision systems) or ordinal (as in a 1–10 severity ranking system). In the approach we recommend, the assessor assigns an integer rating from 0 (highest impact, a very negative evaluation) to 4 (lowest impact, an exemplary evaluation). In essence, what the assessor is doing is providing a figure of merit to represent the estimated result of the more formal LCA inventory analysis and impact analysis stages. She or he is guided in this task by experience, a design and manufacturing survey, appropriate checklists, and other information. The process is purposely qualitative and utilitarian, but does provide a numerical end point against which to measure improvement.

Although the assignment of integer ratings seems quite subjective, experiments have been performed in which comparative assessments of products are made by several different industrial and environmental engineers. When provided with checklists and protocols as guidance, overall product ratings differ by less than about 15 percent among groups of several assessors.

Once an evaluation has been made for each matrix element, the overall Environmentally Responsible Product Rating (R_{ERP}) is computed as the sum of the matrix element values:

$$R_{ERP} = \sum_i \sum_j M_{i,j} \tag{14.1}$$

Because there are 25 matrix elements, a maximum product rating is 100.

Designers who have never performed a product audit may wonder about the relevance of some of the life stage–environmental concern pairs. To aid in perspective, Table 14.2 provides examples for each matrix element. The basis for some of these examples is that the industrial process is responsible (implicitly if not explicitly) for the embedded impacts of the processing of raw materials that are used and for the projected impacts as the products are used, recycled, or discarded.

14.5 TARGET PLOTS

The matrix displays provide a useful overall assessment of a design, but a more succinct display of DfE design attributes is provided by "target plots," as shown in Figure 14.2. To construct the plots, the value of each element of the matrix is plotted at a specific

TABLE 14.2 Examples of Product Inventory Concerns

| Life stage | Environmental concern | | | | |
	Materials choice	Energy use	Solid residues	Liquid residues	Gaseous residues
Resource extraction	Use of only virgin materials	Extraction from ore	Slag production	Mine drainage	SO_2 from smelting
Product manufacture	Use of only virgin materials	Inefficient motors	Sprue, wrap disposal	Toxic chemicals	CFC use
Product delivery	Toxic printing ink use	Energy loss in packing	Polystyrene packaging	Toxic printing ink use	CFC foams
Product use	Intentionally dissipated metals	Resistive heating	Solid consumables	Liquid consumables	Combustion emissions
Recycling, disposal	Use of toxic organics	Energy loss in recycling	Nonrecyclable solids	Nonrecyclable liquids	HCl from incineration

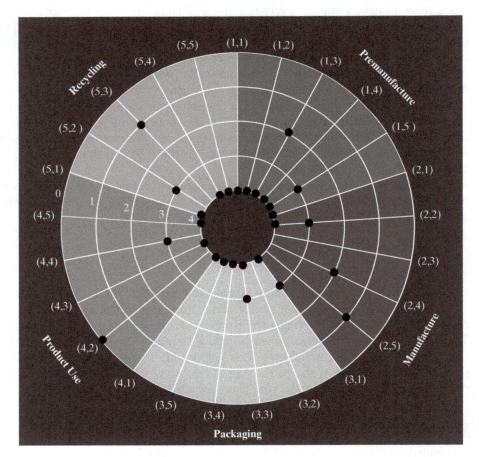

Figure 14.2

The features of the target plot for an environmentally responsible product. The data are for demonstration purposes and do not represent an actual product.

angle. (For a 25-element matrix, the angle spacing is 360/25 = 14.4.) A good product or process shows up as a series of dots bunched toward the center, as would occur on a rifle target in which each shot was aimed accurately. The plot makes it easy to single out points far removed from the bull's eye and to mark their topics out for special attention by the design team. Furthermore, target plots for alternative designs of the same product permit quick comparisons of environmental responsibility. The product design team can then select among design options and can consult the checklists and protocols for information on improving individual matrix element ratings.

14.6   ASSESSING GENERIC AUTOMOBILES OF YESTERDAY AND TODAY

The automobile and its manufacture provide a widely known and widely studied example of how SLCA is accomplished in practice. Automobiles have both manufacturing and in-use impacts on the environment, in contrast to many other products such as furniture or interior hardware. The greatest impacts result from the combustion of gasoline and the release of tailpipe emissions during the driving cycle. However, there are other aspects of the product that affect the environment, such as the dissipative use of oil and other lubricants, the discarding of tires and other spent parts, and the ultimate retirement of the vehicle. To assess these factors, environmentally responsible product assessments have been performed on generic automobiles of the 1950s and 2000s. Some of the relevant characteristics of the vehicles are given in Table 14.3. In overview, the 1950s vehicle was substantially heavier, less fuel efficient, prone to greater dissipation of working fluids and exhaust gas pollutants, and had components such as tires that were less durable.

Premanufacturing, the first life stage, treats impacts on the environment as a consequence of the actions needed to extract materials from their natural reservoirs,

TABLE 14.3   Characteristics of Generic Automobiles (Estimates from
Ward's Automobile Yearbook)

Characteristic materials (kg)	*ca.* 1950s automobile	*ca.* 2000s automobile
Plastics	0	101
Aluminum	0	68
Copper	25	22
Lead	23	15
Zinc	25	10
Iron	220	207
Steels	1290	793
Glass	54	38
Rubber	85	61
Fluids	96	81
Other	83	38
Total weight	1901	1434
Fuel efficiency (mi/gal)	15	27
Exhaust catalyst	No	Yes
Air conditioning	CFC-12*	HFC-134a

* Air conditioning entered the automobile market in the late 1950s on top-of-the-line vehicles.

TABLE 14.4   Premanufacturing Ratings

Element designation	Element value and explanation
1950s auto:	
Materials choice (1,1)	2 (Few toxics are used, but most materials are virgin)
Energy use (1,2)	2 (Virgin material shipping is energy intensive)
Solid residue (1,3)	3 (Iron and copper ore mining generates substantial solid waste)
Liquid residue (1,4)	3 (Resource extraction generates moderate amounts of liquid waste)
Gas residue (1,5)	2 (Ore smelting generates significant amounts of gaseous waste)
2000s auto:	
Materials choice (1,1)	3 (Few toxics are used; much recycled material is used)
Energy use (1,2)	3 (Virgin material shipping is energy intensive)
Solid residue (1,3)	2 (Metal mining generates solid waste)
Liquid residue (1,4)	3 (Resource extraction generates moderate amounts of liquid waste)
Gas residue (1,5)	3 (Ore processing generates moderate amounts of gaseous waste)

transport them to processing facilities, purify or separate them by such operations as ore smelting and petroleum refining, and transport them to the manufacturing facility. Where components are sourced from outside suppliers, this life stage also incorporates assessment of the impacts arising from component manufacture. The ratings assigned to this life stage of generic vehicles from each epoch are given in Table 14.4, where the two numbers in parentheses refer to the matrix element indices. The higher (i.e., more favorable) ratings for the 2000s vehicle are mainly due to improvements in the environmental aspects of mining and smelting technologies, improved efficiency of the equipment and machinery used, and the increased use of recycled material.

The second life stage is product manufacture (see Table 14.5). The basic automobile manufacturing process has changed little over the years but much has been done to improve its environmental responsibility. One potentially high-impact area is the paint shop, where various chemicals may be used to clean the parts and volatile organic emissions can be generated during the painting process. There is now greater emphasis on treatment and recovery of waste water from the paint shop, and the switch from low-solids

TABLE 14.5   Product Manufacture Ratings

Element designation	Element value and explanation
1950s auto:	
Materials choice (2,1)	0 (CFCs used for metal parts cleaning)
Energy use (2,2)	1 (Energy use during manufacture is high)
Solid residue (2,3)	2 (Lots of metal scrap and packaging scrap produced)
Liquid residue (2,4)	2 (Substantial liquid residues from cleaning and painting)
Gas residue (2,5)	1 (Volatile hydrocarbons emitted from paint shop)
2000s auto:	
Materials choice (2,1)	3 (Good materials choices, except for lead solder waste)
Energy use (2,2)	2 (Energy use during manufacture is fairly high)
Solid residue (2,3)	3 (Some metal scrap and packaging scrap produced)
Liquid residue (2,4)	3 (Some liquid residues from cleaning and painting)
Gas residue (2,5)	3 (Small amounts of volatile hydrocarbons emitted)

to high-solids paint has done much to reduce the amount of material emitted. With respect to material fabrication, there is currently better utilization of material (partially due to better analytical techniques for designing component parts) and a greater emphasis on reusing scraps and trimmings from the various fabrication processes. Finally, the productivity of the entire manufacturing process has been improved, substantially less energy and time being required to produce each automobile.

The environmental concerns at the third life stage, product delivery, include the manufacture of the packaging material, its transport to the manufacturing facility, residues generated during the packaging process, transportation of the finished and packaged product to the customer, and (where applicable) product installation (see Table 14.6). This aspect of the automobile's life cycle is benign relative to the vast majority of products sold today, since automobiles are delivered with negligible packaging material. Nonetheless, some environmental burden is associated with the transport of a large, heavy product. The slightly higher rating for the 2000s automobile is due mainly to the better design of auto carriers (more vehicles per load) and the increase in fuel efficiency of the transporters.

The fourth life stage, product use, includes impacts from consumables (if any) or maintenance materials (if any) that are expended during customer use (see Table 14.7). Significant progress has been made in automobile efficiency and reliability, but automobile use continues to have a very high negative impact on the environment. The increase in fuel efficiency and more effective conditioning of exhaust gases accounts for the 2000s automobile achieving higher ratings, but clearly there is still room for improvement.

The fifth life stage assessment includes impacts during product refurbishment and as a consequence of the eventual discarding of modules or components deemed impossible or too costly to recycle (see Table 14.8). Most modern automobiles are recycled (some 95 percent of those discarded enter the recycling system in most countries), and from these approximately 75 percent by weight is recovered as used parts or returned to the secondary metals market. Improvements in recovery technology have made it easier and more profitable to separate the automobile into its component materials.

TABLE 14.6   Product Delivery Ratings

Element designation	Element value and explanation
1950s auto:	
Materials choice (3,1)	3 (Sparse, recyclable materials used during packaging and shipping)
Energy use (3,2)	2 (Over-the-road truck shipping is energy intensive)
Solid residue (3,3)	3 (Small amounts of packaging during shipment could be further minimized)
Liquid residue (3,4)	4 (Negligible amounts of liquids are generated by packaging and shipping)
Gas residue (3,5)	2 (Substantial fluxes of greenhouse gases are produced during shipment)
2000s auto:	
Materials choice (3,1)	3 (Sparse, recyclable materials used during packaging and shipping)
Energy use (3,2)	3 (Long-distance land and sea shipping is energy intensive)
Solid residue (3,3)	3 (Small amounts of packaging during shipment could be further minimized)
Liquid residue (3,4)	4 (Negligible amounts of liquids are generated by packaging and shipping)
Gas residue (3,5)	3 (Moderate fluxes of greenhouse gases are produced during shipment)

TABLE 14.7   Customer Use Ratings

Element designation	Element value and explanation
1950s auto:	
Materials choice (4,1)	1 (Petroleum is a resource in limited supply)
Energy use (4,2)	0 (Fossil fuel energy use is very large)
Solid residue (4,3)	1 (Significant residues of tires, defective or obsolete parts)
Liquid residue (4,4)	1 (Fluid systems are very leaky)
Gas residue (4,5)	0 (No exhaust gas scrubbing; high emissions)
2000s auto:	
Materials choice (4,1)	1 (Petroleum is a resource in limited supply)
Energy use (4,2)	2 (Fossil fuel energy use is large)
Solid residue (4,3)	2 (Modest residues of tires, defective or obsolete parts)
Liquid residue (4,4)	3 (Fluid systems are somewhat dissipative)
Gas residue (4,5)	2 (CO_2, lead [in some locales])

In contrast to the 1950s, at least two aspects of modern automobile design and construction are worse from the standpoint of their environmental implications. One is the increased diversity of materials used, mainly the increased use of plastics. The second aspect is the increased use of welding in the manufacturing process. In the vehicles of the 1950s, a body-on-frame construction was used. This approach was later switched to a unibody construction technique in which the body panels are integrated with the chassis. Unibody construction requires about four times as much welding as does body-on-frame construction, plus substantially increased use of adhesives. The result is a vehicle that is stronger, safer, and uses less structural material, but is much less easy to disassemble.

The completed matrices for the generic 1950s and 2000s automobile are illustrated in Table 14.9. Examine first the values for the 1950s vehicle so far as life stages are concerned. The column at the far right of the table shows moderate environmental stewardship during resource extraction, packaging and shipping, and efurbishment/recycling/disposal. The ratings during manufacturing are poor, and those during customer use are abysmal. The

TABLE 14.8   Refurbishment/Recycling/Disposal Ratings

Element designation	Element value and explanation
1950s auto:	
Materials choice (5,1)	3 (Most materials used are recyclable)
Energy use (5,2)	2 (Moderate energy use required to disassemble and recycle materials)
Solid residue (5,3)	2 (A number of components are difficult to recycle)
Liquid residue (5,4)	3 (Liquid residues from recycling are minimal)
Gas residue (5,5)	1 (Recycling commonly involves open burning of residues)
2000s auto:	
Materials choice (5,1)	3 (Most materials recyclable, but sodium azide presents difficulty)
Energy use (5,2)	2 (Moderate energy use required to disassemble and recycle materials)
Solid residue (5,3)	3 (Some components are difficult to recycle)
Liquid residue (5,4)	3 (Liquid residues from recycling are minimal)
Gas residue (5,5)	2 (Recycling involves some open burning of residues)

TABLE 14.9   Environmentally Responsible Product Assessments for the Generic 1950s and 2000s Automobiles

| Life stage | Environmental concern | | | | | |
	Materials choice	Energy use	Solid residues	Liquid residues	Gaseous residues	Total
Resource extraction						
1950s	2	2	3	3	2	12/20
2000s	3	3	3	3	3	15/20
Product manufacture						
1950s	0	1	2	2	1	6/20
2000s	3	2	3	3	3	14/20
Product delivery						
1950s	3	2	3	4	2	14/20
2000s	3	3	3	4	3	16/20
Product use						
1950s	1	0	1	1	0	3/20
2000s	1	2	2	3	2	10/20
Recycling, disposal						
1950s	3	2	2	3	1	11/20
2000s	3	2	3	3	2	13/20
Total						
1950s	9/20	7/20	11/20	13/20	6/20	46/100
2000s	13/20	12/20	14/20	16/20	13/20	68/100

overall rating of 46 is far below what might be desired. In contrast, the overall rating for the 2000s vehicle is 68, much better than that of the earlier vehicle but still leaving plenty of room for improvement. A more succinct display of DfES design attributes is provided by the target plots of Figure 14.3.

14.7   WEIGHTING IN SLCA

There is an underlying uneasiness in the results of Table 14.9 in that equal importance has been given to, for example, the "packaging and transport" and "in-service" life stages. Because the latter very often has a greater environmental effect, it is appropriate to contemplate some form of matrix element weighting to reflect this information. Thus, instead of the initial approach in which each matrix element had a maximum value of 4 and the overall assessment value was given by summing the matrix element values, a set of weighting factors $\omega_{i,j}$ can be sought to reflect differences in life stage impact such that each matrix element will have a value given by $\omega_{i,j}M_{i,j}$ and for which the overall rating is given by

$$R_{\text{ERP}} = \sum_i \sum_j \omega_{i,j} M_{i,j} \tag{14.2}$$

Determining in a comprehensive way, the appropriate life stage weightings require a complete life cycle assessment and are in conflict with the desire to streamline the LCA process. The assessment can be intuitively improved, however, if the assessment

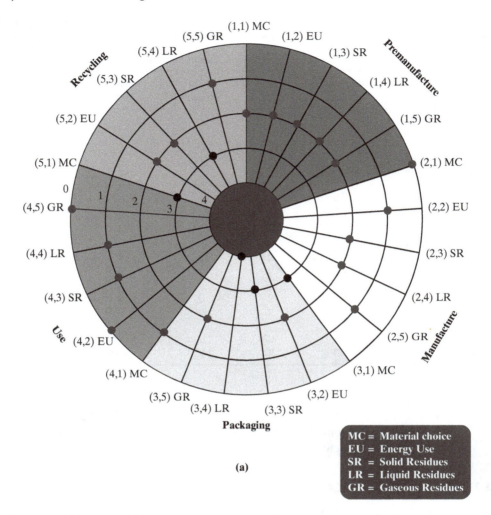

Figure 14.3

Comparative target plots for the display of the environmental impacts of the generic automobile of the 1950s and of the 2000s.

team chooses the life stage likely to produce the most severe environmental impacts and arbitrarily weights that life stage as, say, one half the total assessment value, with the other four life stages arbitrarily weighted at, say, one-eighth the total value. For the automobile example used previously, where the in-service life stage almost certainly results in the most severe of the environmental impacts, that stage is chosen as the dominant one for halftile-octile weighting. The weighting factors are then those shown in Table 14.10.

If the matrix element values are recomputed using these weighting factors, the resultant evaluation for the generic 1950s automobile is shown in Table 14.11, where the increase in influence of life stage 4 is evident and where the overall product rating has dropped from the 44 of Figure 14.3a to 33. This result can be made even more

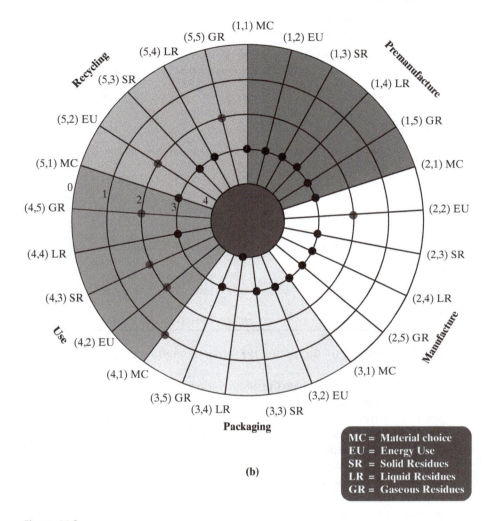

Figure 14.3

Continued

apparent by the target plot. However, the construction of the target plot now requires an additional step in order to demonstrate the degree of departure from optimum of a matrix element value. The procedure is to compute the "deficit matrix," the difference between perfect matrix element scores and actual scores, given by

$$\delta_{i,j} = (\omega_{i,j} M_{i,j})_{\max} - \omega_{i,j} M_{i,j} \tag{14.3}$$

In the 1950s automobile example, the resulting target plot is shown in Figure 14.4a where, by comparison with Figure 14.3a, the deficiencies in environmental responsibility during the in-service life stage are quite obvious.

Although the in-service life stage will be the one most highly weighted for many products, it is not difficult to identify products for which other life stages

TABLE 14.10 Weighting Factors for Singly and Doubly Weighted Matrices

The matrix of weighting factors for a dominant life stage:

$$\omega_{i,j} = \begin{matrix} 0.625 & 0.625 & 0.625 & 0.625 & 0.625 \\ 0.625 & 0.625 & 0.625 & 0.625 & 0.625 \\ 0.625 & 0.625 & 0.625 & 0.625 & 0.625 \\ 2.5 & 2.5 & 2.5 & 2.5 & 2.5 \\ 0.625 & 0.625 & 0.625 & 0.625 & 0.625 \end{matrix}$$

The matrix of weighting factors for a dominant environmentally related attribute:

$$\varphi_{i,j} = \begin{matrix} 0.625 & 2.5 & 0.625 & 0.625 & 0.625 \\ 0.625 & 2.5 & 0.625 & 0.625 & 0.625 \\ 0.625 & 2.5 & 0.625 & 0.625 & 0.625 \\ 0.625 & 2.5 & 0.625 & 0.625 & 0.625 \\ 0.625 & 2.5 & 0.625 & 0.625 & 0.625 \end{matrix}$$

The matrix of weighting factors for a doubly weighted matrix:

$$\omega_{i,j}\varphi_{i,j} = \begin{matrix} 0.39 & 1.56 & 0.39 & 0.39 & 0.39 \\ 0.39 & 1.56 & 0.39 & 0.39 & 0.39 \\ 0.39 & 1.56 & 0.39 & 0.39 & 0.39 \\ 1.56 & 6.25 & 1.56 & 1.56 & 1.56 \\ 0.39 & 1.56 & 0.39 & 0.39 & 0.39 \end{matrix}$$

can be assumed to dominate the environmental impacts. Examples are given in Table 14.12.

Just as some life stages produce larger environmental impacts than others, some environmental impacts are of more concern than others. For a specific product, the priority environmentally related attribute is thus some combination of the magnitude of the product's impacts and the degree to which the impacts meet the high-risk criteria. Determining priorities rigorously requires, as before, a comprehensive and defendable life cycle assessment, but Table 14.12 suggests examples of products for which individual environmental concerns would probably be regarded as dominant by a consensus of experts.

In the case of the 1950s automobile, one could make a case for either energy or gaseous emissions being the dominant impact, but for demonstration purposes energy will be selected here. As with the life stage weightings, we arbitrarily assign the energy

TABLE 14.11 The Singly Weighted Environmentally Responsible Product Assessment for the Generic 1990s Automobile

Life stage	Environmental-related attribute					Total
	Materials choice	Energy use	Solid residues	Liquid residues	Gaseous residues	
Premanufacture	1.25	1.25	1.25	1.25	1.25	6.3/12.5
Product manufacture	0	0.63	1.25	1.25	0.63	3.8/12.5
Product delivery	1.88	1.25	1.88	2.50	1.25	8.8/12.5
Product use	2.50	0	2.50	2.50	0	7.5/20
Refurbishment, recycling, disposal	1.88	1.25	1.25	1.88	0.63	6.8/12.5
Total	7.5/20	4.4/20	8.1/20	9.4/20	3.8/20	33.2/100

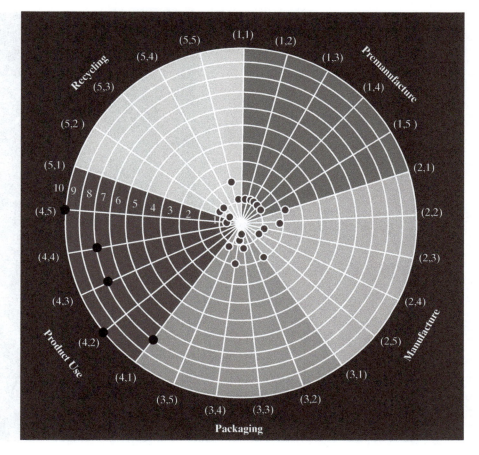

Figure 14.4a

A target plot for the display of the consensus-weighted matrix assessment results for the generic automobile of the 1950s. (From T.E. Graedel, *Streamlined Life-Cycle Assessment*, Upper Saddle River, NJ: Prentice Hall, pp. 204–206, 1998.)

TABLE 14.12   Examples of Products for which Different Life Stages or Environmentally Related Attributes can be Assumed to Dominate

Life stage:	Product
Premanufacture	Laptop computer
Product manufacture	Sheet aluminum from bauxite ore
Product delivery	Triply packaged compact disc
Product use	Automobile
Refurbishment, recycling, disposal	Mercury relays
Environmentally related attributes:	
Materials choice	Medical products with radioactive materials
Energy use	Hair dryers
Solid residues	"Convenience" foods
Liquid residues	Pesticides
Gaseous residues	CFC propellant sprays

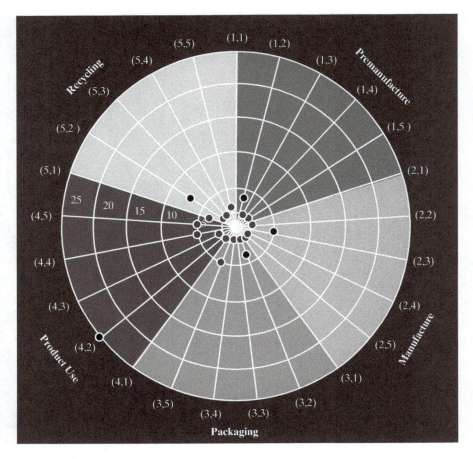

Figure 14.4b

A target plot for the display of the consensus doubly weighted matrix assessment results for the generic automobile of the 1950s. (From T.E. Graedel, *Streamlined Life-Cycle Assessment*, Upper Saddle River, NJ: Prentice Hall, pp. 206–207, 1998.)

use impact half the total assessment value, while the other environmental concerns receive one-eighth each. The second weighting factor $\varphi_{i,j}$ is then as shown in Table 14.5; the doubly weighted matrix element values are given by the product $\varphi_{i,j}\omega_{i,j}M_{i,j}$, and the overall rating by

$$R_{\text{ERP}} = \sum_i \sum_j \varphi_{i,j}\omega_{i,j}M_{i,j} \tag{14.4}$$

The matrix result for the 1950s automobile is shown in Table 14.13, the overall rating now having dropped from 33 to 29 and the impact of high energy use receiving additional emphasis.

The target plot for the doubly weighted matrix is formed in a manner similar to that for the singly weighted matrix, deficits being computed as

$$\Delta_{i,j} = (\varphi_{i,j}\omega_{i,j}M_{i,j})_{\text{max}} - \varphi_{i,j}\omega_{i,j}M_{i,j} \tag{14.5}$$

TABLE 14.13   The Doubly Weighted Environmentally Responsible Product Assessment for the Generic 1990s
Automobile

Life stage	Environmental-related attribute					
	Materials choice	Energy use	Solid residues	Liquid residues	Gaseous residues	Total
Premanufacture	0.78	3.13	0.78	0.78	0.78	6.25/12.5
Product manufacture	0	1.56	0.78	0.78	0.39	3.51/12.5
Product delivery	1.17	3.13	1.17	4	0.78	7.81/12.5
Product use	1.56	0	1.56	1.56	0	4.68/50
Refurbishment, recycling, disposal	1.17	3.13	0.78	1.17	0.39	6.64/12.5
Total	4.68/12.5	10.95/50	5.07/12.5	5.85/12.5	2.34/12.5	28.89/100

The resulting target plot in Figure 14.4b shows dramatically the importance of the
in-service life stage, the energy use environmental concern, and (most importantly) the
failure of the 1950s vehicle to reflect these priorities.

14.8   SLCA ASSETS AND LIABILITIES

As suggested above, when the LCA concept is streamlined it has been thought by some
that part of the legitimacy of a comprehensive LCA may be sacrificed. What is it that is
really lost? Conversely, what are the gains? Having presented the characteristics of
LCAs and semiquantitative SLCAs, it is useful to compare their characteristics one by
one, as done in Table 14.14.

To begin, we ask if both approaches cover the full life cycle of the product being
evaluated. Assuming that the SLCA is the semiquantitative, full life cycle approach
described earlier in this chapter, the answer is yes. So far as quantification is
concerned, LCAs definitely generate detailed numerical results, and SLCAs generate

TABLE 14.14   Attributes of LCAs and SLCAs

Attribute	LCA	SLCA*	Preference
Covers entire life cycle	Yes	Yes	—
Quantitative	Yes	Somewhat	LCA
Comprehensive	No	No	—
Creates awareness of environmental issues	Yes	Yes	—
Usable early in the design phase	No	Yes	SLCA
Relative ranking of environmental impacts	Doubtful	No	Undetermined
Valuation/weighting	Possible	Possible	—
Generates recommendations	Yes	Yes	—
Prioritizes recommendations	Possible	Possible	—
Time requirements	Long	Short	SLCA
Cost	Expensive	Inexpensive	SLCA

* The attributes refer to a semiquantitative, full life cycle approach, rather than to other methods of LCA streamlining.

approximations. Neither method is comprehensive, as complex environmental concerns such as biodiversity and land use are generally not directly addressed in either.

From the product designer's perspective, both methods create an awareness of environmental issues that otherwise might not be part of the design process. SLCAs, however, can be employed early in the design stage, while LCAs cannot; here the LCA requirement for numerical specificity is a disadvantage, not an advantage.

Can the environmental impacts of the product be rigorously compared? This is not possible in an SLCA. It is possible in theory in an LCA—this is the normalization step—but in practice is rarely attempted. In both cases, however, valuation or weighting can be employed.

A list of recommendations for product improvement in environmental performance is generated by both methods, and prioritizing the recommendations is possible in both cases.

Two final attributes rank quite differently between the two approaches: LCAs are much more time-consuming and much more expensive than SLCAs.

The last column of Table 14.14 indicates that for six of the eleven attributes, the methods are essentially equivalent, and SLCAs are superior in three categories. The LCA advantage shows up clearly in only one: It is extensively quantitative. This instinctively appealing attribute is only decisive if the quantitation is such that it can provide results of high confidence. As discussed above, there are many arguments to the effect that it seldom does so.

14.9   THE LCA/SLCA FAMILY

To some degree, the tools presented in Chapters 12–14 can be regarded as a family expressible on a two-by-two matrix, as shown in Figure 14.5. The type of assessment (streamlined or not) is the abscissa, the focus of the assessment (retrospective [sometimes termed "attributional"] or prospective [sometimes termed "consequential"]) is the ordinate. Let us examine the quadrants in turn.

> *Quadrant I: The Academician's Quadrant:* The goal of users in this quadrant is to evaluate a type of product for its environmental attributes, where the necessary information is readily available. Achieving as much precision as possible is the intent. An example is the recent effort to link physical and economic input–output analyses.

	LCA	SLCA
Prospective/ Retrospective	I Academician	II Technological historian
Prospective/ Consequential	III Policy maker	IV Designer

Figure 14.5

The family of life cycle assessment methodologies and suggestions as to appropriate users in each category.

Quadrant II: The Historian's Quadrant: The goal of users in this quadrant is to efficiently create a generic analysis of a type of product or service, especially where the available information is approximate or must be estimated. Historians of technology, historical or recent, are typical users. An example is the telephone service analysis of Graedel and Saxton (2002).

Quadrant III: The Policy Maker's Quadrant: Users in this quadrant wish to have detailed information on the environmental consequences of a policy decision, as in the leaded/nonleaded solder analysis of Ekvall and Andrae. Some large corporations also employ LCAs to evaluate completed designs, as in the Mercedes-Benz S-Class evaluations described by Finkbeiner and colleagues.

Quadrant IV: The Designer's Quadrant: Users in this quadrant want to efficiently evaluate the environmental attributes of new designers, often early in the design process. Because this information is generally proprietary, case studies are seldom published. As a result, it is easy to underestimate the level of activity in this quadrant. Among the few published examples is that of Bennett and Graedel (2000) for home air conditioners.

There is little question that LCA/SLCA is generally effective at raising environmentally related issues for product, process, and service designers to address, and in separating the more important of those issues from the less important. It is less certain that LCA/SLCA can reliably enable the user to say with certainty that "Product A is environmentally preferable to Product B." The most utilitarian approach to these tools for designers may be to use LCAs to perform retrospective analyses to frame the design approaches for new products, processes, and services, and then to perform prospective SLCAs to identify and address specific design choices of environmental interest. The result will invariably be a product more environmentally superior than would have been the case had some version of LCA/SLCA approaches not been taken. For policy use, carefully crafted and restricted LCAs are likely to serve as useful aids in decision making.

FURTHER READING

Bennett, E.B., and T.E. Graedel, Conditioned air: Evaluating an environmentally preferable service, *Environmental Science & Technology, 34*, 541–545, 2000.

Finkbeiner, M., et al., Application of life cycle assessment for the environmental certificate of the Mercedes-Benz S-Class, *International Journal of Life Cycle Assessment, 11*, 240–246, 2006.

Graedel, T.E., and E. Saxton, Improving the overall environmental performance of existing telecommunications facilities, *International Journal of Life Cycle Assessment, 7*, 219–224, 2002.

Graedel, T.E., B.R. Allenby, and P.R. Comrie, Matrix approaches to abridged life cycle assessment, *Environmental Science & Technology, 29*, 134A–139A, 1995.

Hochschorner, E., and G. Finnveden, Evaluation of two simplified life cycle assessment methods, *International Journal of Life Cycle Assessment, 8*, 119–128, 2003.

Hur, T., J. Lee, J. Ryu, and E. Kwon, Simplified LCA and matrix methods in identifying the environmental aspects of a product system, *Journal of Environmental Management, 75*, 229–237, 2005.

EXERCISES

14.1 The SLCA approach presented in this chapter is essentially qualitative in nature. Discuss the degree to which this limitation influences the usefulness of the final result.

14.2 Product inventories have been prepared for two different designs of a high speed widget. The matrices are reproduced below—the figure on the left side of each matrix element referring to Design 1, that to the right to Design 2. Select the better product from a DfE viewpoint. What features of each design would you address if improvement were needed?

Life Stage	MC	EU	SR	LR	GR
PM	1/1	4/3	4/3	2/2	3/2
M	2/1	1/2	1/2	2/1	2/4
PD	3/2	1/1	2/3	1/1	1/1
PU	1/2	1/2	1/3	1/1	1/3
RD	2/1	2/2	2/1	1/2	1/2

14.3 Evaluate the legitimacy and utility of the SLCA weighting approach described in Section 14.7.

14.4 Select a product of moderate complexity, such as a toaster, a desktop telephone, or an overhead projector. Conduct an SLCA on the product. Prepare a report that summarizes your findings, comments on where it was difficult to assign ratings because of lack of information, and proposes design changes that would improve the environmental responsibility of the product.

PART IV   Analysis of Technological Systems

C H A P T E R   1 5

Systems Analysis

15.1   THE SYSTEMS CONCEPT

Perhaps the most important operational feature of the industrial ecology approach is its ability to focus not solely on the product or process itself, but also on the related systems and their behavior. Identifying the appropriate system level and properly relating it to its technological context can be a critical step in any successful industrial ecology assessment. Accordingly, it is useful to introduce and discuss the general concept of technological systems.

A system may be thought of as a group of interacting, interdependent parts linked together by exchanges of energy, matter, and/or information, and subject to a common plan or serving a common purpose. There are two types of systems: simple and complex. The terms are used here not in the sense of classical physics (e.g., the "simple" harmonic oscillator), but of the dictionary definitions, "complex" meaning "consisting of interconnected parts so as to make the whole difficult to understand"; "simple" is thus "not combined or compound." Some of the implications of this distinction are:

1. Simple systems tend to behave in a linear fashion: the output of the system is linearly related to the input. That is, a system with stressors A and B will have an impact Y given by

$$Y = A + B \tag{15.1}$$

or, to give a common example, if you combine two quarters, you get $0.50, no more and no less. With complex systems, on the other hand, emergent behaviors may not be additive: We cannot predict the characteristics of an anthill by summing up the observed behavior of individual ants, because we have omitted interactive effects. In a complex system with stressors A and B, impact Y may be dictated by these interactive AB terms, the magnitude and behavior of which may be far from obvious.

$$Y = A + B + f(AB) \tag{15.2}$$

For example, a salt marsh may be relatively resistant to chemical pollution until a threshold is exceeded, after which even a small increment of additional insult will cause it to suddenly degrade precipitously. It is a complex system.

2. Simple systems can generally be evaluated in terms of cause and effect: An action is easily traceable through the system to its predictable effect. With complex systems, on the other hand, feedback loops and other response properties that often make the linkage between cause and effect difficult to establish.

3. Complex systems, unlike simple systems, are characterized by significant time and space discontinuities. A problem for most people in comprehending the issues surrounding global climate change is that there are large time lags and spatial discontinuities between the forcing function (e.g., driving automobiles or using electricity), and the resulting shifts in regional and global climate patterns.

4. Simple systems tend to be characterized by a stable and known equilibrium point to which they return in a predictable fashion if perturbed. Many complex systems, on the other hand, operate far from equilibrium, in a state of constant adaptation to changing conditions. Complex systems often evolve; simple systems generally stay more or less as they are.

A second attribute of systems that helps define and classify them is whether or not they are purposeful, or guided by a definite aim. Consider Figure 15.1, which characterizes four examples. The automobile is a planned system, predictable when operated according to its design, and developed with a purpose in mind—transportation. The traffic flow on an interstate highway reflects independent human agents with a common purpose—reaching their destinations—but with unpredictable changes in speed and traffic jams that emerge from the system as a whole as a result of interactions among individual decisions. Nature is a contrast to these purposeful actions involving technology. It can be predictive, in the sense that an experienced naturalist can anticipate with great certainty the actions of a beehive, which has no more plan than survival. Nature can also be unpredictable, as in the

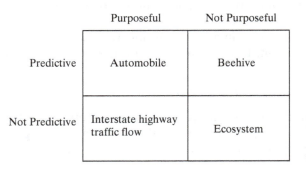

Figure 15.1

A two-dimensional classification of systems, with examples (see text for discussion).

behavior of an ecosystem with many different species each trying to find its own niche. In many cases, natural systems will tend toward homoeostasis, or a dynamic stability; while it may appear to be purposeful, this is an emergent characteristic of the system over time rather than the result of an overall plan.

Fundamentally, industrial ecology systems are purposeful. To the degree that they restrict their scope to strictly technological products and processes, they are also predictive. Once independent human actors are added, however, complexity sets in. The results cannot be studied solely by natural science approaches, but inextricably involve specialists from the social sciences. This is another example of the fact that industrial ecology is inherently multidisciplinary, and, given the coupling in such systems between the social and environmental, an appropriate framework for sustainable engineering.

Case Study: An Interfirm Trade Network

The complexity of even fairly simple systems in industrial ecology is exemplified by W. Ashton's study of interfirm trade in the Barceloneta region of Puerto Rico. She interviewed 39 firms to establish their trading partners in the immediate region and plotted the results on a cluster diagram. The network turned out to be highly centralized, with a small number of firms establishing most of the trading connections, and most of these were connected with the chemical industry. Waste management firms were peripheral members of the network. Food products, electronic equipment, and traders were scattered about the diagram. The results permit a highly informed perspective on opportunities for resource sharing (see Chapter 16), social networking (see Chapter 7), and adaptation to changing conditions (see Section 15.2) (see Figure CS-1).

15.2   THE ADAPTIVE CYCLE

Many industrial ecology systems of interest are not only complex, they are evolving. As this occurs, a system may flip from one metastable state to another. As we often do in this book, we turn to biological ecology for an example, and this brings us to shallow lakes. In a number of cases, such lakes are known to be bistable (Figure 15.2a): if low in nutrients the water is generally clear; if high in nutrients it is generally turbid. The transition is not gradual, however, but rapid once a bifurcation point is crossed. The behavior is related to the biological communities involved. Some nutrient conditions favor algae feeders that reduce turbidity, while some favor bottom feeders that increase it. The turbidity, and especially the unanticipated flip from one state to another, results both from the general conditions of the system (temperature, water depth, etc.) and from the particular types and number of organisms that comprise it and whose populations evolve with it.

Transitional behavior of systems is expressed in another way in C.S. Holling's "adaptive cycle," illustrated in Figure 15.2b. Holling identifies four system functions, expressed on a plot of "potential" (a function of accumulated economic and social capital) and the degree of connectedness of the system. An early-stage system in the exploitation (r) phase gradually increases connectedness and stability. At a certain level of connectedness, the potential stabilizes; new opportunities for exploitation decrease;

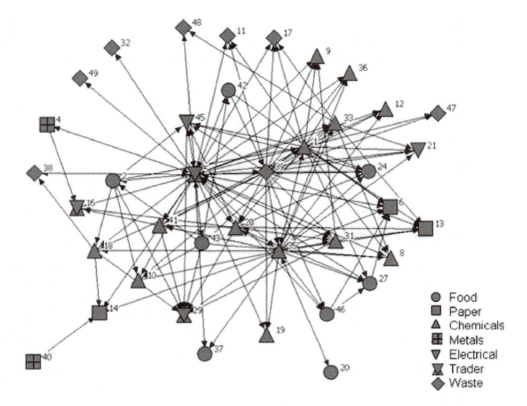

Figure CS-1

The interfirm trade network in Barceloneta, Puerto Rico. Each node represents a firm (numbered nonsequentially), and the lines indicate the sale of a product or service between two firms (arrows point to buyers). The firms with the highest number of buyer–seller relationships are located in the center of the diagram. A firm's position depends on the number of others to which it is directly tied, and the geodesic (shortest) distances to others in the network. (Reproduced with permission from W. Ashton, Understanding the organization of industrial ecosystems: A social network approach, *Journal of Industrial Ecology, 12*, 34–51, 2008.)

and the system (*K*) becomes stable and conserves its capital. A major disturbance is then needed to release the system; in this Ω phase the potential and connectedness are rapidly transformed. The reorganization (*α*) phase that follows is unpredictable; it can lead to a regeneration of the original system, probably with variations, or an escape to another type of system entirely.

The adaptive cycle provides considerable perspective on the interpretation of industrial ecology systems. Consider the Barceloneta, Puerto Rico, industrial ecosystem, described more fully by Ashton (2008). The collapse and reorganization of this system occurred in the 1940s and 1950s, when sugar industry exports declined markedly. From the mid-1950s through 1970, a shift to manufacturing-based industry signaled the exploitation phase. In the following 20 years, pharmaceutical industries were added, and the system entered the conservation phase. By 2008, manufacturing was contracting, perhaps signifying the beginning of the collapse phase of the cycle.

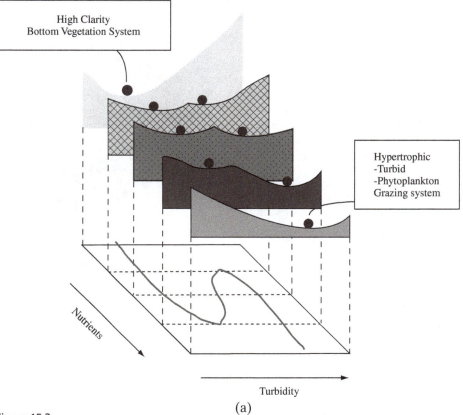

High Clarity
Bottom Vegetation System

Hypertrophic
-Turbid
-Phytoplankton
Grazing system

Nutrients

Turbidity

(a)

Figure 15.2

(a) The bistability pattern in a shallow lake. The response of turbidity to nutrients is shown at the bottom, and conceptual stability diagrams at the top. (Adapted from M.S. Scheffer, et al., Alternative equilibria in shallow lakes, *Trends in Ecology and Evolution, 8*, 275–279, 1993; courtesy of J.J. Kay.) (b) The adaptive cycle for a complex system, connecting four system functions (r, K, Ω, α) and the flow of events among them. The arrows show the speed of the cycle—long (rapidly changing) or short (slowly changing). The exit from the cycle in the r box indicates the stage where a flip into a different stability regime is most likely. (Reproduced with permission from C.S. Holling, Understanding the complexity of economic, ecological, and social systems, *Ecosystems, 4*, 390–405, 2001.)

15.3 HOLARCHIES

Industrial ecology has technology at its core, but also recognizes technology's tight links to human actions (the social system) and to nature (the environmental system). In other words, industrial ecology deals with systems of systems. Nature does, too. Figure 15.3 sketches a few examples. Hierarchies in natural systems are familiar to biological ecologists, who analyze what they observe at different spatial levels, that turn out to be different types of organizational structures as well. The industrial product (automobile) as a feature of the comprehensive built environment, already presented in Chapter 7, emphasizes stocks of materials in use. The example of an industrial

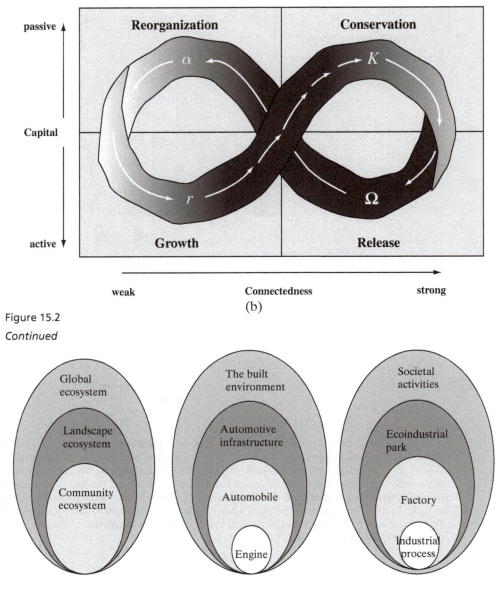

Figure 15.2

Continued

Figure 15.3

Examples of holarchies: (left) a classical holarchy of nature; (center) a technological holarchy based on stocks of material in use; (right) a technological holarchy based on flows of materials and energy.

process as a component of societal action emphasizes flows of materials and energy. All these examples are systems of systems.

In complex systems theory, an individual entity in a system of systems is termed a "holon," and the network of holons a "holarchy." Because the behavior of some or all of the holons may not be predictive, a holarchy cannot in general be the product of a comprehensive, anticipatory design. Rather, a holarchy tends to be a self-organizing, hierarchical, open (SOHO) entity; it is hierarchical in the sense of "a conceptually or causally linked system of grouping objects or processes along an analytical scale," and open in the sense of requiring external inputs of materials and energy. Hence, the SOHO entity of which the automobile is a part is properly termed an "industrial ecology holarchy."

All holarchies have a number of properties in addition to their hierarchical structure that are worth noting. Perhaps the most obvious is that they are nonlinear and thus not strictly predictive. A second is that they are dynamic systems that exist in quasi-steady states rather than at equilibrium. They often exhibit feedback loops involving materials or energy cycling. Finally, there is the potential for chaotic and catastrophic behavior if they find themselves removed from their quasi-steady state.

The study of holarchies is a vigorous activity within biological ecology; its incorporation into technologically related fields is more recent and less accomplished. Unlike the nature-centered perspective, however, holarchies related to technology can be regarded as purposeful. This has enabled Mark Maier of the Aerospace Corporation to list five characteristics of technological holarchies:

- Operational independence of the holons, each holon being useful in its own right.
- Managerial independence of the holons, they exchange information but not control.
- Evolutionary development, a result of the incorporation of new resources, technologies, and ideas.
- Emergent behavior rather than planned behavior (as discussed below in more detail).
- Geographic distribution, rendering close linkages of material and energy flows, and perhaps information flows, problematic.

Maier goes on to define and exemplify three types of technological holarchies, as follows:

The Directed Technological Holarchy. Example—integrated air defenses of modern military forces. In such a system, the holons (the networks of surveillance radars, missile launch batteries, fighter aircraft) operate independently, but their normal operational mode is subordinated to the centrally managed purpose as required.

The Collaborative Technological Holarchy. Example—urban transport systems. In these systems, the holons (private vehicles, taxis, subways, railroads, aircraft) voluntarily collaborate to fulfill the agreed-upon central purpose.

The Virtual Technological Holarchy. Example—World Wide Web. Virtual holarchies lack any central authority, other than the basic enabling protocols.

Technological holarchies are obviously very difficult entities to study, but undeniably meet the industrial ecology criterion of involving human transformations of mass and energy, and thus can be addressed from a systems perspective.

15.4 THE PHENOMENON OF EMERGENT BEHAVIOR

A feature of holarchies that frequently confounds analysts is that of emergent behavior, in which even a detailed knowledge of one holon is insufficient to predict behavior at a different holonic level. An obvious example from the realm of nature is the beating heart. The heart at its lowest level consists of cells, of course, which can be described extensively from physical and chemical perspectives. Nothing at the cell level suggests electrical activity leading to rhythmicity, however. Rather, rhythmicity arises as a consequence of the electrical properties of numerous intracellular gap junctions, and as modified by the three-dimensional architecture and structure of the organ itself.

It is useful to think of the cardiac holarchy from the perspective of Figure 15.4. This diagram suggests a property of holarchies not previously mentioned: that size tends to increase as one moves from lower holons to higher, and speed tends to decrease. In the cardiac case, cells are small and their physical and chemical processes are rapid; the opposite is true of the organ itself. The result is that the organ level sets constraints and boundaries for the cell coupling level. The cell level itself provides a mechanistic foundation for the cell coupling level.

In any holarchy in a metastable state, the details of the –1 holon's processes are too small and fast to be anything but background noise to level 0. When the system is disrupted, however, the rapid dynamics of the –1 holon are what overcomes the constraints imposed by the +1 holon, and the system adopts a new metastable state. In the cardiac example, damage to an artery of the wall of the heart can lead to an irregular heartbeat, or perhaps to no heartbeat at all.

Emergent behavior is also a feature of industrial ecology holarchies. Consider the example of cellular telephony. This complex technology was developed in the 1980s and 1990s. The fixed-location base stations that were originally needed were few, and the telephones expensive and briefcase sized. The cell phone holon and the infrastructure holon that supported it were largely predictable, and users were anticipated to be a modest number of physicians, traveling salespeople, and others not having convenient access to a landline phone.

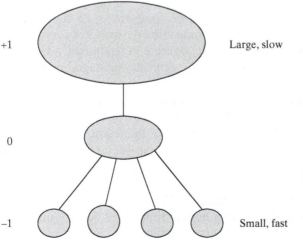

+1   Large, slow

0

–1   Small, fast

Figure 15.4

Relationships between holons in a holarchy. The level of interest (level 0) is in the center of the diagram. The dynamics of level 0 are constrained and bounded by the larger and slower-moving level +1. Interactions at the small, fast-moving level –1 generate the signatures felt in level 0. (Adapted from a diagram in R.V. O'Neill, Hierarchy theory and global change, in *Scales and Global Change*, T. Rosswall, R.G. Woodmansee, and P.G. Risser, Eds., New York: John Wiley, 1988.)

Around 2000, improved technology made cell phones much smaller and much cheaper. Parents began to buy cell phones for their children, and firms for their employees. Suddenly it was possible to call anyone from anywhere. Demand skyrocketed, especially in developing countries where the technology made it possible to avoid installing landline phones almost completely. An entirely new pattern of social behavior had emerged, unpredicted and certainly unplanned.

This story is relevant to industrial ecology because it ultimately involves humans, resources, energy, and the environment. The hundreds of millions of cell phones demand an incredible diversity and quantity of materials for optimum functionality. At one point in their rapid evolution, tantalum came into short supply, and the mineral coltan was mined in Africa by crude technological approaches in order to fill this supply gap, doing significant environmental damage in the process. This challenge appears to be past, but the cell phone network worldwide is now trying to address a new emergent behavior: the recovery of precious metals from discarded cell phones by primitive "backyard" technologies. This social/technological activity did not exist when cell phones were few; as they became abundant, the recycling networks flipped into a new and unanticipated state.

15.5   ADAPTIVE MANAGEMENT OF TECHNOLOGICAL HOLARCHIES

Humans desire to control their surroundings for the benefit of themselves and others, of course. As technology becomes more and more deeply ingrained and integrated into human life, the challenges faced by control-minded humans increase. This is particularly true when the challenges are recognized as holarchical. As Table 15.1 demonstrates, the distinctions between the management of simple systems and of holarchies are great.

Because of the human element, holarchies display dynamics that are characteristic of social rather than engineered systems, such as:

1. Problems cannot be definitively described or defined. Any boundaries drawn around such systems necessarily oversimplify other aspects of the system, making even basic steps such as system definition somewhat arbitrary.
2. There is nothing like an indisputable public benefit. In other words, not only is there no definitive prioritization of values that can guide design and analysis,

TABLE 15.1   Distinctions Between Unitary Systems and Holarchies

Area	Unitary system	Holarchy
Focus	Single complex system	Multiple complex integrated holons
Objective	Achieve a purpose	Optimize holarchic functions
Expectation	Solution	Appropriate response to situation
Boundaries	Fixed	Fluid
Problem	Defined	Emergent
Goals	Unitary	Pluralistic
Approach	Intellectually rigorous	Intellectually adaptive

Source: Adapted from C. Keating, et al., System of systems engineering, *Engineering Management Journal*, *15* (3), 36–45, 2003.

but every "benefit" identified by the systems analyst necessarily carries with it a "disbenefit." For example, policies that encourage foreign direct investment in underdeveloped countries necessarily also change the culture and institutions of that country, which some people regard as "cultural imperialism."

3. There are no objective definitions of equity. Even in the case of sustainability, the assumption of egalitarian values necessarily means that libertarian values are de-emphasized (i.e., equality of outcome is elevated over equality of opportunity).

4. Policies for social problems cannot be meaningfully correct or incorrect. They may or may not achieve their stated goals, but even if they do they will frequently create perturbations elsewhere in the complex fabric of technological, environmental, economic, and social networks.

5. There is no optimality. One can certainly design better or worse solutions, but a single perfect state is neither definable nor achievable.

The automotive holarchy that was shown in Figure 7.1 exemplifies many of these points. Even a cursory evaluation of the automotive system indicates that attention is being focused on the wrong holon, and illustrates the fundamental truth that a strictly technological solution is unlikely to fully mitigate a culturally influenced problem. The engineering improvements of the vehicle, its energy use, its emissions, its recyclability, and so forth, on which much attention has been lavished, are truly spectacular. Nonetheless, and contrary to the usual understanding, the greatest attention so far as the system is concerned should probably be directed to the highest levels—the infrastructure technologies and the social structure. Consider the energy and environmental impacts that result from just two of the major system components required by the use of automobiles. The construction and maintenance of the "built" infrastructure—the roads and highways, the bridges and tunnels, the garages and parking lots—involve huge environmental impacts. The energy required to build and maintain that infrastructure, the natural areas that are perturbed or destroyed in the process, the amount of materials demanded, from aggregate to fill to asphalt—all required by the automobile culture, and attributable to it. In addition, the primary customer for the petroleum sector and its refining, blending, and distribution components—and, therefore, causative agent for much of its environmental impacts—is the automobile. Efforts are being made by a few leading infrastructure and energy production firms to reduce their environmental impacts, but these technological and management advances, desirable as they are, cannot in themselves begin to compensate for the increased demand generated by the cultural patterns of automobile use.

The final and most fundamental effect of the automobile may be in the geographical patterns of population distribution for which it has been a primary impetus. Particularly in lightly populated and highly developed countries such as Canada and Australia, the automobile has resulted in a diffuse pattern of residential and business development which is otherwise unsustainable. Lack of sufficient population density along potential mass transit corridors makes public transportation uneconomic within many such areas, even where absolute population density would seem to augur otherwise (e.g., in densely populated suburban New Jersey in the United States). This transportation infrastructure pattern, once established, is highly resistant to change in the short term, if for no other reason than that residences and commercial buildings last for decades.

The story of the automobile and its holarchy contains a rich lesson for society: that management of these entities must be adaptive rather than prescriptive. Brian Walker of the Commonwealth Scientific and Industrial Research Organization in Australia, together with colleagues in the United States, suggests that three related aspects of the technological holarchies determine their future trajectories: resilience, adaptability, and transformability. They define these aspects as follows (we have done some minor rewording):

> *Resilience.* The capacity of a technological holarchy to absorb disturbances and reorganize while undergoing change, so as to retain essentially the same function, structure, identity, and feedbacks.
> *Adaptability.* The capacity of actors in a technological holarchy to have influence (i.e., to manage resilience to at least a modest degree).
> *Transformability.* The capacity to create a fundamentally new technological holarchy when technological, environmental, or social conditions make the existing holarchy untenable, or make alternatives preferable.

Ludwig and colleagues (1993) suggest five principles for adaptive management; they are worth listing here:

- Include human motivation and responses as part of the system to be studied and managed
- Rely on scientists and engineers to recognize problems, but not to remedy them
- Act before scientific consensus is achieved
- Distrust claims of sustainability
- Confront uncertainty

In the industrial ecology sense, resilience, adaptability, and transformability are centered on resources, energy, and the environment. They cannot be rigorously managed. Nonetheless, flexible adaptive management, properly informed, and conducted with a few clearly stated goals in mind, has the potential to much improve our approaches to these complex, intertwined entities that are so important to sustainability.

FURTHER READING

Holarchies:

Geels, F.W., Processes and patterns in transitions and system innovations: Refining the co-evolutionary multi-level perspective, *Technological Forecasting & Social Change, 72,* 681–696, 2005.

Gunderson, L.H., and C.S. Holling, Eds., *Panarchy: Understanding Transformations in Human and Natural Systems,* Washington, DC: Island Press, 2002.

Kay, J.J., On complexity theory, exergy and industrial ecology: Some implications for construction ecology, In *Construction Ecology: Nature as a Basis for Green Buildings,* C.J. Kibert, J. Sendzimer, and G.B. Guy, Eds., Washington, DC: Spon Press, pp. 72–107, 2002.

Maier, M.W., Architecting principles for systems-of-systems, *Systems Engineering, 1,* 267–284, 1998.

Systems and Networks:

Ashton, W.S., *Coordinated Resource Management in Regional Industrial Ecosystems*, PhD Dissertation, Yale University, New Haven, CT, 2008.

Levin, S.A., The problem of pattern and scale in ecology, *Ecology, 73*, 1943–1967, 1992.

Strogatz, S.H., Exploring complex networks, *Nature, 410*, 268–276, 2001.

Metastable States and Adaptive Management:

Keating, C., et al., System of systems engineering, *Engineering Management Journal, 16* (3), 36–45, 2003.

Ludwig, D., R. Hilborn, and C. Walters, Uncertainty, resource exploitation, and conservation: Lessons from history, *Science, 260*, 17 and 36, 1993.

Scheffer, M., and F.R. Westley, The evolutionary basis of rigidity: Locks in cells, minds, and society, *Ecology and Society, 12* (2), article 36, December, 2007.

Scheffer, M.S. et al., Alternative equilibria in shallow lakes, *Trends in Ecology and Evolution, 8*, 275–279, 1993.

Walker, B., C.S. Holling, S.R. Carpenter, and A. Kinzig, Resilience, adaptability and transformability in social-ecological systems, *Ecology and Society, 9* (2), 5, 2004.

EXERCISES

15.1 Is a group of automobiles on a multilane highway a simple system or a complex system? Why?

15.2 Characterize the following entities according to three attributes—purposeful or not purposeful; predictive or not predictive; and a system, a holarchy, or neither. If a holarchy, identify the holons. Defend your answers: (1) a wetland, (2) public transportation in Switzerland, (3) a cellular telephone manufacturing plant, (4) a toaster, (5) nested "Russian" dolls, (6) the chemistry and physics of Earth's atmosphere.

15.3 Identify the holons in a cellular telephone system.

15.4 Evaluate integrated air defense systems, urban transport systems, and the World Wide Web by the five characteristics of technological holarchies.

CHAPTER 16

Industrial Ecosystems

16.1   ECOSYSTEMS AND FOOD CHAINS

In Chapters 4 and 5 we explored the relevance of biological ecology (BE) to technology from the perspective of organisms (known in industrial ecology as manufacturing facilities). We demonstrated that tools of BE could enhance an understanding of industrial ecology (IE) in such areas as studies of resource utilization, linkages, and recycling. The next hierarchical level to consider is the ecosystem. In BE, an ecosystem consists of the interacting parts of the physical and biological worlds. By analogy, an industrial ecosystem consists of the interacting parts of the technological and nontechnological worlds. The interactions among participants in an ecosystem generally involve the transfer of resources from one participant to another, or the sharing of resource acquisition or disposal, sometimes in ways or along paths that would not be intuitively anticipated. Such a structure constitutes a food chain.

A simplified marine biological food chain is shown in Figure 16.1, the boxes indicating the principal elements of the hierarchy, called "trophic levels" (from the Greek word for "food"). In addition to the dominant trophic-level actors, the extractor is included, because minerals obtained from inorganic reservoirs are essential to life. Note that the food chain is not completely sequential: Decomposers receive egested material from several trophic levels, for example. Bacteria act as both extractors and decomposers, receiving carbon in one activity, minerals in the other. Omnivory (i.e., feeding on more than one trophic level) is common in nature, but it complicates the food chain diagram without adding conceptual insight, so it is not incorporated into the figure.

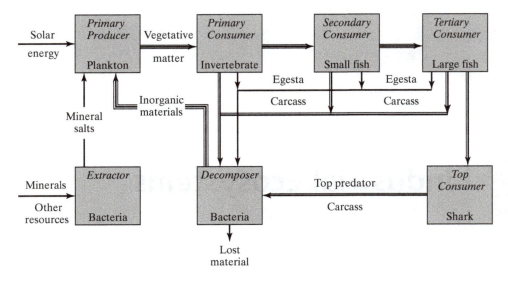

Figure 16.1

The biological food chain. The cycle begins with primary producers, who use energy and materials to generate resources usable to higher trophic levels. The number of consumer trophic levels varies in different ecosystems, and it is in some degree a matter of definition: The final consumer stage is that of the top predator or ultimate consumer. Decomposers return materials to the primary producers, thus completing the cycle. At the bottom of each box is an example of an organism in an aquatic ecosystem that plays the trophic level role. Types of resources are indicated along the flow arrows. The widths of the arrows are very rough indications of typical relative quantities of materials flow.

Similarly, a simplified industrial food chain is pictured in Figure 16.2, again neglecting omnivory. It appears that the biological trophic levels and industrial trophic levels can be described with essentially the same terminology. What is most interesting when comparing the two diagrams is perhaps not the similarities, but the differences:

1. The IE food chain equivalent of the BE decomposer is the recycler. Unlike the biological decomposer who furnishes reusable materials to primary producers, recyclers can often return resources one trophic level higher, to primary consumers. In practice, of course, the system is seldom this prescribed; many recyclers send intermediate materials to the same smelters used for primary production and are thus more like biological decomposers.

2. The IE food chain has an additional actor, the disassembler. The disassembler's goal is to retain resources at high trophic levels, passing as little as possible on to the recycler.

3. Little in the way of resources is lost in the overall biological food chain, but much is lost in the industrial food chain. As a consequence, the industrial ecosystem must extract a substantial portion of its resources from outside the system, the biological ecosystem only a small amount.

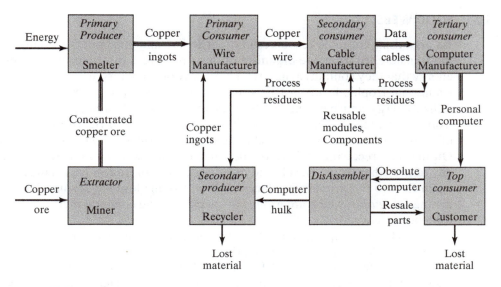

Figure 16.2

The industrial food chain, drawn by analogy with the biological food chain of Figure 16.1. Examples of organisms that play specific trophic-level roles for the example of the use of copper in personal computers appear at the bottom of each box. The widths of the arrows are very rough indicators of typical relative quantities of materials flow.

A bit of reflection on these diagrams point up some of the other distinctions between biological and industrial ecosystems:

1. *Speed in adapting to change.* Consider what happens if, for reason of disease in BE or regulatory inhibition in IE, the nutrient supply from an intermediate trophic level is restricted or disrupted. The industrial organism that is affected can, in all probability, quickly develop an alternative nutrient supply, perhaps by process or product redesign or by negotiation with new suppliers. The biological species can do the same, but generally over timescales of decades to millennia, rather than days to weeks.

2. *Response to increased need for resources.* Suppose a species in an intermediate trophic level finds conditions favorable for multiplication (due to a lack of competition, for example), provided suitable nutrients are available. In the IE case, increased activity on the part of extractors and primary producers can generally supply the needed materials, perhaps with lag times of days to a few years. In the BE case, little systemic capacity exists for substantially increasing the flow of nutrients, and the likely result is that mean individual growth and/or population increase does not take place, or does so at the expense of another species at the same trophic level.

3. *Approach to initiative.* Biological organisms are expert at working within the environment in which they find themselves. In contrast, industrial organisms strive to define the environment for themselves. BE systems are responders; IE systems are initiators.

16.2   FOOD WEBS

As organisms consume resources in order to live and go about their daily functions, the transference from one organism of these resources (nutrients and the embodied energy that they contain) generates a hierarchy of predators and prey. The first of the traditional trophic levels in BE is that of the primary producers (e.g., plants), who utilize energy and basic nutrients to produce materials usable at the next higher trophic level (seeds, leaves, and the like). The next several BE trophic levels consist of herbivores and carnivores (including the "top carnivore," who is prey for no one). On the final receiving end are the detritus decomposers, who receive residues from any of the other trophic levels and regenerate from those residues materials that can again flow to the primary producers. Within trophic levels it is sometimes useful to distinguish *guilds*, which are groups of species having common methods, location, or foraging practice.

Food chains imply a linear flow of resources from one trophic level to the next, as in Figure 16.3a. In such a construct, the interspecies interactions are straightforward. No resource flow systems in BE really follow this simple structure, however; they resemble much more the web structure of Figure 16.3b. Here species at one trophic level prey upon several species at the next lower level, and omnivory is common, as in Figure 16.3c. Finally, a fully expressed food web can exhibit all the various features: multiple trophic levels, predation, and omnivory (Figure 16.3d).

Food web analysis has two important goals: to study the flows of resources in ecosystems and to analyze ecosystems for dynamic interactions related to those resources. Consequently, constructing the food web diagram is a prelude to developing the source and sink budget for the resource of interest. In concept, organisms are identified, their trophic levels assigned, and their interactions specified. The food web is then diagrammed, and resource flows and stabilities are analyzed.

BE provides several statistics for food web analysis. The simplest is *species richness* (S): the number of different types of organisms contained within the system. The second is *connectance* (C). C is derived by constructing a community matrix, as shown in Table 16.1 for the web of Figure 16.3b. On the table, a numeral one is entered for

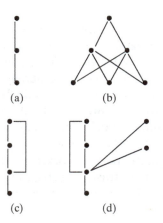

(a)   (b)

(c)   (d)

Figure 16.3

Diagrams of food chains and food webs. Filled circles represent species; lines represent interactions. Higher species are predators of lower species, so resource flow is from bottom to top. Species and rates of interactions vary with time. (a) A food chain in a three-level trophic system. (b) A food web in a three-level trophic system. (c) Omnivorous behavior in a food chain. (d) A food web encompassing multiple trophic levels, predation, and omnivorous feeding behavior.

TABLE 16.1   Community Matrix for the Ecosystem of Figure 16.1b.

Prey species*	Consuming species					
	1	2	3	4	5	6
1	X	0	0	1	1	0
2	0	X	0	1	1	0
3	0	0	X	1	1	0
4	0	0	0	X	0	1
5	0	0	0	0	X	1
6	0	0	0	0	0	X

*Species at the lowest trophic level are numbered 1, 2, and 3, those at the intermediate level 4 and 5, and that at the top level 6.

each consumer that receives resources from a given producer. If no resource transfer occurs, a zero is entered. C is then calculated by

$$C = 2L/(S[S-1]) \tag{16.1}$$

where L is the number of nonzero interaction coefficients in the community matrix. L is multiplied by two because there are two interactions: the effect of the predator on the prey and that of the prey on the predator. For the ecosystems of Table 16.1 and Figure 16.3b, $S = 6$ and $C = 16/30 = 0.53$. A handy reference point is to think of connectance in terms of industrial ecosystem types (Figures 4.1 and 4.2). For Type I systems, $C = 0$. In a Type II industrial ecosystem, $0 \leq C < 1$. In a Type III system, all possible connections are accomplished, so C approaches 1.

In BE, many (but certainly not all) food webs have large numbers of primary producers, fewer consumers, and very few top predators. Omnivores may be scarce in these systems, whereas decomposers are abundant. Food web models have provided a potential basis for fruitful analyses of resource flows in both BE and IE. Difficulties arise, however, when one wishes to quantify the resource flows and subject the web structure and stability properties to mathematical analysis. Many of the needed data turn out to be difficult to determine with certainty, particularly in the case of organisms that appear to function at more than one trophic level. This characteristic may not be a major complication for resource flow studies, but it seriously complicates stability analysis. The contention that more complex communities are more stable—because disruptions to particular species or flow paths merely shunt energy and resources through other paths rather than putting a roadblock on the entire energy or resource flow—remains a hotly debated topic.

Food web studies in BE are often centered around the question, "Given a perturbation of a certain type and size, how will the ecosystem respond?" The perturbation may be an infestation, the loss of a routine food source, or a severe climatic event. One view is that an ecosystem encountering modest frequencies and intensities of disturbance may be the most stable, a view that seems in harmony with what we know of industrial ecosystems as well.

Food webs are formed as opportunistic responses to local resource availability and desirability. An example of a biological system is shown in Figure 16.4a. In this

Figure 16.4a

A food web for the Benguela marine ecosystem off the southwest coast of South Africa. (Reproduced with permission from P.A. Abrams, B.A. Menge, G.A. Mittelbach, D.A. Spiller, and P. Yodzis, The role of indirect effects in food webs, in *Food Webs*, G.A. Polis and K.O. Winemiller, Eds., New York: Chapman & Hall, pp. 371–395, 1996.)

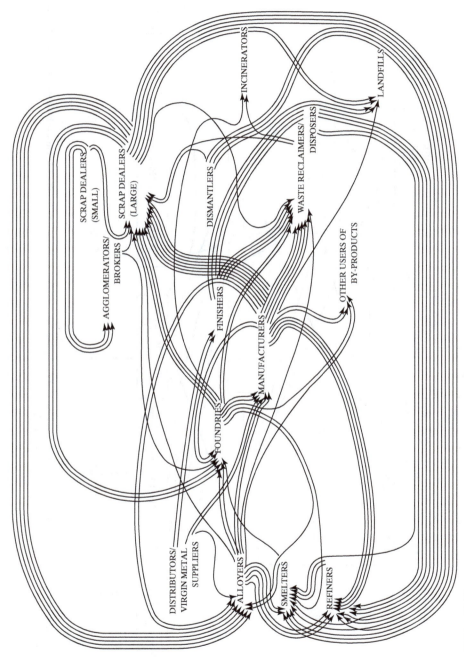

Figure 16.4b

A diagrammatic representation of an industrial food web. In this diagram, each line represents one set of transactions between categories of interviewed firms. For example, the transactions of a firm that might send its wastes to four different waste disposal companies appears as only one line. Similarly, a firm that might buy from three different alloyers also receives a single line for these transactions. The magnitudes of the flows are not captured on this diagram. Also, flows to and from the consumers of the industrial goods are not shown in this diagram, nor are fugitive emissions to air, water, and land. (Reproduced with permission from A.D. Sagar, and R.A. Frosch, A perspective on industrial ecology and its application to a metals-industry ecosystem, *Journal of Cleaner Production, 5, 39–45,* 1997.)

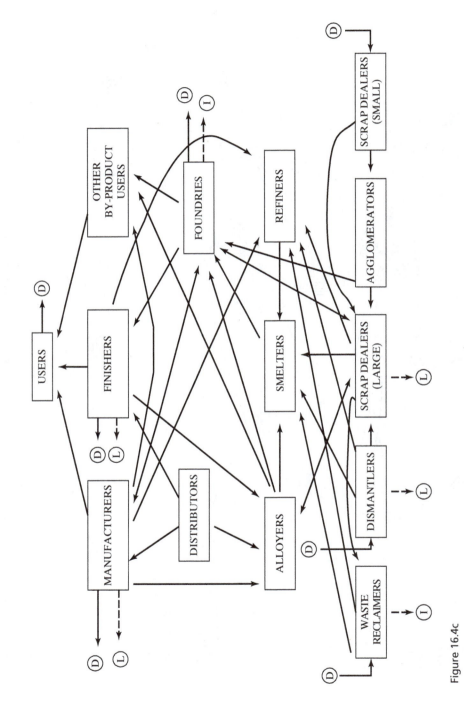

Figure 16.4c

The industrial food web of Figure 16.2b, redrawn as is customary in biological ecology with the lowest trophic level at the bottom and the highest at the top. D = detritus, L = loss to landfill, I = loss to incineration.

diagram, the position of the populations indicates their normal location with respect to the ocean surface. The connecting arrows indicate the flow of resources from one population to another. Diagrams such as this one are often drawn with the lowest trophic level at the bottom and the top predator at the top; were the present system so drawn it would, by the dictates of marine environments, be rather similar. Even without that presentation, it is clear that omnivory is common—the anchovy is fed upon by everything from mackerel to fur seals, for example.

Figure 16.4b shows an industrial food web for the flow of copper among industries in the Boston area. Termed a "spaghetti diagram" by its author, it appears so only because of the approach used to create the display; it could also be drawn to have much the same format as the BE example, as we do in Figure 16.4c.

It is instructive to compare the forms and characteristics of typical biological and industrial ecosystems. Those for biological ecosystems are sketched in Figure 16.5a; it tends to be pyramidal in shape, with richness, geographic stasis, and the ability to evolve rapidly increasing downward, size, longevity, and vulnerability to habitat fragmentation

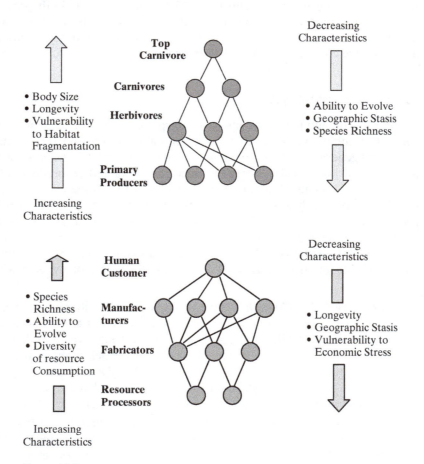

Figure 16.5

The generic structure and attributes of (top) biological and (bottom) industrial food webs.

increasing upward. In the industrial ecosystem (Figure 16.5b), the pyramid is inverted: There are few extractors (mining companies, crude oil extractors, etc.), many more fabricators (of rolls of steel, cables of copper, etc.), and many more manufacturers still. The characteristics tend to follow the inversion of the pyramid in terms of longevity, ability to evolve, vulnerability to stress (economic in this case), and geographic stasis. The top predator in the industrial system, the human customer, is not a good match for the top predator in biology, because he or she does not transform resources but only receives and "stores" them. Thus, the concept of trophic levels carries over from one system to the other, but the food webs that result are almost mirror images.

The concept of trophic levels is, of course, a convenient simplification of reality. Strictly speaking, only the initial user of materials from an extractor is an unambiguous primary producer, but organisms from many trophic levels are consumers, and many organisms operate at multiple trophic levels. In biological systems, for example, trees use nutrients to produce nuts, which are eaten by squirrels, and the nut resources are (among other things) used to produce baby squirrels. Some of those babies become food for foraging mammals and birds. The squirrel is thus both prey and predator, secondary producer and consumer. A similar situation exists in industry, where a factory acting as a consumer may receive disk drives, housings, and keyboards as a consumer and assemble computers as a secondary producer.

16.3 INDUSTRIAL SYMBIOSIS

Symbiosis is the intimate association of two species, either for the benefit of one of them (parasitic symbiosis) or for both of them (mutualistic symbiosis). It is a recurring situation in BE and generally arises from a long period of coevolution. The concept can be applied as well to technological systems, in which the unwanted by-products of one industry become the new materials for another. Variations on this concept are the sharing of utilities or infrastructure such as wastewater treatment plants, or the joint provisioning of services such as transportation or worker training. Industrial symbiosis (IS) would thus appear to offer the promise of developing environmentally superior industrial ecosystems.

Marian Chertow of Yale University has designated industrial symbiotic systems as being divisible into five categories. We describe these types below, each with an example.

Category 1 IS—Through waste exchanges. In these situations, recovered materials are donated or sold to other organizations. Some of the exchanges are informal or opportunistic, while others are formalized through waste exchange networks. A common example is the automobile scrap yard, which recovers and sells automobile components and prepares the bulk metal body and chassis for recycling. These interactions are essentially unplanned, however, so that the exchange of resources is insufficient to regard Category 1 IS exchanges as examples of industrial symbiosis.

Category 2 IS—Within a facility, firm, or organization. In this type of IS, materials or products are exchanged within the boundaries of a single organization, but among different organizational entities. This is a common approach to the design of petrochemical complexes, for example, where a by-product from one process serves as the feedstock to another.

Category 3 IS—Among colocated firms in a defined industrial area. In this type, corporations or other entities located close together, perhaps in an industrial park, organize themselves to exchange energy, water, materials, and/or services. An example of such a system is where brewery waste can be used as a resource for mushroom, pig, fish, and vegetable farming.

Category 4 IS—Among nearby firms not colocated. Category 4 systems are exemplified by that at Kalundborg, Denmark, in which a number of firms within a 3-km radius exchange steam, heat, fly ash, sulfur, and a number of other resources (Figure 16.5). Not designed with IS in mind, Kalundborg became an example of industrial symbiosis as a consequence of a series of "green twinning" trades, each economically advantageous.

Category 5 IS—Among firms organized across a broader region. The final EIP type consists of exchanges across a broad spatial region. In principal, it can incorporate any or all of the types described above. A successful Category 5 IS (and none have yet been fully realized anywhere) would probably require an active management organization to identify new twinning opportunities and recruit new participants.

As can be seen from this discussion, IS typology designations are evolutionary, not static. Simple Category 1 systems can become Category 4 or even Category 5 over time, for example. Indeed, IS systems can, in principle, be consciously developed, often around a key tenant such as a power plant that can readily begin the resource exchange process with a wide variety of potential industrial partners.

16.4 DESIGNING AND DEVELOPING SYMBIOTIC INDUSTRIAL ECOSYSTEMS

How might one design, or at least recognize, an efficient industrial ecosystem? It seems clear that such a system would need to involve a broad sectoral and spatial distribution of participants, and be flexible and innovative. It is likely that several IS types (see above) would be included. A key to an effective broad-scale IS is a high degree of synergy between input and output flows of resources. Few industrial food webs have been analyzed for connectance, but those that have include the ones pictured in Figures 16.4c and 16.6. When those and 16 others had their connectance computed, the median turns out to be very similar to that of biological systems (Figure 16.7). This is a somewhat surprising result, in view of the possible reasons why we might expect $C_{IE} < C_{BE}$:

- Industrial ecosystems are in earlier evolutionary states than biological ecosystems, and connectance generally increases as an ecosystem matures.
- BE organisms all exchange organic nutrients of rather similar chemical form. IE organisms, however, have much broader feeding habits—some desire petrochemicals, some metals, some forest products. In such a system, mismatches between output flows and input needs act against high connectance.

Analyses of a large number of industrial food webs may thus start to reveal characteristics not demonstrated by other approaches—a missing sector or type of industrial

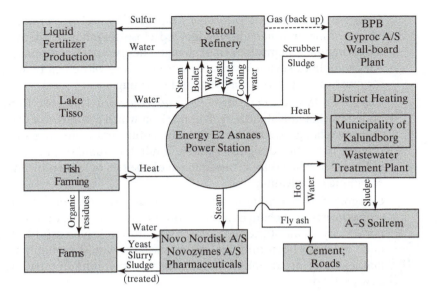

Figure 16.6

Flows of resources in the eco-industrial system at Kalundborg, Denmark. (Reproduced with permission from M.R. Chertow, "Uncovering" industrial symbiosis, *Journal of Industrial Ecology, 11* (1), 11–30, 2007.)

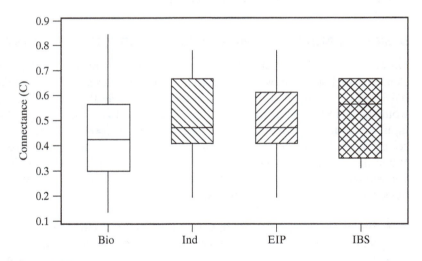

Figure 16.7

Box plots of connectance in different types of food webs. BIO = 113 biological food webs; IND = 19 industrial food webs, consisting of 15 ecoindustrial parks (EIP) and 4 integrated biosystems (IBS). (Reproduced with permission from C. Hardy and T.E. Graedel, Industrial ecosystems as food webs, *Journal of Industrial Ecology, 6* (1), 29–38, 2002.)

activity that could increase connectance, for example. Conversely, such studies might help address a question that remains as difficult for BE as for IE: What will be the response of a particular ecosystem to a particular type of perturbation? These topics provide a rich area for detailed investigation.

16.5   UNCOVERING AND STIMULATING INDUSTRIAL ECOSYSTEMS

How can one decide whether several actors who exchange resources constitute an industrial ecosystem? Category 1 IS would seem not to qualify, because the exchanges are unplanned and episodic. Category 2 IS probably fails as well, because symbiosis involves more than a single actor and is not preplanned. Chertow adopts the "3–2 heuristic" shown in Figure 16.8 as the minimum criterion: At least three different entities must be involved in exchanging at least two different resources. None of the three entities may be primarily engaged in a recycling-oriented business. Chertow characterizes Category 1 IS activities as "kernels" (having bilateral or multilateral exchanges) and "precursors" (having resource exchanges with a public goods component) that do not (yet) meet the minimum industrial ecosystem requirement, but have the potential to evolve into that status.

There is substantial debate over whether industrial ecosystems can be *planned,* but it is clear that existing ecosystems can be *discovered.* The latter frequently begin by sharing the acquisition, exchange, or disposal of the most universal of resources—water and energy. Processed water from one facility may be used as cooling water to another, or excess steam in one may provide heat or power to another. Shared construction of a water treatment facility is an often-encountered example.

Surveys of existing or potential industrial ecosystems reveal a number of drivers that tend to stimulate them, as well as a number of barriers that tend to inhibit them. We summarize some of the more important drivers and barriers in Table 16.2. What is evident from the table is that industrial ecosystems are not creations of technology; rather they result from the interplay of technology, economics, government, and society. It is this complexity that probably accounts for the difficulty in developing planned ecosystems. It is also clear, however, that triggering events may sometimes initiate

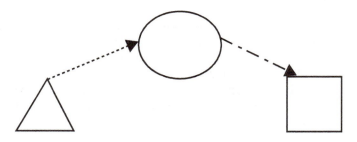

Figure 16.8

The 3–2 symbiosis defining a minimal industrial ecosystem. (Reproduced with permission from M.R. Chertow, "Uncovering" industrial symbiosis, *Journal of Industrial Ecology, 11* (1), 11–30, 2007.)

TABLE 16.2 Drivers for and Barriers to Industrial Ecosystems

Drivers	Barriers
Financial opportunities. Most IS exchanges make good business sense in terms of lower input costs, lower operating costs, and/or increased revenues.	*Informational.* A lack of understanding of the process inputs and outputs of potential symbionts often inhibits exchange possibilities.
Resource scarcity. Water is the most common example.	*Economic.* Exchanges or joint approaches to acquisition or disposal may carry with them excessive economic costs.
Reduced liability. Potentially problematic discards or by-products that are exchanged become the responsibility of others.	*Technical.* The set of potential symbionts may not "fit together" so far as inputs and outputs are concerned.
Sustainability focus. Industrial symbiosis is a natural component of increasing corporate attention to sustainability.	*Regulatory.* In some cases, regulations may prevent or inhibit the exchange of potentially hazardous resources.
Staff Mobility. Technically trained people moving from one facility to another, especially in a different industry, often see symbiotic opportunities not visible to others.	*Motivational.* Firms, regulators, and others must be willing to commit to symbiotic relationships.

Adapted from D. Gibbs, P. Deutz, and A. Procter, Sustainability and the local economy: The role of eco-industrial parks, Paper presented at *Ecosites and Eco-Centres in Europe*, Brussels, June 19, 2002; and D. van Beers, G. Corder, A. Bossilkov, and R. van Berkel, Industrial symbiosis in the Australian minerals industry: The cases of Kwinana and Gladstone, *Journal of Industrial Ecology, 11* (1), 55–72, 2007.

ecosystem behavior—a regional water shortage, the technical obsolescence of a facility, or a new environmental regulation, for example.

If governments find it difficult to create industrial ecosystems do they have any role at all? Chertow suggests three policy initiatives that have the potential to stimulate symbolic activities:

- Bring to light kernels of cooperative activity that are still hidden.
- Assist the kernels that are taking shape.
- Provide incentives to catalyze new kernels by identifying and supporting precursors to symbiosis.

Notwithstanding all of the above, industrial ecosystems can be fragile creatures. One of the symbionts may go out of business, or a resource flow valued by one symbiont may no longer be produced by another, or a symbiosis champion may retire without a suitable successor on hand. When they succeed, however, industrial ecosystems are financially beneficial, environmentally sound, and socially satisfactory. These attributes make it most worthwhile to discover, stimulate, and propagate industrial ecosystems of all kinds, in all kinds of places.

16.6 ISLAND BIOGEOGRAPHY AND ISLAND INDUSTROGEOGRAPHY

Islands have been the foci of evolution and ecosystem studies since Charles Darwin's work in the latter part of the nineteenth century. Islands have well-delineated boundaries; they are isolated to greater or lesser degrees; and they are generally limited in size. These characteristics act to limit resource availability, as well as to make analytical studies considerably easier than is generally the case elsewhere. The result has been a rich set of investigations and results in the field of "island biogeography."

Islands inhabited by modern-day humans retain many of the features of Darwinian locales. It is true that modern ships and planes have increased the connections of islands to the rest of the planet. Nonetheless, limitations remain—on energy sources, on water, on waste disposal sites, and on the availability of most intermediate and final products of modern technology.

The industrial ecology tools discussed in previous chapters can be quite valuable in an "island industrogeography" setting. An example is a material flow analysis performed for the Island of Hawai'i (Figure 16.9a). The diagram demonstrates clearly that two-thirds of all material employed in a year is imported from off the island, including virtually all food, energy resources, consumer products, and machinery. Half the inputs are used for new construction, and a quarter of them are released to the environment. In a similar study in Puerto Rico, Chertow found that different entities on the island were importing glass for glass manufacturing while discarding twice as much to landfills, and importing and simultaneously discarding similar amounts of boxboard. Once such mismatches are revealed, opportunities for improved efficiency can be readily discerned.

Similarly, comprehensive energy systems analysis on islands can demonstrate dependencies and opportunities. Figure 16.9b, also for the Island of Hawai'i, shows that virtually all the energy for the island is supplied by imported petroleum products, and that vehicle transport directly consumes more than half of all imports. Such analyses provide policy makers with food for thought, and with options to explore in increasing detail the sustainability of their islands.

Island industrogeography is quite a new field, but one that already can demonstrate its usefulness. It may indeed be possible that such research can provide for industrialists and governments insights to rival those of Darwin for biology. In the semi-isolated systems which islands exemplify, constraints and responses are more vivid, and solutions potentially more innovative, than is often the case in more complex and more connected locales.

FURTHER READING

Chertow, M.R., Industrial symbiosis: Literature and taxonomy, *Annual Reviews of Energy and the Environment, 25*, 313–337, 2000.

Chertow, M.R., "Uncovering" industrial symbiosis, *Journal of Industrial Ecology, 11* (1), 11–30, 2007.

Deschenes, P.J., and M. Chertow, An island approach to industrial ecology: Towards sustainability in an island context, *Journal of Environmental Planning and Management, 47*, 201–217, 2004.

McElroy, J.L., and K. Albuquerque, Managing small-island sustainability: Towards a system design, *Nature and Resources, 26*, 23–31, 1990.

Pimm, S.L., J.H. Lawton, and J.E. Cohen, Food web patterns and their consequences, *Nature, 350*, 669–674, 1991.

Sagar, A.D., and R.A. Frosch, A perspective in industrial ecology and its application to a metals-industry system, *Journal of Cleaner Production, 5*, 39–45, 1997.

Zhu, Q., E.A. Lowe, Y. Wei, and D. Barnes, Industrial symbiosis in China: A case study of the Guitang Group, *Journal of Industrial Ecology, 11* (1), 31–42, 2007.

(Unit: gigagrams)

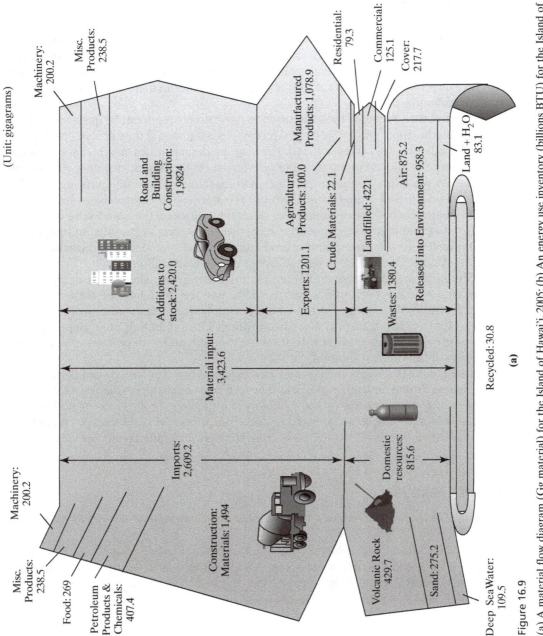

Machinery: 200.2

Misc. Products: 238.5

Road and Building Construction: 1,9824

Additions to stock: 2,420.0

Material input: 3,423.6

Machinery: 200.2

Misc. Products: 238.5

Food: 269

Petroleum Products & Chemicals: 407.4

Imports: 2,609.2

Construction: Materials: 1,494

Domestic resources: 815.6

Volcanic Rock 429.7

Sand: 275.2

Manufactured Products: 1,078.9

Residential: 79.3

Commercial: 125.1

Cover: 217.7

Agricultural Products: 100.0

Crude Materials: 22.1

Exports: 1201.1

Landfilled: 4221

Air: 875.2

Wastes: 1380.4

Released into Environment: 958.3

Land + H₂O 83.1

Recycled: 30.8

Deep SeaWater: 109.5

(a)

Figure 16.9

(a) A material flow diagram (Gg material) for the Island of Hawai'i, 2005; (b) An energy use inventory (billions BTU) for the Island of Hawai'i, 2005. (Both diagrams courtesy of M. Chertow.)

238

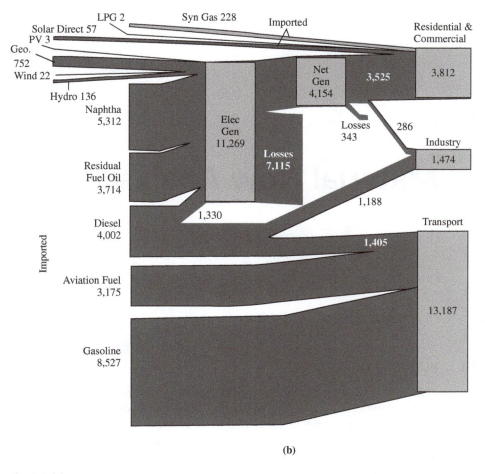

(b)

Figure 16.9

Continued

EXERCISES

16.1 Compute the richness and connectivity of the industrial food web of Figure 16.5.

16.2 Compute the richness and connectivity of the industrial food web of Figure 16.4c.

16.3 Select a local industrial park for study and identify the principal input and output flows. Determine any existing "green twinning" relationships and propose others for consideration.

16.4 The food web of Figure 16.4c pictures many types of actors at several different trophic levels. For any single actor type (smelters, users, etc.) do input flows equal output flows? Why or why not?

16.5 Does island industrogeography require an island? Dubai in the United Arab Emirates has some of the characteristics of an island. Investigate its geographical situation and write an essay on why or why not it might be suitable for such an analysis.

CHAPTER 17

Material Flow Analysis

17.1 BUDGETS AND CYCLES

Industrial ecology is a concept with cycles at its very heart, and cycles are analyzed by means of materials and energy budgeting. Nearly everyone is familiar with the concept of a household or personal financial budget, whether or not she or he is conscientious about making and sticking to it. An approach very similar to that of financial budgeting is used to fashion budgets in industrial ecology. The situation can be appreciated with the aid of the diagram in Figure 17.1a, which shows a tub receiving water from several faucets and having a number of drains of different sizes. When the water is supplied at constant (but probably different) rates by all the faucets and is removed at an equal total rate by drains with (probably) different capacities, the water level remains constant. When the tank is very large, however, and has some wave motion that makes it difficult to tell whether the absolute level is changing, an observer may have difficulty telling whether the system is in balance or not. In that case, he or she may try instead to measure the rate of supply from each of the faucets and the rate of removal in each of the drains over a period of time to see whether the sums are equivalent. A part of this technique involves the determination of the pool size (the quantity of water in the tank) and either the rate of supply or the rate of removal. Determination of changes in the pool size then gives information about rates that are difficult to measure. The process of estimating or measuring the input and output flows and checking the overall balance by measuring the amount present in the reservoir constitutes the budget analysis.

Suppose the input from one of the sources is increased—that is, in our analogy, the flow from one of the faucets increases—will the water level keep increasing? The answer depends on whether one of the drains can accommodate the additional supply, as can the

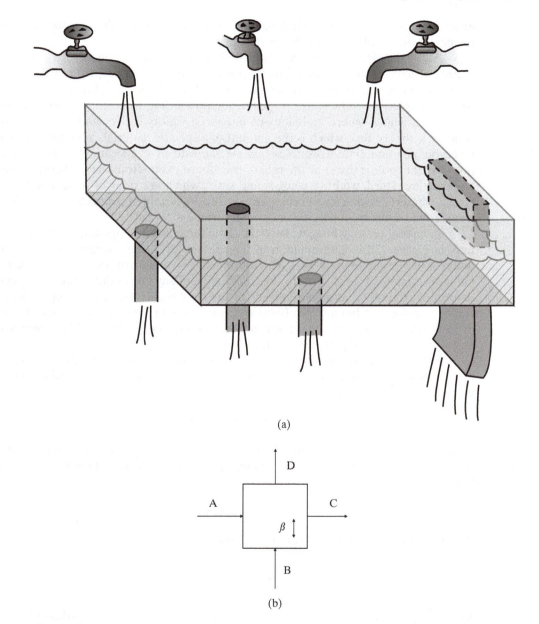

(a)

(b)

Figure 17.1

(a) A simple conceptual system for budget calculations. The water level in the tub is determined by the water flows in and out, as discussed in the text; (b) A schematic diagram of the flows to (i.e., A and B) and from (i.e., C and D) a reservoir. β indicates the "stock" of the reservoir (i.e., its contents).

"trough drain" at the right side of the tank in the figure. If no such drain is present, then the water level will indeed increase. Conversely, if the flow into a drain is enhanced for some reason, such as the removal of an obstruction, then the water level will decrease in the absence of a corresponding increase in the supply. Such a process will continue in this manner unless the flows through the drains adjust themselves to this new factor or unless the new factor results in other drains changing their functioning.

All of the circumstances mentioned above occur in budgets devised for various industrial ecology studies, and all budgets involve the same concepts. One is that of the reservoir, in which material is stored. Examples include the shipping department where completed products are prepared for forwarding to customers, or the atmosphere as a whole, where emissions of industrial vapors collect and react. A second concept is that of flux, which is the amount of a specific material entering or leaving a reservoir per unit time. Examples include the rate of evaporation of water from a power plant cooling tower or the rate of transfer of ozone from the stratosphere to the troposphere. Third, we have sources and sinks, which are rates of input and loss of a specific material within a reservoir per unit time. A system of connected reservoirs that transfer and conserve a specific material is termed a "cycle."

Industrial ecology budgets have the same three basic components as those for the tank in Figure 17.1a: determination of the present level (the concentration of a single material or a group of materials), a measurement or estimate of sources, and a measurement or estimate of sinks. A perfect determination of any two of these three components determines the other, as a consequence of the *conservation of mass* principle: Material can be transformed, but not lost. Because any material of interest in an industrial facility or in the environment may have several sources and sinks (Figure 17.1b), each source and sink must generally be studied individually.

For a system chosen with a boundary surrounding the entire Earth, and assuming no significant loss to interplanetary space, the overall quantity of material (the content β) will not change with time:

$$\Sigma\beta = \text{Constant} \tag{17.1}$$

Similarly, for any geographically limited system in a steady state, the source and sink fluxes into each reservoir exactly balance, that is, $A + B = C + D$ in Figure 17.1b, so that in each case

$$\beta_r = \text{Constant} \tag{17.2}$$

that is, the contents of each reservoir will not change with time:

$$\Delta_r = \frac{d(\beta_r)}{dt} = 0 \tag{17.3}$$

In a changing system, however, the source and sink fluxes are not equal and Δ_r is real-valued. Over time, the result for any reservoir will be a change in the content (also termed the "stock") of the element in question, the change being given by

$$\Delta_r = \int_{t_2}^{t_1} (\Sigma F_i - \Sigma F_o)dt \tag{17.4}$$

where F_i ($=A + B$ in Figure 17.1b) is an input flux, F_o ($=C + D$ in Figure 17.1b) is an output flux, and t_2-t_1 the time interval. Δ_r can be either positive or negative.

Another useful parameter is the *residence time*, τ_r, which is the average time spent in the reservoir by a specific material. The average τ residence time is composed of all forms of the substance, weighted with appropriate probability factors. For example, when one evaluates nitrogen flow into and out of an animal, one finds that some of the nitrogen rapidly flows through the animal whereas other nitrogen flows much more slowly, a reflection of animal lifetimes and foraging and excretion characteristics. We can define the average residence time in a situation with a variety of residence times as

$$\tau_r = \int \tau \psi(\tau) d\tau \tag{17.5}$$

where $\psi(\tau)$ indicates the fraction of the constituent having a residence time between τ and $\tau + d\tau$. This probability fraction $\psi(\tau)$ is a function of the reservoir processes (the rate of breakdown of nitrogen-containing molecules in food in biological ecology, or the rate of product obsolescence in industrial ecology, for example). An alternative route to determining residence time is the use of Markov chain modeling, as developed by Eckelman and Daigo (2008).

The *age* is the time elapsed since a particle entered a reservoir. The average age of all particles of a specific kind within a reservoir is thus given by

$$\tau_a = \int \tau \Omega(\tau) d\tau \tag{17.6}$$

where $\Omega(\tau)$ is the age probability function.

An industrial ecology resource analysis may deal with any spatial scale and with many reservoirs. The anthropogenic nickel cycle, shown in Figure 17.2, is an example. There are seven principal processes (or reservoirs or life stages). First the ore (rock containing nickel minerals) is *mined* and the minerals separated from the rock, followed by *smelting* (to separate the metal minerals from the rock debris) and *refining* (to purify the nickel). At the *fabrication* stage, the nickel is fashioned into intermediate products such as rolls and sheets, which are then employed in the *manufacture* of products such as nickel-containing appliances or vehicles. Following *use* (generally for a number of years), the products are discarded to *waste management*, where they may be recycled, sold as scrap (presumably for reuse), or lost.

The cycle in Figure 17.2 is characterized by processes that are linked through markets, each indicating trade with other regions at the respective life stages. The scrap market plays a central role in that it connects *waste management* with *production* and *fabrication*. The cycle is surrounded by entities lying outside the system boundary: trade partners (other regions), the lithosphere from which ore extraction takes place, and repositories for nickel in tailings and slag (i.e., production wastes), and in landfilling.

No matter what the cycle, the careful construction of a flow diagram such as this is essential if an accurate picture is to be obtained. In some cases, as in Figure 17.2, the flows crossing the boundary may be well known, and the challenge is to quantify them (and the related stocks as well). In others, as in the atmospheric chlorofluorocarbons (CFCs), the importance of some of the sources and sinks may not be realized and must be deduced by quantifying those for which information is available. In all cases, however, the process begins by specifying the metabolic cycle, its metabolites, and its enzymes, in as much detail as possible.

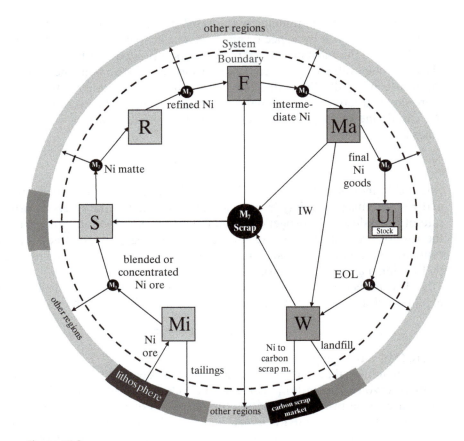

Figure 17.2

A diagrammatic representation of the resource analysis cycle for nickel. Mi, mining; S, smelting; R, refining; F, fabricating; Ma, manufacturing; U, use; W, waste management; IE, industrial waste; EOL, end of life. The economic markets through which nickel may be traded are indicated as M_i. (Reproduced by permission from B. Reck, et al., The anthropogenic nickel cycle: Insights into use, trade, and recycling, *Environmental Science & Technology, 42*, 3394–3400, 2008.)

In industrial ecology, the concepts of budgets and cycles are applied to the anthropogenic use of resources such as nickel, or sometimes to the combination of anthropogenic and natural use. The results help us evaluate present metabolic needs and to estimate those that may be required in the future. Similarly, we can study specific resources as they pass through various technological organisms and thus evaluate resource supply, use, and loss. Substance-centered cycles complement the product cycles addressed by life cycle assessment (Chapters 12–14).

17.2 RESOURCE ANALYSES IN INDUSTRIAL ECOLOGY

Resource analyses occur at several levels that may be distinguished by the chemical state of the resource. If one follows a resource in either atomic or molecular form, the study may be termed "substance analysis." The simplest case is where an element's rate of transition from reservoir to reservoir is the subject of analysis. The next most complex approach deals with molecules or alloys, that is, with well-defined chemical entities. Finally, if a resource is followed in various chemical states through a series of natural and anthropogenic collective reservoirs such as ores or automobiles, the exercise may be titled "material analysis." For any of these, the approach may involve determination of flows only, or of both flows and stocks. "Material flow analysis" (MFA) is a commonly used term that includes all of these approaches.

17.2.1 Elemental Substance Analyses

In an elemental analysis, the emphasis is on the atom. This may be because the supply of the atom is highly limited (gold, for instance), or because it is a biotoxicant (cadmium, for instance). It does not mean that the subject of the analysis is necessarily present in atomic form, but that the chemical form in which the atom exists is not considered in the analysis. The advantage of this approach is that its data are unambiguous—a flow of sulfur is expressed as mass of sulfur per unit time, for instance, regardless of whether the actual entity being transferred is flowers of sulfur, sulfur dioxide, or sulfuric acid.

Let us return to the cycle diagram for nickel (Figure 17.2). It turns out that data exist for many of the flows, but not all. Those that are not immediately quantifiable from extant data include the export of final nickel goods (poor statistics and uncertain nickel content), flow from use (little data), and amount landfilled (landfill information rarely addresses individual metals). To deal with these issues, it is necessary to consider the amount of nickel in stocks of nickel-containing products in use.

In-use stocks may be estimated by either a "top-down" or a "bottom-up" method. The bottom-up method begins with inventories of the different service units that contain the resource in question, such as buildings, factories, or vehicles. The content of the resource per service unit obtained from engineering data is combined with census information on the number of units in a given geographic area to determine the metal stock in use. Mathematically, the computation is expressed as

$$S(t) = \sum_{i=1}^{n} N_i(t) \cdot \frac{M_i(t)}{N_i(t)} \tag{17.7}$$

where S is the in-use stock, N the number of units of a particular product (automobiles, say) that use the resource to provide services of some sort, and M_i/N_i is the material intensity per unit of product, all summed over the i types of products addressed by the analyst.

The bottom-up method allows determination of the spatial distribution of stocks in particular localities, but yields less useful data on wastes, because we lack extensive information on the content and extent of landfills.

The top–down method computes the mass balance between the flow of new resources into use and the flow out of use arising from products that reach the end of their service lives. Some of the outflow of end-of-life resources is recycled into new products and so remains in use. The rest enters a steadily increasing stock placed in waste repositories and is found by integrating the balance between the discard rate and the recycling rate. Integration of the mass balance year-by-year determines the cumulative amount of stock that remains in use and the amount accumulated in wastes.

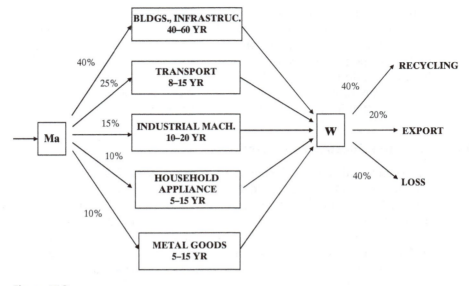

Figure 17.3

Flows and fates for nickel from manufacturing to the use and waste management life stages.

The nickel flow into stock in use can be divided into five product groups, as shown in Figure 17.3. Each has a different average lifetime. Given the input flow proportions and the lifetimes, the flow each year to waste management can be calculated from a top–down perspective. This leads to quantification of other previously unspecified flows as well. The results for four countries with rather different nickel cycles are shown in Figure 17.4. Russia and the United States are almost mirror images of each other—the former mines nickel ore, processes it, exports it, and uses it very little, while the latter mines none, imports lots of refined nickel, and then uses it extensively. China is a miner as well as an importer (mostly of semi-products). Japan imports ore, carries it through the cycle, and exports quite a bit of both semi-products and final products. Results like these inform industries and governments about import and export flows, rates of recycling and reuse, loss to the environment, additions to stock, and areas in which more investigation is needed to improve the characterization of the cycle.

Case Study: The "Hidden" Trade of Metals

The balance of trade in various metals between countries has traditionally been determined from data on trade in metal ores and refined metal. This is an important part of the picture, but only a part. Metal also flows from country to country as a constituent of finished products, which have not customarily been evaluated as part of the metal trade statistics. A recent study looked at the case of comprehensive metal trade and the United States in 2000. For chromium, no ores were imported, but the United States did import large amounts of the intermediate refinery product ferrochromium (top diagram). U.S. manufacturing is thus dependent on Russia and Africa for chromium imports. For the country as a whole, however, a different picture is provided by the flows of chromium in products (bottom diagram). Here the major unbalanced chromium flow is in products from Asia.

The analysis demonstrates, among other things, that the United States is less dependent on Russia and Africa for chromium than was thought to be the case, provided the chromium entering the country in products can be efficiently recycled and reused (Figure CS-1).

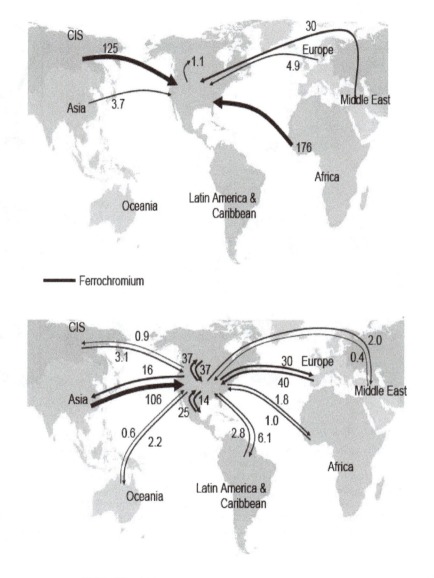

Figure CS-1

United States trade in chromium (Gg Cr/yr) in ferrochromium (top) and finished products (bottom).

Source: J. Johnson, and T.E. Graedel, The "hidden" trade of metals in the United States, *Journal of Industrial Ecology, 12,* 739–753, 2008.

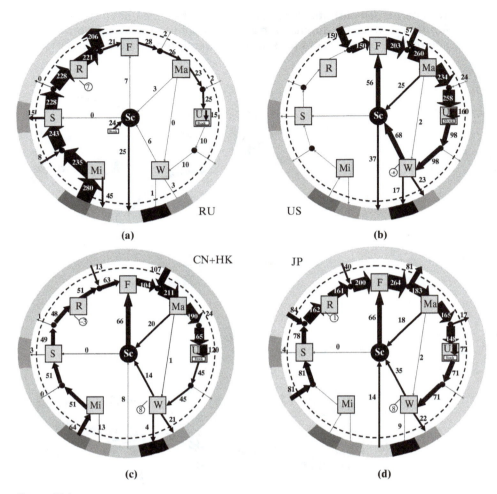

Figure 17.4

Circular nickel diagrams for year 2000 for the largest extractor, (a) Russia, and the three principal users: (b) United States; (c) China; and (d) Japan. The units are gigagram per year nickel (Gg/a Ni, 1 gigagram = 1000 metric tons). The widths of the arrows are proportional to the flow magnitudes. If no data are provided, nickel flows are 0 Gg/a Ni. The data are in units of Tg Ni/yr for 2000. The dotted circle at life stage W indicates a "phantom flow" required for mass balance but not reflected in the available statistics or calculations. (Reproduced by permission from B. Reck, et al., The anthropogenic nickel cycle: Insights into use, trade, and recycling, *Environmental Science & Technology, 42*, 3394–3400, 2008.)

17.2.2 Molecular Analyses

One can make the case that the first molecular resource analysis in industrial ecology was the study of the atmospheric cycle of CFCs by Mario Molina and F. Sherwood Rowland of the University of California-Irvine. The CFCs, created by industry as refrigerants and propellants, had been detected in the atmosphere. Molina and Rowland deduced by an analysis of their sources and sinks that the eventual fate of the CFCs

TABLE 17.1   Molecular Flow Analysis for Atmospheric Methane

Budget item	Flux (Tg C/yr)
Natural sources	
Wetlands	160
Termites	20
Ocean	15
Wild animals	15
Geologic, wildfires	10
Anthropogenic sources	
Coal, gas mining	100
Rice fields	60
Ruminants	80
Waste treatment	80
Biomass burning	40
Sinks	
Reaction with hydroxyl radical	490
Removal by soils	30
Transport to the stratosphere	40
Reaction with chlorine atoms	10

Source: The estimates are from T.E. Graedel, and P.J. Crutzen, Atmospheric *Change: An Earth System Perspective*, New York: W.H. Freeman, 1993; and D.F. Ferretti, et al., Stable isotopes provide revised global limits of aerobic methane emissions from plants, *Atmospheric Chemistry and Physics, 7,* 237–241, 2007.

was to be broken apart by high-energy solar radiation in the upper atmosphere, the molecular fragments subsequently reacting with and removing the ozone that screens harmful ultraviolet radiation from Earth's surface. The discovery of the "ozone hole" over Antarctica several years later demonstrated the accuracy of their analysis, and Rowland and Molina were awarded the Nobel Prize in Chemistry in 1995 for their discovery.

Molecular approaches can be powerful tools to study species that have both natural and anthropogenic sources. Atmospheric methane is a good example; it is important as a molecule because in that form it plays a significant role as a "greenhouse gas." The methane budget is given in Table 17.1. It shows that the flux from anthropogenic sources pretty clearly exceeds that from natural sources, and that no single anthropogenic source is dominant.

17.3   THE BALANCE BETWEEN NATURAL AND ANTHROPOGENIC MOBILIZATION OF RESOURCES

Only a few of the elements in the periodic table have been the subject of extensive budget and cycle analyses. Nonetheless, it is readily apparent that natural flows are dominant for some elements, while anthropogenic flows are dominant for others. Figure 17.5 presents a comparison of anthropogenic and natural mobilization of the elements of the periodic

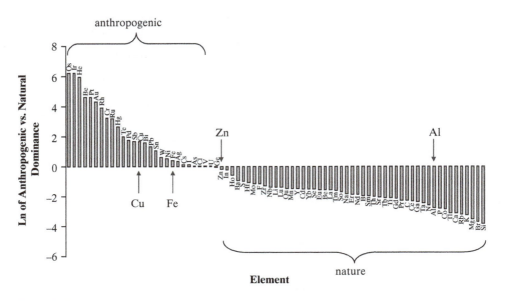

Figure 17.5

An assessment of the degree to which the global cycles of the elements are dominated by nature or by human activity. The arrows indicate the four principal metals in magnitude of use. (Adapted by Jason Rauch from R.J. Klee and T.E. Graedel, Elemental cycles: A status report on human or natural dominance, *Annual Reviews of Environment and Resources, 29*, 69–107, 2004.)

table. It can be seen that natural flows are generally more important for the halogens, alkali metals, and alkaline-earth metals, while anthropogenic flows dominate the mobilization of the metals and semi-metals.

What is the cause of these patterns of mobilization? It appears to relate to the solubility in water of the principal compound in which an element occurs in nature. Those that are highly soluble—sodium in sea salt, for example—are efficiently mobilized and redeposited by natural processes. These elements are used as building blocks by living organisms—calcium in shells, chlorine in cells, and so forth. Elements in the center of the periodic table, however, tend to occur as highly insoluble oxides or sulfides. Natural processes have no effective way to isolate these elements, nor do they know how to deal with them when they are isolated by human industrial processes. As a result, it is these elements that form corrosion-resistant structures, that provide high-strength materials, and that are often biotoxic. For the majority of the elements, though not all, aqueous solubility determines whether nature or humans will control their mobilization and flow.

The balance between resource mobilization and use by humans and nature is illustrated for copper in Figure 17.6. This "anthrobiogeochemical" copper cycle is for the region from Earth's core to the Moon and combines natural biogeochemical and human anthropogenic stocks and flows. As would be anticipated from Figure 17.5, the anthropogenic mining, manufacturing, and use flows (on the order of 10^4 Gg Cu/year) clearly dominate the material flows in the cycle. In fact, the anthropogenic rate of copper extraction is about a million times the natural rate of renewal. In contrast, the natural

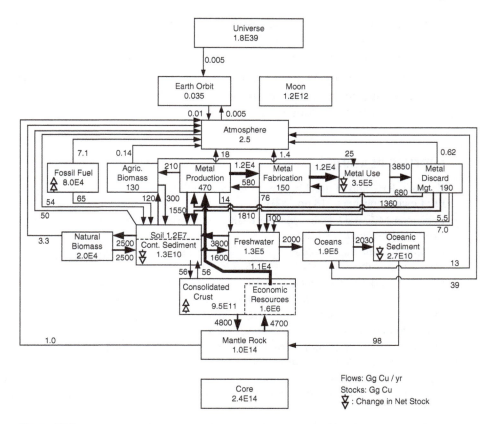

Figure 17.6

Earth's anthrobiogeochemical copper cycle, ca. 1994. The thickness of arrows represents a rough approximation of magnitude to aid visualization. The box arrow symbol indicates those reservoirs that are not in a state of mass balance, and are either accumulating or losing copper. Input–output flows may not equilibrate precisely in those reservoirs that are in a state of mass balance due to associated error ranges. (Reproduced from J. Rauch and T.E. Graedel, Earth's anthrobiogeochemical copper cycle, *Global Biogeochemical Cycles, 21*, GB2010, doi:10.1029/2006GB002850, 2007.)

repositories of Earth's core, mantle, and crust, and of the Moon, hold much higher stocks of copper ($>10^{10}$ Gg) than do anthropogenic repositories ($<10^6$ Gg).

17.4   THE UTILITY OF MATERIAL FLOW ANALYSIS

Material flow analyses have many uses, some realized, some proposed. Any list is idiosyncratic, but the following provides a sense of SFA utility:

- Establish and monitor the material and energy requirements for activities related to the maintenance and growth of economies.
- Identify and evaluate trends in discards and the potential for resource conservation and recycling.

- Identify and monitor the losses of substance to the environment and assess the environmental implications.
- Assess the status and trends of mineral and energy systems.
- Track the evolving amounts of material stocks in use, reuse, and disposal.
- Respond to issues related to materials demand and the potential for materials scarcity over the long term.

FURTHER READING

Brunner, P.H., and H. Rechberger, *Practical Handbook of Material Flow Analysis*, Boca Raton, FL: Lewis Publishers, 318 pp., 2004.

Eckelman, M.J., and I. Daigo, Markov chain modeling of the global technological lifetime of copper, *Ecological Economics, 67*, 265–273, 2008.

Journal of Industrial Ecology, Special issue on materials use across world regions: Inevitable pasts and possible futures, *12* (5/6), 629–798, 2008.

Klee, R.J., and T.E. Graedel, Elemental cycles: A status report on human or natural dominance, *Annual Reviews of Environment and Resources, 29*, 69–107, 2004.

Molina, M. J., and F.S. Rowland, Stratospheric sink for chlorofluoromethanes: Chlorine atom catalyzed destruction of ozone, *Nature, 249*, 810–812, 1974.

Murakami, S., M. Yamanoi, T. Adachi, G. Mogi, and J. Yamatomi, Material flow accounting for metals in Japan, *Materials Transactions, 45*, 3184–3193, 2004.

OECD (Organisation for Economic Co-operation and Development), *Measuring Material Flows and Resource Productivity: The OECD Guide*, Paris, FR: OECD Environment Directorate, 2007.

Rauch, J., and T.E. Graedel, Earth's anthrobiogeochemical copper cycle, *Global Biogeochemical Cycles, 21*, GB2010, doi:10.1029/2006GB002850, 2007.

Sörme, L., B. Bergbäck, and U. Lohm, Goods in the anthroposphere as a metal emission source, *Water, Air, and Soil Pollution Focus, 1*, 213–227, 2001.

EXERCISES

17.1 You are assigned to analyze a reservoir generally similar to that of Figure 17.1a. The reservoir has three input flows ($I_1 = 6$ l/s, $I_2 = 13$ l/s, $I_3 = 9.5$ l/s) and two output flows ($O_1 = 16$ l/s and $O_2 = 10$ l/s). Is the system in steady state? If not, at what rate is the reservoir contents changing?

17.2 The reservoir of Exercise 17.1 contains 1500 liters of water. A third output flow, $O_3 = 2.5$ l/s, has been added. Compute the turnover time for the water in the reservoir.

17.3 The nickel cycles of Figure 17.4 are not at steady state. What does this imply for global in-use reservoirs of nickel? Which reservoirs are changing most rapidly? What will be the change in the contents of each of the in-use reservoirs after 10 years, given constant flow rates?

17.4 Choose an element from the fifth row of the periodic table and research its substance flow aspects. Describe its principal uses. For a country designated by you, construct a diagram of the form of Figure 17.2 and add quantitative flow information where available or where you can make estimates. What information do you need to complete your cycle, and how might you acquire it? (Supplementary material you may find helpful includes the U.S. Geological Survey's annual *Mineral Commodity Summaries* [http://minerals.usgs.gov/minerals/pubs/mcs/] and the United Nations Comtrade database [http://comtrade.un.org/].)

17.5 Consider a scenario in which the concentration of CO_2 in the atmosphere gradually rises to 400 parts per million by volume (ppmv), about 8% higher than the level in 2000, then stabilizes by the year 2100, as shown in the top graph. The bottom graph shows anthropogenic CO_2 emissions from 1900–2000, and current net removal of CO_2 from the atmospheric by natural processes. Sketch:

(a) Your estimate of likely future net CO_2 removal, given the scenario above.

(b) Your estimate of likely future anthropogenic CO_2 emissions, given the scenario above.

(Reproduced with permission from J.D. Sterman, Risk communication on climate: Mental models and mass balance, *Science,* 322, 532–533, 2008.)

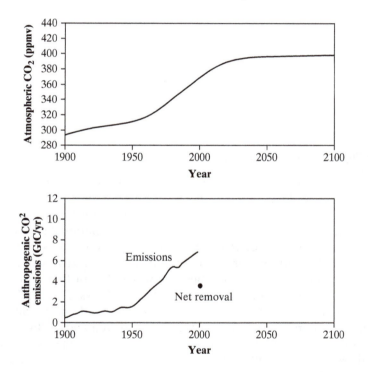

CHAPTER 18

National Material Accounts

18.1 NATIONAL-LEVEL ACCOUNTING

Metabolic studies are not limited to organisms, natural or industrial. In the late 1990s, country-level metabolisms were found possible to assemble for several countries, at least for the major flows. This effort, coordinated by the World Resources Institute (WRI), essentially defined the form of national material accounts (NMAs), as shown in Figure 18.1. An important contribution of this approach was the identification and quantification of "hidden flows"—materials moved or mobilized in the course of providing commodities, but which do not themselves enter the economy. There are two components to these hidden flows: (1) ancillary material removed along with the target material and later separated and discarded, such as the rock matrix containing a metal ore, and (2) excavated and/or disturbed material, such as the soil removed to gain access to an ore body, and they are measured or estimated both within the country and (for imported resources) at the extracting country as well. This inclusion means that all material flows, whether obvious to the user or not, are included in the NMA. NMAs also represent an expansion of the concept of metabolism to include flows that occur beyond the boundary of the organism; the biological analogy would be to include the material in a bird's nest as part of the bird's metabolism.

Notwithstanding their benefits, NMAs have some significant limitations. They attempt in principle to count all flows (with the usual exception of water) crossing national borders, as well as those crossing the boundary between the ecosphere and the anthroposphere. In practice, however, minor flows are usually excluded in the

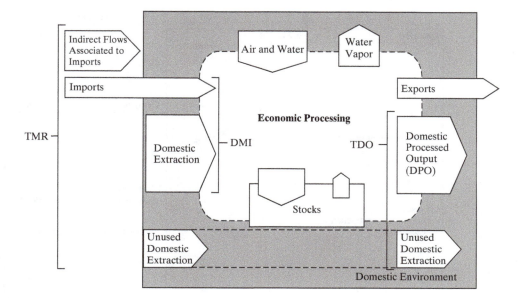

Figure 18.1

The schematic representation of an economy-wide material balance. The acronyms are: TMR, total material requirements; DMI, domestic material inputs; and TDO, total domestic outputs. (Reproduced with permission from *Measuring Material Flows and Resource Productivity: The OECD Guide,* Report ENV/EPOC/SE(2006)1/REV3, Paris: OECD Environment Directorate, 2007.)

interest of tractability even if they have significant meaning for sustainability, environmental impact, or some other important parameter. Additionally, because the focus is on material crossing the boundaries, transformations occurring within the country (e.g., the fabrication of automobiles from domestic steel) are not part of the analysis.

18.2 COUNTRY-LEVEL METABOLISMS

Country-level assessments of material entering use invariably show that the largest flows by far are related to construction materials and to energy resources. This was illustrated by the example in Figure 11.1, which characterized U.S. material use in the twentieth century. When features such as hidden flows are added, the typical result is that nonsaleable extraction (e.g., of waste rock in mining) is much larger even than minerals entering use, as shown in Figure 18.2 for China in 2004. The total amount of material entering the system in that year was about 28 Pg (thousand million metric tons), including 17 Pg of extraction and excavation that were unused. Several significant emissions flows are shown as well. An important statistic is the 5.4 Pg added to stock (in buildings, appliances, etc.).

Construction materials, and the other major components shown in Figure 11.1, may be thought of as the *elephants* of material flows—very large, but perhaps not of

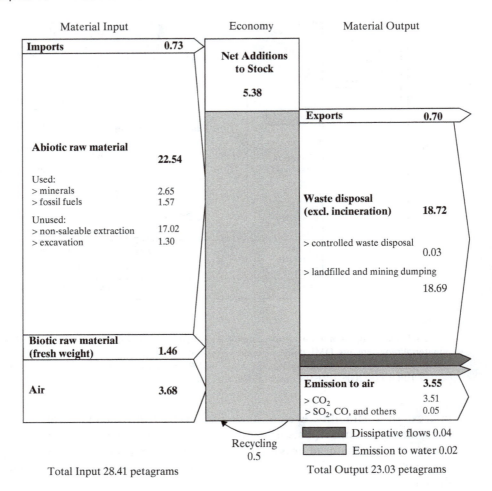

Figure 18.2

The NMA for China in 2002, with flows given in Pg/yr. (Reproduced with permission from M. Zu, and T. Zhang, Material flows and economic growth in developing China, *Journal of Industrial Ecology, 11* (1), 121–140, 2007.)

too much concern, either from an environmental or sustainability perspective. The contrasting flows, the *scorpions*, are exemplified by that for mercury in the United States since 1975 (Figure 18.3)—low in total amount, but of high concern because of mercury's toxicity. The figure shows a dramatic reduction in mercury use in the last decade of the twentieth century as a result of legislation limiting or banning mercury's use in a variety of products. (Note that the figure does not include residual flows that might be important in specific cases, such as the mercury emitted from coal-fired power plants.)

Comparisons of NMA parameters among countries indicate both the level of a country's wealth and the abundance and use of its natural resources. Figure 18.4a compares the domestic material input (DMC, defined as TMR minus exports minus export

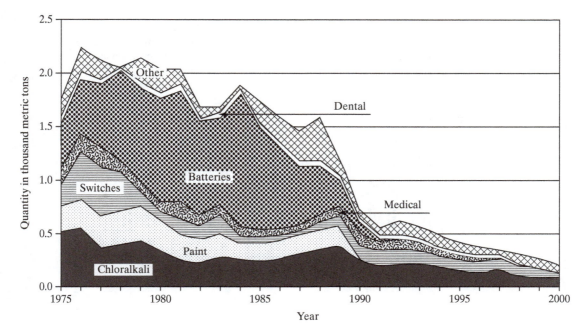

Figure 18.3

The use of mercury in the United States for the period 1975–2004. (W.E. Brooks, and G.R. Matos, *Mercury Recycling in the United States in 2000,* Circular 1196-U, Reston, VA: U.S. Geological Survey.)

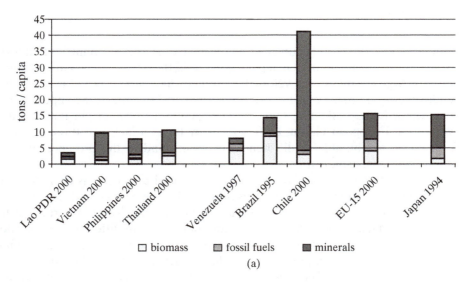

Figure 18.4

(a) Domestic material consumption (DMC) per capita for a number of countries and for EU-15. (Courtesy of Marina Fischer-Kowalski, Klagenfurt University.) (b) DMC per capita in European Union countries in 2000. (Reproduced with permission from H. Weisz, et al., The physical economy of the European Union: Cross-country comparison and determinants of material consumption, *Ecological Economics, 58,* 676–698, 2006.)

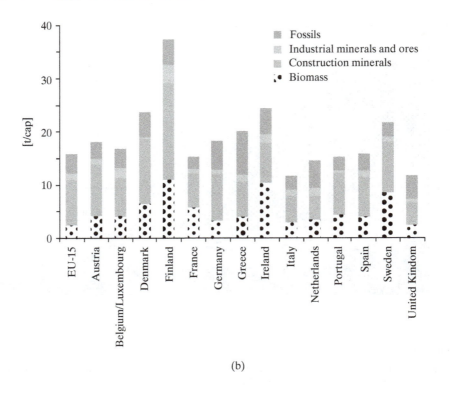

(b)

Figure 18.4

Continued

hidden flows) for several countries and for the EU-15 (the 15 countries that formed the original European Union). On a per capita basis, the developing countries have DMCs of 5–10 Mg/yr, while the developed economies have about twice as much. Resource endowments are readily visible, especially the very large mineral component of Chile's per capita DMC, reflecting the very large amount of mining in that country. Even within the European Union, DMC differs by up to a factor of three (Figure 18.4b), a property attributable to differences in wealth and to the presence or absence of material-intensive industries.

Table 18.1 compares NMA results for several countries, as characterized by an international team under the auspices of the WRI. The data are a decade old, but these flows change slowly, so the perceptions remain relevant. The top portion of the table consists of magnitudes of flows to the environment from different sectors. The results are a combination of country size and level of activity—the U.S. flows are the highest in all categories. Japan's output from the industrial sector is all out of proportion to its size, however, as is that of The Netherlands for agriculture. The differences are markedly decreased when these flows are expressed on a per capita basis, however. Household-related flows are similar, for example, except in Japan, where they are much lower. Construction-related flows are similar as well, except in Austria, where they are much higher. Environmental flows from agriculture in The Netherlands and transport in the United States present obvious possible targets for public policy initiatives.

TABLE 18.1   Domestic Processed Outputs to the Environment in Five Countries, 1996

	Austria	Germany	Japan	The Netherlands	United States
	National totals (Tg)				
Agriculture	9.5	44.8	30.6	44.6	231.6
Construction	5.4	26.2	25.4	3.0	110.0
Energy supply	2.9	133.1	119.7	12.4	723.9
Industry	11.0	35.3	171.3	18.1	653.0
Household	6.5	56.4	35.2	14.0	280.4
Transport	5.5	57.9	77.8	17.1	668.9
	Per capita averages (Mg)				
Agriculture	1.2	0.6	0.2	2.9	0.9
Construction	0.7	0.3	0.2	0.2	0.4
Energy supply	0.4	1.6	1.0	0.8	2.7
Industry	1.4	0.4	1.4	1.2	2.4
Household	0.8	0.7	0.3	0.9	1.0
Transport	0.7	0.7	0.6	1.1	2.5

Among the interesting and useful results presented by the WRI-led team for country-level NMAs are the following:

- Industrial economies are becoming more efficient in their use of materials, but waste generation continues to increase because of overall economic growth.
- One-half to three-quarters of annual resource inputs to industrial economies are returned to the environment as wastes within a year.
- The extraction and use of fossil energy resources dominate output flows in industrial countries.
- On average, about 10 metric tons of materials per capita are added each year to domestic stocks in industrial countries, mostly in buildings and infrastructure.

In recent years, as more countries have completed NMAs, comparisons have become more detailed. An interesting example is the relationship between per capita TMR and per capita GDP, shown in Figure 18.5 for 11 countries and the EU-15. In most cases, there is a weak positive dependence of TMR on economic growth. Nonetheless, TMR is at a reasonably high level even for low-income countries such as China and Poland. For most of the higher-income countries, TMR grows slowly with GDP, but in Germany and the United States, the trend is opposite. Special circumstances apply in the latter two cases, however, and illustrate the way the TMR index is defined. In the United States, efforts to reduce soil erosion in agriculture were responsible for almost all the TMR decrease. In Germany (and in the low-income Czech Republic), the closure of lignite mines (as alternative energy sources were utilized) caused substantial decreases in the amount of mobilized material that was required.

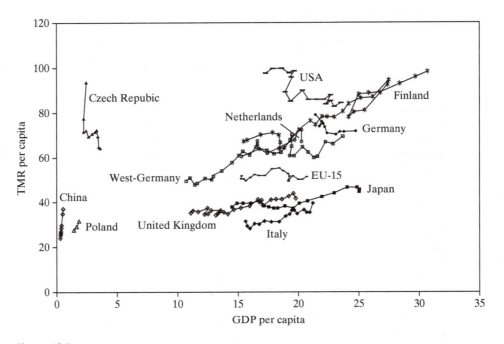

Figure 18.5

Total material requirements (TMR, the sum of domestic extraction, imports, and domestic and foreign hidden flows) as a function of GDP, both on a per capita basis, for a number of countries and for EU-15. The time periods vary, but most are for 20–25 years beginning in the 1970s. (Reproduced with permission from S. Bringezu, H. Schütz, S. Steger, and J. Baudisch, International comparison of resource use and its relation to economic growth, *Ecological Economics, 51*, 97–124, 2004.)

18.3   EMBODIMENTS IN TRADE

When a product made in one country is exported for use in another, it can be viewed as trailing behind it what has been termed its "ecological rucksack," that is, the resources extracted and consumed in the manufacture and transport of the product and the emissions therefrom, many of which occur in countries other than where the product is used.

In the life cycle inventory presentation of Chapter 12, we discussed the "1.7 kg computer chip," a chip weighing only 2 g, but requiring some 1600 g of fossil fuels, 72 g of chemicals, and 20 kg of water, and generating kilogram-sized amounts of wastewater and solid waste, as well as air emissions. In many cases, these flows, and their implications, do not occur in the importing country, but that country is, in a sense, responsible for them. They are thus "hidden" or "virtual flows" from the importing country's perspective.

The concept of virtual flows began with the development of national materials accounts. It was found in every case that the hidden flows exceeded the visible flows. In addition, for many countries the majority of the material flows sustaining their economies did not take place within the country.

Embodied energy is perhaps the best known of the hidden flows. This is because most energy is generated from fossil fuels, and fossil fuels produce climate-altering carbon dioxide. We discuss this topic in some detail in Chapter 19. Water is another widely analyzed example, discussed in Chapter 20.

TABLE 18.2   Methane Embodied in 1990 Imports from Developing Countries (GgCH$_4$)

	Meat products	Milk products	Biomass burning	Rice	Total
Canada	6	1	4	—	11
France	119	11	24	22	177
Germany	112	21	65	85	283
Japan	120	—	51	—	172
UK	238	31	77	9	355
USA	84	—	63	34	182
Total	679	64	283	150	1178

Reproduced from S. Subak, Methane embodied in the international trade of commodities: Implications for global emissions, *Global Environmental Change, 5*, 433–446, 1995.

Emissions other than CO_2 can be important virtual flows considerations. An example is methane (CH_4), a greenhouse gas much more potent than CO_2 on a molecule-for-molecule basis. Methane related to international trade is generated during cultivation of rice, production of meat and milk products, and the periodic burning of grasslands. An analysis done for six importing countries, shown in Table 18.2, indicated that in 1990 more than 1 TgCH$_4$ was embodied in imports into those countries, more methane than was emitted from all sources in many countries around the world. As with CO_2, this embodied CH_4 has become a discussion item in international climate negotiations.

The final embodied flow we discuss is that of toxics and air pollutants. Industrial processes tend to release these entities, of course, and they are released where manufacturing occurs. To the extent that the products of the manufacturing are exported, the importing country has, in a sense, transferred its pollution to the exporting country. The situation is exacerbated if the exporting country has weaker environmental regulations than the importing country. In the case of the United States, a study in 2004 (Ghertner and Fripp 2007) found average emissions in exporting countries to be two to four times higher than was the case domestically. The result was the avoidance of around 20 percent of emissions for most air pollutants and more than 80 percent for lead.

The topic of embodied flows is not a simple one. These flows are not necessarily bad; indeed, some provide food and employment to those who need it. In many cases, however, these flows burden the exporting countries to a significant degree. For the greenhouse gases, embodied flows have become a policy topic. None of these issues would be raised without the careful analyses encompassed in NMAs.

18.4   RESOURCE PRODUCTIVITY

Resource Productivity is a measure of the quantity of output produced by a corporation, an industrial sector, or a political entity relative to one or more inputs used in the production of the output. A common metric used for this purpose is GDP divided by the natural resource input, both domestic and imported. Japan's performance by this metric is shown in Figure 18.6a. A more stringent measure is to divide GDP by total material input, thus including the hidden flows in the calculation.

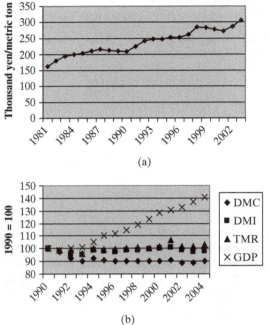

(a)

(b)

Figure 18.6

(a) Changes in resource productivity in Japan, 1780–2003, with the target for 2010. (Source: Tokyo, Ministry of the Environment, 2006.) (b) The decoupling of economic growth and resource use in the United Kingdom, 1790–2004. DMI = Direct material input, DMC = Direct material consumption, TMR = Total material requirements, GDP = gross domestic product. (Source: UK Environmental Accounts, Office of National Statistics, 2005.)

A goal in the evaluation of resource productivity is the measurement of the degree of *decoupling* of economic growth and the use of resources. Such an effect is shown for the United Kingdom in Figure 18.6b. Three different resource use indicators are shown, and substantial decoupling has occurred in each case. The decoupling is weaker in the case of TMR than for DMI or DMC, demonstrating that hidden flows are significant and suggesting that a significant portion of the environmental impacts associated with United Kingdom economic growth are being felt outside the country.

18.5   INPUT–OUTPUT TABLES

(Note: This section is advanced material that is optional.)
The NMAs described and discussed above deal largely with flows that cross national boundaries. Internal trades and transformations are excluded, a characteristic that limits the utility of NMAs. This issue was addressed from the perspective of economics by Wassily Leontief in the 1930s. His approach, termed "economic input–output (I/O) analysis," was to construct industrial sector transaction tables, as shown in Figure 18.7a for a two-sector economy. The variables are

$z_{i,j}$ = flows from one sector to another
y_i = flows entering use (final demand)
x_i = total production (output of all components of a sector)

In this arrangement, the columns (e.g., $z_{11} + z_{21}$) represent the inputs to a sector, while the rows (e.g., $z_{11} + y_1 = x_1$) represents the outputs. In I/O, all values are expressed in monetary units, and the number of sectors (depending on the country) may be as many as several hundreds.

From \ To	Producers		Final Demand	Total Output
	Sector 1	Sector 2		
Sector 1	z_{11}	z_{12}	y_1	x_1
Sector 2	z_{21}	z_{22}	y_2	x_2

From \ To	Producers		Export	Final Demand	Total Output
	Sector 1	Sector 2			
Sector 1	z_{11}	z_{12}	e_1	y_1	x_1
Sector 2	z_{21}	z_{22}	e_2	y_2	x_2
Import	i_1	i_2			

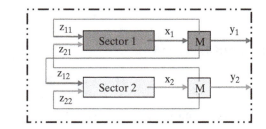

Figure 18.7

(a) A two-sector input–output transaction table; (b) A two-sector input–output transaction table including imports and exports; (c) A representation of an input–output table as a material flow diagram.

Import and export flows (also expressed in monetary units) can be included by adding an additional row and an additional column, as shown in Figure 18.7b.

Each of the $z_{i,j}$ factors contributes a portion of the total production, so we can write

$$z_{11} = a_{11}*x_1, z_{12} = a_{12}*x_2, \ldots \tag{18.1}$$

where the $a_{i,j}$ terms are called technical coefficients.

For the two-sector model of Figure 18.7a, the production equations then form a system of linear equations:

$$x_1 = a_{11}*x_1 + a_{12}*x_2 + y_1$$

$$x_2 = a_{21}*x_1 + a_{22}*x_2 + y_2 \tag{18.2}$$

or, expressed in matrix notation,

$$X = \mathbf{A}*X + V \tag{18.3}$$

where

$$X = \begin{bmatrix} x_1 \\ x_2 \end{bmatrix} \qquad V = \begin{bmatrix} y_1 \\ y_2 \end{bmatrix} \qquad \mathbf{A} = \begin{bmatrix} a_{11} & a_{12} \\ a_{21} & a_{22} \end{bmatrix}$$

production   final demand   technical coefficient
vector   vector   matrix

The inverse of a square matrix $\mathbf{A}$ is a matrix $\mathbf{A}^{-1}$, so that

$$\mathbf{A}*\mathbf{A}^{-1} = \mathbf{A}^{-1}*\mathbf{A} = \mathbf{I} \tag{18.4}$$

Where $\mathbf{I}$ is the identity matrix:

$$\mathbf{I} = [10.,01.,\ldots] \tag{18.5}$$

To solve equation (18.3), subtract $\mathbf{A}*X$ from both sides:

$$X - \mathbf{A}*X = y \tag{18.6}$$

and by equations (18.4) and (18.5),

$$(\mathbf{I} - \mathbf{A})X = y \tag{18.7}$$

Multiply both sides of equation (18.7) by $(\mathbf{I} - \mathbf{A})^{-1}$ to give

$$(\mathbf{I} - \mathbf{A})^{-1}(\mathbf{I} - \mathbf{A})X = (\mathbf{I} - \mathbf{A})^{-1}* y \tag{18.8}$$

which then becomes

$$X = (\mathbf{I} - \mathbf{A})^{-1}*y \tag{18.9}$$

where the term $(\mathbf{I} - \mathbf{A})^{-1}$ is designated the *Leontief inverse*.

The use of equation (18.9) enables the calculation of the effects of a single change in a single sector (say, on demand or supply) on all other sectors. This formulation of the monetary inputs and outputs of the sectors of a society is one of the central pillars of modern economics.

Notwithstanding the utility of this approach for the study of monetary linkages among sectors, there are important aspects of modern economies not addressed by I/O analysis. The basic one for our purposes is that traditional I/O tracks flows of resources only very indirectly. This is because the z_{bj} values include not only the cost of materials but also the cost of labor, the cost of capital, and so on. Thus, questions such as "What is the consequence of producing an additional metric ton of steel?" cannot be addressed in this way.

Industrial ecologists approach this challenge by recognizing that an input–output table can be represented as a material flow system, or in Figure 18.7c, where the matrix element values now refer to mass of material rather than monetary value. There are two implications of this formulation. The first is that conservation of mass holds across markets, that is, that the equivalence

$$x_1 = z_{11} + z_{12} + y_1 \tag{18.10}$$

implies that no mass is unaccounted for. The second implication is that the technical coefficients $a_{i,j}$ are constant. If numerical values can be derived for x_i, y_i, and $a_{i,j}$ parameters, the result is a *physical* input–output (PIO) table analogous to the I/O tables of the economists.

PIO analyses are at a rather early stage as this is written. A major challenge is generating the necessary numerical values, which has been achieved only in a few cases because data are sparse, frequently old, and may not be collected in comparable formats. In addition to data difficulties, a fundamental limitation of input–output models is that they assume linear behavior. While this is necessary to simplify the mathematics, it means that major technological or material shifts will almost certainly be mischaracterized if only viewed through the use of such methods. Another challenge arises from the desire to include the generation of waste, in order to satisfy the conservation of mass requirement, or to calculate embodied emissions in international trade. Still another is approaches that combine I/O and PIO in hybrid analyses that allow researchers to ask questions such as "What are the sectoral implications in terms of metal mass for a certain increase in monetary investment?" Unfortunately, PIO-related data that are available are often quite old—useful for demonstrations but not as guidance for decision makers. Nonetheless, PIO approaches are quite promising, and this area of industrial ecology is likely to develop rapidly in the future.

Hybrid Input–Output Analysis and Lead-Free Solder

Because metal deposits and the subsequent processing of ore typically involve more than a single metal, metal subsectors and metal import–export flows are strongly linked, as demonstrated by a study of metal flows in Japan under a transition to lead-free solder.

The top diagram shows these linkages, which result in single-metal demand changes influences the budgets of several other metals. A hybrid EIO–PIO analysis was conducted for two solder alternatives: the traditional lead-tin solder and the newer tin-silver-copper (SAC) product. In the former case, the majority of Japan's silver supply for all purposes came from by-product recovery in copper and lead smelters (bottom diagram). In the latter case, this by-product flow could not be significantly increased, and the demand for silver as a solder constituent could only

(continued)

(continued)

be met by an increase in silver import flows. This result could not have been derived from a simple EIO model, but the hybrid version allows the material-sensitive questions to be posed and resolved.

Source: S. Nakamura, et al., Hybrid input–output approach to metal production and its application to the introduction of lead-free solders, *Environmental Science & Technology, 42,* 3843–3848, 2008.

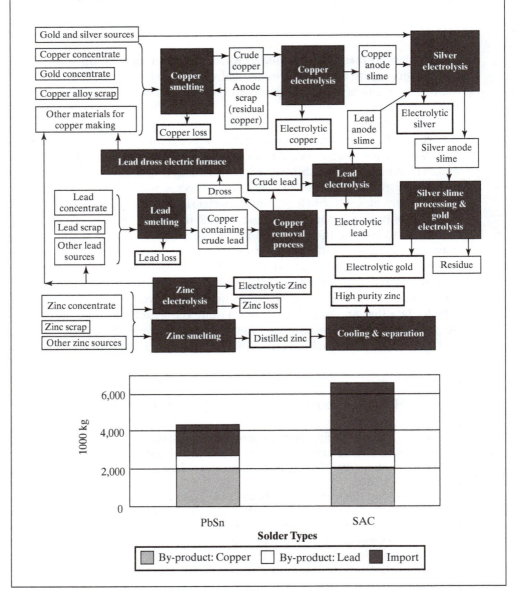

18.6   THE UTILITY OF METABOLIC AND RESOURCE ANALYSES

The types of analyses discussed in this chapter have proven themselves to be quite useful as industrial ecology tools. Organismal metabolic analyses permit us to examine resource use by individuals, factories, and by countries, and provide insights into the cultural and technological patterns that are revealed. Country-level NMAs and the insights that arise from them have the potential to be useful for a variety of public policy purposes — traffic planning, recycling incentives, and waste treatment plant construction, to name a few possibilities. NMAs are also related to environmental impacts, which result from actions taken to stimulate flows of resources or from the flows themselves. However, NMAs to date have not had a strong influence on policy, for several reasons: there is little supporting theory, there are problems with incomplete data, no suitable measure of reliability is at hand, the level of detail is often insufficient to support policy initiatives, and the use of weight at the common metric is at odds with the contrasting implications of "elephant" and "scorpion" flows.

A different situation holds for physical input–output analyses. These methods have a strong theoretical grounding, and the policy uses are quite obvious. However, the available data are much poorer than even the data used for NMAs, and much is seriously out of date. To the extent that the potentials of PIO are recognized, and that recognition leads to the generation of relatively current data (as is the case with I/Os), national accounts generated on a physical basis have the potential to provide a very useful supplement to national economic accounts; such a situation is indeed beginning to take hold in Europe and Japan.

FURTHER READING

Adriaanse, A., et al., *Resource Flows: The Material Basis of Industrial Economies,* Washington, DC: World Resources Institute, 1997.

Bringezu, S., H. Schütz, S. Steger, and J. Baudisch, International comparison of resource use and its relation to economic growth, *Ecological Economics, 51,* 97–124, 2004.

Ghertner, D.A., and M. Fripp, Trading away damage: Quantifying environmental leakage through consumption-based life-cycle analysis, *Ecological Economics, 63,* 563–577, 2007.

Hashimoto, S., and Y. Moriguchi, Proposal of six indicators of material cycles for describing society's metabolism from the viewpoint of material flow analysis, *Resources, Conservation, and Recycling, 40,* 185–200, 2004.

Hawkins, T., C. Hendrickson, C. Higgins, H.S. Matthews, and S. Suh, A mixed-unit input–output model for environmental life cycle assessment and material flow analysis, *Environmental Science & Technology, 41,* 1024–1031, 2007.

Hendrickson, C.T., L.B. Lave, and H.S. Matthews, *Environmental Life-Cycle Assessment of Goods and Services: An Input-Output Approach,* Washington, DC: Resources for the Future, 2006.

Mass Balance UK, *The Mass Balance Movement: The Definitive Reference for Resource Flows Within the UK Environmental Economy,* http://www.massbalance.org/index.php, accessed March 1, 2008.

Matthews, E., et al., *The Weight of Nations: Material Outflows From Industrial Economies,* Washington, DC: World Resources Institute, 2000.

Nakamura, S., K. Nakajima, Y. Kondo, and T. Nagasaka, The waste input-output approach to material flow analysis, *Journal of Industrial Ecology, 11* (4), 50–63, 2007.

Weber, C.L., and H.S. Matthews, Embodied environmental emissions in U.S. international trade, 1997–2004, *Environmental Science & Technology, 41,* 4875–4881, 2007.

Weisz, H., and F. Duchin, Physical and monetary input-output analysis: What makes the difference?, *Ecological Economics, 57,* 534–541, 2006.

EXERCISES

18.1 Do you think the concept of "hidden flows" in Figure 18.1 is useful? How?

18.2 What policy initiatives are suggested by the Chinese national material flow analysis of Figure 18.3?

18.3 Is resource productivity a useful measure of environmental performance? Why or why not?

18.4 Assume that the following transactions are observed: Suppose that the final demand of Sector 1 doubles. By what percentages will x_1 and x_2 increase?

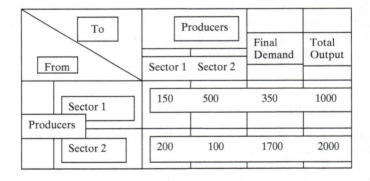

From \ To	Producers		Final Demand	Total Output
	Sector 1	Sector 2		
Sector 1 (Producers)	150	500	350	1000
Sector 2 (Producers)	200	100	1700	2000

C H A P T E R   1 9

Energy and Industrial Ecology

The first and second laws of thermodynamics dictate that all biological and industrial processes require energy conversion and result in entropy production. Biological processes receive their energy from the sun. Industry's energy sources are more varied, but rely heavily on fossil fuels. In Chapter 11, we discussed industrial energy generation and distribution. In this chapter, the focus is on energy use.

19.1 ENERGY AND ORGANISMS

Biological organisms expend energy to transform materials into new forms suitable for use (growing muscle or babies, for example). They also release waste heat and material residues. Excess energy is released by biological organisms into the surroundings, as are material residues (feces, urine, expelled breath, etc.). The energy flow through any organism, which reveals many features of its physiology, can be diagrammed as shown in Figure 19.1. In a somewhat similar fashion, industrial organisms expend energy for the purpose of transforming materials of various kinds into their own new forms suitable for use—their products. Energy residues (heat, exhaust gases, etc.) are emitted by industrial organisms into the surroundings, as are material residues (solid waste, liquid waste, etc.). By analogy with biological organisms, the energy flow of an industrial organism can be diagrammed as shown in Figure 19.2.

It is often the case in industry that the amount of energy required to operate a facility is well monitored (because it must be paid for), but the energy required for each individual operation or set of operations within a facility is not known. In such

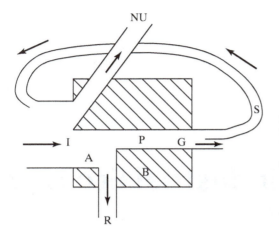

Figure 19.1

A model of an organism's energy flow in biological ecology. I = ingestion, A = assimilation, P = production, NU = not used, R = respiration, G = growth, S = storage (i.e., as fat for future use). B = the biomass in the organism, is stable only if inputs and outputs exactly balance. (Reproduced with permission from R.E. Ricklefs, *The Economy of Nature*, 3rd ed., New York: W.H. Freeman, 1993.)

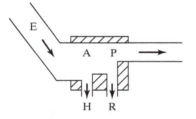

Figure 19.2

A model of an industrial organism's (i.e., a manufacturing facility's) energy flow. E = energy ingestion, A = assimilation, H = heat loss, R = respiration (i.e., energy used to run motors, etc.), P = production. Unlike the biological analog, there is no storage (at least not by design), and the organism's biomass is static.

cases, an energy audit is advisable to show where the opportunities for gains might lie, as well as to provide data for "green" accounting systems. Figure 19.3 shows such an audit for a facility that uses oil, coal, and electricity to provide energy for three different industrial processes, as well as for lighting and heating. The diagram demonstrates that more than enough energy is available in losses from the Process A energy stream to operate Processes B and C, and to heat and light the entire factory in the bargain. The diagram also suggests that boiler losses would be the highest priority target for improvement, and that steam losses also appear to constitute a substantial opportunity.

For a particular process, one wishes to audit the energy use at each stage of material extraction, processing, and manufacturing. In the production of aluminum cans, for example (Figure 19.4), the major energy use is in the separation and purification of aluminum contained in the ore. Production of sheet and of cans is also significant, but at a much-reduced level. The transport of material between stages is a minor contributor to total energy use. With this information as a basis, one might choose to increase the amount of recycled material used to produce metal products rather than extract metals directly from ore. Although aluminum presents the greatest opportunity for energy savings through recycling, the use of many other kinds of scrap material can also result in energy savings of 30 percent or more.

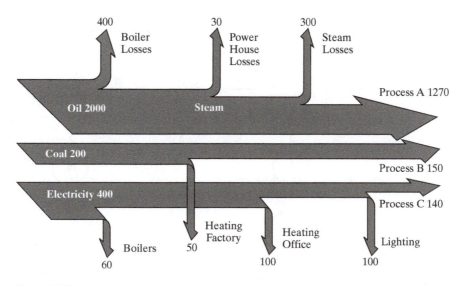

Figure 19.3

A "Sankey diagram" of energy sources, uses, and losses for a typical industrial facility. The units are arbitrary. (Reproduced with permission from *Climate Change and Energy Efficiency in Industry.* Copyright 1992 by International Petroleum Industry Environment Conservation Association.)

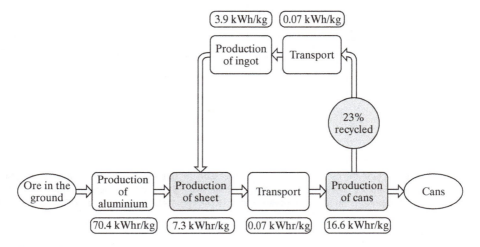

Figure 19.4

A process energy use diagram for the production of aluminum cans. (Reproduced with permission from *Climate Change and Energy Efficiency in Industry.* Copyright 1992 by International Petroleum Industry Environment Conservation Association.)

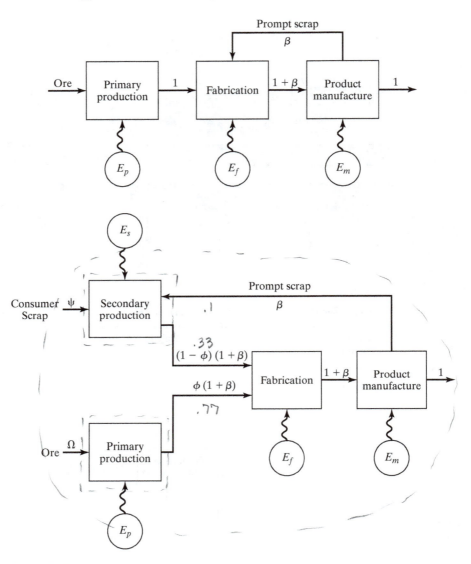

Figure 19.5

(a) Schematic diagram of a metals processing system using only virgin material; (b) Schematic diagram of a metals processing system using both virgin material and consumer scrap. (Adapted from P.F. Chapman and F. Roberts, *Metal Resources and Energy*, Boston, MA: Butterworths, 1983.)

To examine the energy use implications of virgin material and recycled material from a rigorous perspective, consider the process sequence shown in Figure 19.5(a). Each processing step has associated with it an energy per unit of throughput. For simplicity, we choose the amount of output material to be 1 kg. β is the fraction of throughput that is immediately reused as "prompt scrap" rather than output material: rejected material, sprues, runners, lathe turnings, and so forth. The energy consumed per kilogram of output material is then given by

$$\Phi = E_p + E_f(1 + \beta) + E_m(1 + \beta) \qquad (19.1)$$

$$= E_p + (E_f + E_m)(1 + \beta)$$

It is obvious from this equation that manufacturing operations that produce a smaller fraction of scrap will require less energy per unit of output than those where a large fraction of material must be refabricated.

A more relevant case for industrial ecology is a manufacturing sequence that uses both virgin and consumer recycled material. The latter need undergo only secondary production, which is generally much less energy-intensive than primary production. The situation is illustrated in Figure 19.5b, where ϕ is the fraction of output material from primary production, Ω is the amount of material entering the process in the ore, and Ψ is the amount of the material entering the process as consumer scrap. The energy consumed by this system per kg of output material is given by

$$\Phi = E_p(\phi)(1 + \beta) + E_s(1 - \phi)(1 + \beta) + E_f(1 + \beta) + E_m(1 + \beta) \qquad (19.2)$$

$$= (\phi E_p + (1 - \phi)E_s + E_f + E_m)(1 + \beta)$$

Because $E_p \gg E_s$, total energy use is minimized by making ϕ and β as low as possible. In this connection, it should be noted that product designers who specify virgin materials in their products may not be directly paying the high energy cost that results, but the virgin material specification forces the cost to be borne at some point within the in dustrial system.

19.2   ENERGY AND THE PRODUCT LIFE CYCLE

In Chapters 10–14, we mentioned that energy use is an important feature of the product life cycle. That property is assessed by calculating energy use and resultant emissions at each life stage. The results of such a calculation are shown in Figure 19.6 for a study related to the potential substitution of magnesium or aluminum parts (body, doors, hoods, wheels, wheel hubs, sheet frames, and steering) for steel parts in a midsize automobile. Figure 19.6a demonstrates that the magnesium parts require somewhat more energy to manufacture, but produce a lighter vehicle that saves considerable energy in use. Aluminum parts can be made with slightly less energy than magnesium parts, but the former are heavier than the latter so the energy savings in use are not as great.

A somewhat different picture emerges if the emphasis is on CO_2 emissions rather than energy (Figure 19.6b). This is because CO_2 emissions for producing magnesium ingots from virgin material are quite high. If a high proportion of recycled magnesium is used, however, the CO_2 emissions drop by more than 10 percent and the resulting vehicle is preferable to the aluminum equivalent. The final choice is thus dictated in part by regulations, in part by the manufacturer's desire for environmental superiority, by the availability of appropriate metal scrap, by recycling technology, and by other

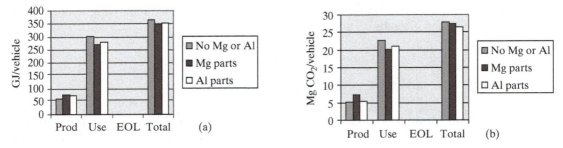

Figure 19.6

(a) Energy consumption at different life stages for vehicles with steel, magnesium, or aluminum parts; (b) CO_2 emissions for the vehicles. (The data are from M. Hakamada, et al., Life cycle inventory study on magnesium alloy substitution in vehicles, *Energy, 32*, 1352–1360, 2007.)

familiar concerns such as safety (although a metal, magnesium can burn rapidly under crash conditions). None of these factors could have been evaluated properly without the product energy study.

The energy use picture is more complex when dealing with long-lived products that require maintenance and periodic renovation in addition to initial manufacturing. Buildings are a prime example of this situation. As shown in Table 19.1 for a residence, the energy needed for construction is by far the largest quantity, but some aspects of renovation requires substantial energy expenditures in manufacture, and demolition carries with it an energy penalty as well. Even this is not the whole picture, of course: Figure CS-1 in Chapter 12 demonstrated that operating energy dwarfs the energy needs of construction and renovation, even for energy-efficient buildings.

TABLE 19.1   Energy Requirements for the Construction, Maintenance, and Renovation of a Typical New Zealand House

Process	Energy requirement (MJ)	Cycle (yr)
New build and replacement construction	255,109	—
Annual maintenance	357	1
Renovation		
External		
Replace gutters, downspouts (PVC)	3,119	25
Replace aluminum windows	9,761	40
Replace galvanized steel roofing	54,143	50
Internal		
Replace wool carpeting	14,029	10
Kitchen upgrade	13,479	25
Replace electrical wiring	2,242	40
Demolition	13,378	—

Source: Abridged from I.M. Johnstone, Energy and mass flows of housing, *Building and Environment, 36*, 27–41, 2001.

Case Study: Energy Use by Refrigerators

Attention to energy use in product design can result in some quite startling achievements, as shown by a history of energy use in household refrigerators over more than half a century (D. Kammen, University of California, Berkeley, 2007). The refrigerators in use in the late 1940s and 1950s in the United States were rather small in volume and used modest amounts of energy. The volume grew to more than twice as large in the 1960s and 1970s, and energy consumption quadrupled. In the late 1970s the volume stabilized, and attention to energy-efficient design began to rapidly reduce consumption. By 2000, refrigerators with volumes two and one-half times larger than those of 1947 were consuming only slightly more energy than those early units. Energy was consumed in manufacturing, of course, but life-cycle inventories invariably find that any product that requires energy during its use phase has an energy consumption profile in which all other phases are small to negligible by comparison (Figure CS-1).

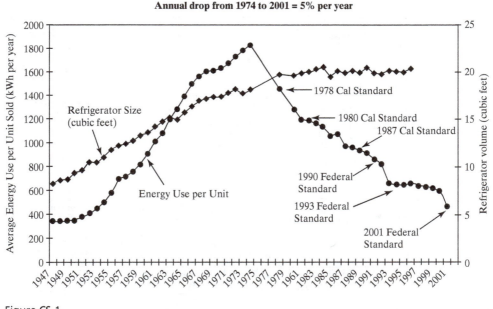

Figure CS-1

United States Refrigerator Use vs. Time.

19.3 THE ENERGY CYCLE FOR A SUBSTANCE

Just as in material flow analysis, it is possible to do energy flow analysis for a substance, as opposed to a facility. We illustrate that approach by the analysis of the energy required to produce one metric ton of stainless steel. The energy used at each production process is underpinned by the amount of material required to be produced by that process, so it is necessary to begin by quantifying the material flows of chromium, nickel, and iron (the constituents of the type of stainless steel that was analyzed). The material flows shown in Figure 19.7a represent the contained mass of the element in each flow. For example, the flow into chromium mining is 146 kg of contained chromium in ore.

Combining the results of the material flows with the unit energy use of each production process, the energy required at each stage for transportation, and the country-level electricity generation profiles resulted in the quantification of energy use. The result (Figure 19.7b) was that 53,000 MJ/t of stainless steel was required. The authors of the paper went on to compare the energy and carbon dioxide changes resulting from different fractions of recycled material usage.

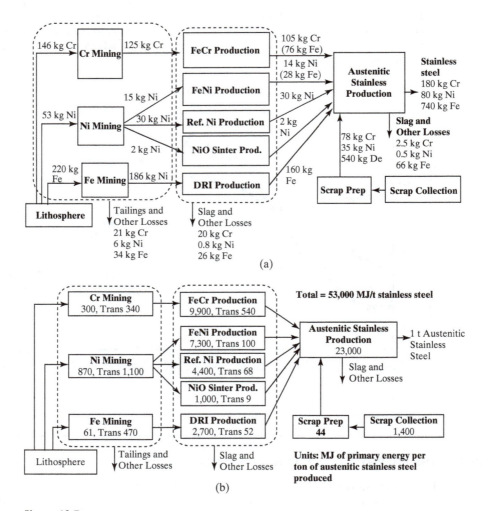

(a)

Figure 19.7

(a) The material flows of chromium, nickel, and iron required to produce one ton of austenitic stainless steel, *ca.* early 2000s, under current operating conditions; (b) The primary energy needed to produce the stainless steel. (Reproduced by permission from J. Johnson, et al., The energy benefits of stainless steel recycling, *Energy Policy, 36*, 181–192, 2008.)

19.4   NATIONAL AND GLOBAL ENERGY ANALYSES

Another approach to energy analysis that parallels material flow analysis is to generate an assessment on a country basis. Such an analysis for Canada is shown in Figure 19.8a. Among the results are the following: (1) Most of Canada's energy comes from domestic fossil fuels; (2) Several different sectors use similar amounts of energy, so reducing energy consumption would be a complex process; (3) About half of the Canadian energy that is produced is exported (to the United States).

Similar analyses have been carried out at the global level. The current sectoral distribution of greenhouse gas emissions from fossil fuel use (Figure 19.8b) shows that some 20 percent is directly charged to industry, and technological products or activities are related to much more. Given long-term concerns about climate, constraints on industrial energy use are likely in the future. Diagrams like those of Figure 19.8 are very important in suggesting policy decisions related to energy use, and to where those policies might be most effective.

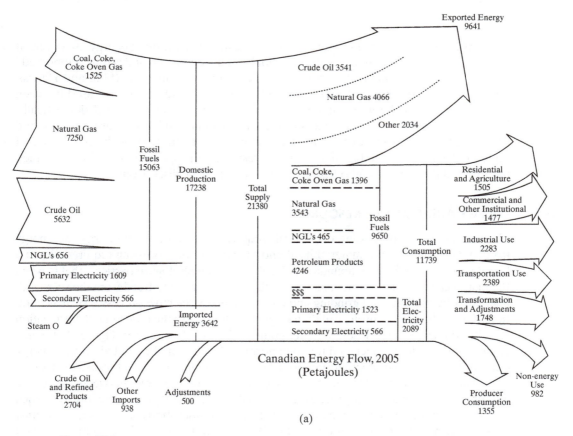

(a)

Figure 19.8

The energy cycle for Canada in 2005. (Reproduced from Statistics Canada, *Report on Energy Supply-Demand in Canada*, Ottawa, ON, 2006); (b) Sector shares of global greenhouse gas emissions in 2004. (Reproduced from *Climate Change 2007: Synthesis Report*, Intergovernmental Panel on Climate Change, Geneva, Switzerland, 2007.)

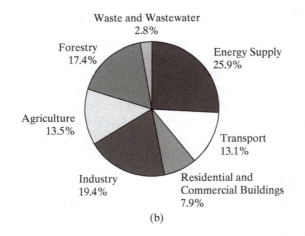

Figure 19.8

Continued

As mentioned in the previous chapter, a substantial fraction of energy use occurs at locations other than where the material is finally being employed and is thus "embodied" or "virtual" energy. One of the most dramatic reports on this topic (Shui and Harriss, 2006) finds that during 1997–2003, U.S. CO_2 emissions would have increased between 3 and 6 percent if the goods imported from China had been produced domestically. By the same token, between 7 and 14 percent of China's emissions during that period resulted from producing exports for U.S. buyers. The realization of the magnitude of these hidden flows, in combination with CO_2 restrictions proposed in the climate treaties, has injected the topic of embodied flows into global policy discussions.

19.5   ENERGY AND MINERAL RESOURCES

The extraction and processing of minerals require large amounts of energy, the energy needed depending on the material itself and the amount of processing required. When a target mineral is sufficiently abundant to be above the mineralogical barrier—that is, when the matrix can be called an ore and not a rock—the minerals are freed from the surrounding matrix by crushing and grinding and concentrated by selective processes such as flotation. The resulting concentrate can then be purified to acquire the target metal. Purification is rather directly related to the melting point of the metal, as seen in Figure 19.9 (some engineering plastics are included for comparison). The chart makes it obvious that energy optimization is much more important for titanium and aluminum than for lead or steel.

The situation for copper presents clearly the relationship between ore grade and energy. Extraction and processing are relatively efficient for copper ore (the rock matrix containing copper sulfide minerals), though considerations of energy are certainly present. At mineral concentrations below 0.1 percent, however, the copper is thought to be dispersed in solid solution within silicate minerals rather than concentrated into mineral form. To recover the dispersed copper, the silicate minerals themselves must be separated and processed. The energy costs of so doing are very large, because the chemical bonds holding the atoms together in a silicate mineral are much stronger than those holding the atoms of a copper sulfide mineral together. Metallurgical experience

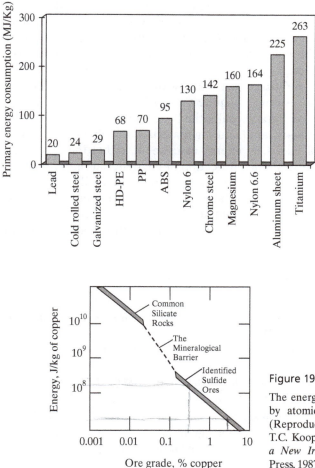

Figure 19.9

The primary energy consumption required to produce one kilogram of various materials. (Adapted from M. Schuckert, H. Beddies, H. Florin, J. Gediga, and P. Eyerer, Quality requirements for LCA of total automobiles and its effects on inventory analysis, in *Proceedings of the Third International Conference on Ecomaterials,* Tokyo: Society of Non-Traditional Technology, pp. 32–329, 1997.)

Figure 19.10

The energy necessary to recover copper from its ores and by atomic substitution from common silicate minerals. (Reproduced with permission from R.B. Gordon, T.C. Koopmans, W.D. Nordhaus, and B.J. Skinner, *Toward a New Iron Age?* Cambridge, MA: Harvard University Press, 1987.)

suggests that the overall energy needs will be about 10 times as great per recovered copper atom if recovery from silicate minerals is required. Figure 19.10 shows the ore grade–energy relationship for copper. This is a log-log plot, indicating that recovery from ore of grade 0.01 percent will require roughly 100 times the energy needed for ore of grade 0.1 percent.

19.6   ENERGY AND INDUSTRIAL ECOLOGY

Because almost every action of industry involves the use of energy, energy analysis is inherently a part of industrial ecology. Historically, this analysis has been primarily a focus of the most energy-intensive industrial sectors: metal mining and processing, pulp and paper, cement, petrochemicals, and glass manufacture. With energy use becoming a more general concern, energy analysis is now nearly universal and is moving to encompass products as well as manufacturing facilities. In the case of products, there is now a general realization that energy use occurs at all life stages, is a natural part of life cycle analysis and is one of the principal ways in which designers find their work related to environmental impacts and sustainability. For products that use energy during their service lifetimes—engines, computers, buildings, and so on—there is a particular premium on designs that minimize energy consumption.

Energy intensity differs with industrial sector and with life stage. It is usually the case that the initial extraction of resources is quite energy intensive because of the separation and purification involved. Paradoxically, the benefits of modern technology often require very pure materials, and manufacturing materials at high purity requires lots of energy. Another aspect of industrial energy assessment is that the extraction stage for materials often occurs far from the fabrication of the materials into products, perhaps on another continent. This is a "hidden energy flow," essentially invisible to those using the material, but no less present for all that. In general, the energy needed to provide materials through recycling is substantially less than that needed for virgin materials, so the use of recycled materials is desirable from both a material and an energy perspective.

FURTHER READING

Arthur D. Little, Inc., *Overview of Energy Flow for Industries in Standard Industrial Classifications 20–39*, Report 71563-00, Cambridge, MA, 2000.

Ayres, R.U., L.W. Ayres, and V. Pokrovsky, On the efficiency of U.S. electricity usage since 1800, *Energy, 30*, 1092–1145, 2005.

Glicksman, L.R., Energy efficiency in the built environment, *Physics Today, 61* (7), 35–40, 2008.

Hu, S.-C., Power consumption of semiconductor fabs in Taiwan, *Energy, 28*, 895–907, 2003.

Lampert, M., et al., Industrial energy flow management, *Applied Energy, 84*, 781–794, 2007.

Ramirez, C.A., M. Patel, and K. Blok, How much energy to process one pound of meat? A comparison of energy use and specific energy consumption in the meat industry of four European countries, *Energy, 31*, 2047–2063, 2006.

Schaefer, C., C. Weber, and A. Voss, Energy usage of mobile telephone services in Germany, *Energy, 28*, 411–420, 2003.

Shui, B., and R.C. Harriss, The role of CO_2 embodiment in US—China trade, *Energy Policy, 34*, 4063–4068, 2006.

EXERCISES

19.1 You have been asked to make energy reduction recommendations for the facility whose current energy budget is shown in Figure 19.3. What are your recommendations, and why?

19.2 Assume a material processing system as shown in Figure 19.5, with $E_p = 31$ GJ/t, $E_f = 5$ GJ/t, $E_m = 5$ GJ/t, and $\beta = 0.1$. Compute Φ.

19.3 To the system of the previous problem, add a secondary production component to reprocess consumer scrap with $E_p = 9$ GJ/t and $\phi = 0.7$. Find Ψ, Ω, and Φ.

19.4 In the system of Exercise 19.3, a fraction λ of the material entering the primary production process is irretrievably lost to slag. Reformulate equation (19.2) to take this loss into account. If $\lambda = 0.2$, compute Ψ, Ω, and Φ.

19.5 A city has 100,000 homes, each with one refrigerator. The 2001 electricity cost was \$0.15/kWh. Referring to the figure in the refrigerators case study, compute how much money was saved in 2001 if all homes had just replaced a 1974 refrigerator with a new refrigerator.

19.6 An office building in your community has 50 offices, each with an average of four desks. Each desk has a desk lamp that can use either a 60-watt incandescent bulb or a 13-watt fluorescent unit. The average use of a lamp is seven hours per day. How much power is required for the building per year for each of the two options? Given your local energy cost, what is the annual cost of each of the two options? If the price of an incandescent bulb is 88 cents and that of a fluorescent unit is 12 dollars, how long will it take to justify the purchase of fluorescent units, assuming everything is newly purchased?

CHAPTER 20

Water and Industrial Ecology

20.1 WATER: AN OVERVIEW

Water is necessary for life, for industry, for agriculture, and for the continuing function of many coupled natural systems such as the carbon cycle. Yet it is frequently overlooked in lists of sustainability concerns, even though the United Nations estimates that by 2025, two-thirds of the human population will experience shortages of water, and almost two billion will experience severe shortages. Water scarcity already affects most of the Middle East, Malta, Singapore, Poland, and other countries. Agriculture is involved in an estimated 70 percent of all global water use, with industrial use, municipal use, and reservoir losses accounting in roughly equal amounts for the remainder.

From an industrial ecology perspective, the use of water is a classic subject for sustainable engineering. It is a critical material, often regulated and used in a highly politicized environment, and subject to very different laws depending on jurisdiction. (For example, water use in the eastern United States is governed by a very different legal framework than common in the western United States.) Water is closely coupled to public health, food production, industrial manufacture, national security, and social stability, and water infrastructures are often among the most important elements of a society's built environment.

Water quantity and conservation is not the only concern of the industrial ecologist, who also needs to consider water quality. For example, there is plenty of water for drinking in Bangladesh, but much of it suffers from high arsenic levels, making it toxic if consumed over long periods of time. Quantity, in other words, is adequate, but quality is not, suggesting the need for treatment technology rather than more wells and delivery infrastructure.

20.2   WATER AND ORGANISMS

Biological organisms are, we are told, 70–90 percent water. As with energy, the flows of this resource within and through an organism can be measured and diagrammed. An example, for the least shrew, is shown in Figure 20.1a. The analysis accounts for direct water ingestion, indirect ingestion as a consequence of extracting water from food, and losses to evaporation and excretion.

Water is important for most industrial organisms as well, where it plays a central role in such processes as washing, plating, cooling hot machinery, and serving as a chemical reaction medium. By analogy with biological organisms, the water flow of an industrial organism can be diagrammed as shown in Figure 20.1b. Either analysis, of course, can show much more detail—the water budget of a shrew's kidney or of a wastewater treatment facility, for example.

Industry may be aware of the overall amount of water required to operate a facility (because it must be paid for), but often the water required for each individual operation or set of operations within a facility is not known. In such cases, a water audit can show where the opportunities for gains might lie, as well as providing data for "green" accounting systems. Figure 20.2a shows the water budget for a semiconductor manufacturing facility. The diagram demonstrates that the majority of the water must

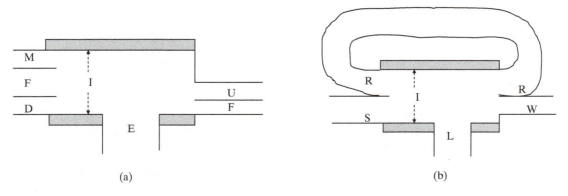

(a)   (b)

Figure 20.1

(a) A model of the water flow of a least shrew. F = food, D = drinking, M = metabolic water (i.e., water extracted from food by metabolic transformation), I = ingestion, E = evaporation, U = urine, F = feces. The shrew's daily water ingestion is 5.1 ml. (From D.L. Goldstein and S. Newland, Water balance and kidney function in the least shrew (*Cryptotis parva*), *Comparative Biochemistry and Physiology—Part A: Molecular and Integrative Physiology, 139*, 71–76, 2004.) (b) A model of the water flow of an industrial facility. S = water supplied, R = recycled water, I = ingestion, W = wastewater, L = loss. The common term "water use" applies to flow I, "water consumption" to flow S.

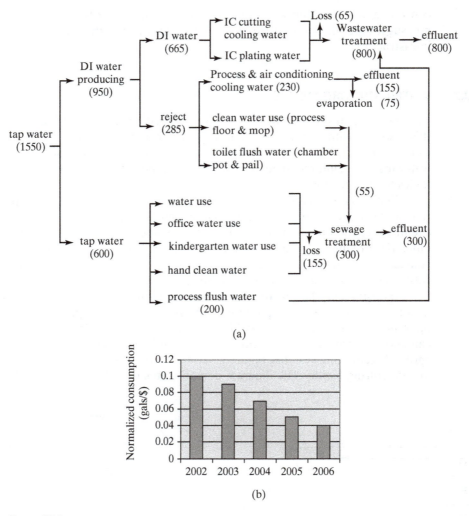

(a)

(b)

Figure 20.2

(a) A mass balance for water in a Taiwanese semiconductor manufacturing facility. The units are m^3/day. (Reproduced with permission from S.-H. You, D.-H. Tseng, and G.-L. Guo, A case study on the wastewater reclamation and reuse in the semiconductor industry, *Resources, Conservation, and Recycling, 32*, 73–81, 2001.); (b) A five-year history of water use per dollar of sales by the United Technologies Corporation, 2002–2006. (From http://www.utc.com/responsibility/key_performance_ indicators.htm, accessed December 26, 2007.)

be deionized, and two-thirds of that water is used for integrated circuit (IC) cutting and plating. Other water is used for process cooling, for cleaning, and for washing. This inventory of uses and losses shows many activities and processes where the potential for improved water use might be investigated. A common opportunity is to "cascade" the use of water—perhaps once for high-purity manufacturing, a second time for toilets, and a third time (after some filtration) for irrigation of the property.

Similar analyses can be undertaken for an entire corporation, and the results are often reported to the public. Figure 20.2b shows the five-year history of water use by

Water Requirements for Electric Cars

Realization of the water requirements of actual or contemplated activities often results in new perspectives. An example is an analysis of the water requirements related to electric vehicles. These vehicles are often promoted as improving air quality and benefiting trade balances for the country in which they are manufactured, which they do. However, the generation of electricity requires substantial amounts of cooling water, some of which is lost to evaporation. An analysis by King and Webber quantified the water requirements, as shown in the chart below. The result is that displacing the U.S. gasoline-powered automotive fleet with electric vehicles would result in withdrawing 17 times more water and consuming 3 times more. This does not necessarily mean that such an action is undesirable, but that water requirements need to be a part of the planning process in determining the optimum approach to vehicle propulsion (Figure BX-1).

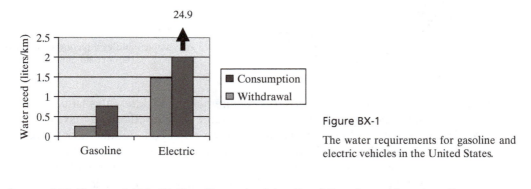

Figure BX-1

The water requirements for gasoline and electric vehicles in the United States.

Source: C.W. King, and M.E. Webber, The water intensity of the plugged-in automotive economy, *Environmental Science & Technology, 42,* 4305–4311, 2008.

United Technologies Corporation (UTC), normalized to corporate sales (e.g., gallons of water used per dollar of sales). By that metric, UTC's water consumption (flow S on Figure 20.1b) was cut by more than half in four years—a commendable performance.

Water use within a facility or a corporation is itself a component of a multilevel holonic system, as suggested in Figure 20.3 (a mining operation is the example here). Worldwide, more than 70 percent of water use is for agriculture and 10 percent for human activities, so the industrial holons are minor entities in the overall water cycle.

20.3 WATER AND PRODUCTS

In principle, it should be possible to construct a life-cycle water analysis for any commercial substance or product of choice, quantifying the water used at each substance or product life stage and the resulting environmental impacts. In practice, few such studies have been done, except perhaps for products such as washing machines that require water for everyday use, and fewer still have been published. Additionally, it is appropriate to include water used in the generation or processing of materials, just as the hidden flows of

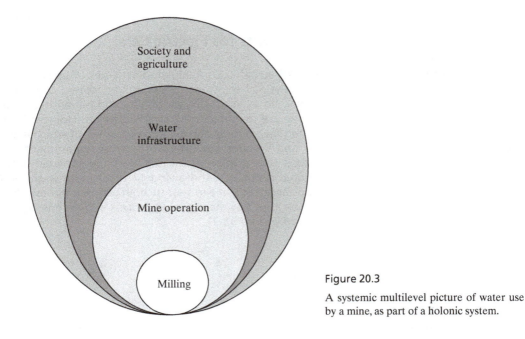

Figure 20.3

A systemic multilevel picture of water use by a mine, as part of a holonic system.

waste rock are included in total material requirements. To address this latter challenge, Hoekstra and Chapagain of the University of Twente in The Netherlands have publicized *virtual water* as a metric, that is, the volume of water required to produce a commodity or service rather than that contained within it. For an agricultural crop, for example, the virtual water would include that used for irrigation or washing. For a machine, it would include the water used in the separation of the metal minerals from their ores, the amount used as cooling water in the cutting and shaping of metals, and so forth. The virtual water contents of a small selection of products are given in Table 20.1. The values for the few nonagricultural products should be regarded as very preliminary, as only a handful of studies of this type have been performed.

An industrial water study restricted to the mining and processing of metal ores is that of Norgate and Lovel, who have analyzed the water required to produce one metric ton of various metals at contemporary Australian ore grades, as shown in Figure 20.4. Their approach was to include both the water directly consumed in the production of refined metal from the ore, as well as the water consumed in producing other new materials used in the process stages and in the generation of electricity. (The sum of these latter terms is the "indirect water consumption.") The large indirect consumption for aluminum and titanium reflects the very large amount of energy required in their processing, because energy production is often water intensive.

In mining, water use largely reflects the grade of the ore used to produce the metal. Norgate and Lovel have developed a relationship between the water required for metal mining and processing and the grade of the ore being mined. It is

$$W = 167.7\ G^{-0.9039} \tag{20.1}$$

where W = embodied water (m³/metric ton of refined metal) and

G = grade of ore used to produce metal (in percent metal).

TABLE 20.1   Global Average Virtual Water Content of Some Selected Products, Per Unit of Product

Product	Virtual water content (liters)
1 microchip (2 g)	32
1 sheet of A4-size paper (80 g/m^2)	10
1 slice of bread (30 g)	40
1 potato (100 g)	25
1 cup of coffee (125 ml)	140
1 bag of potato crisps (190 g)	185
1 hamburger (150 g)	2,400
1 cotton T-shirt (250 g)	1,900
1 pair of shoes (bovine leather)	8,000

Source: A.Y. Hoekstra and A.K. Chapagain, Water footprints of nations: Water use by people as a function of their consumption pattern, *Water Resources Management, 21*, 35–48, 2007.

It is important to note, however, that different ore processing technologies can have significantly different water usage (see Schüller, et al., 2008).

Many corporations are beginning to report their water performance publicly. A particularly interesting example is the Coca-Cola Company, because the product of the company is largely water. Over a four-year period, the "water efficiency" (water use per unit of product) declined worldwide by nearly 20 percent—an impressive performance.

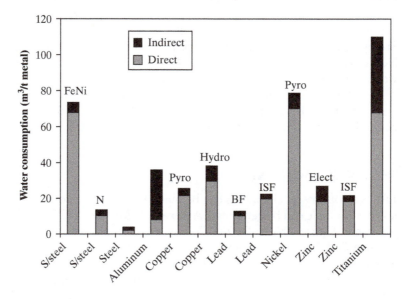

Figure 20.4

Embodied water for a sampling of major industrial metals for contemporary Australian mines. FeNi = ferronickel process, Ni = nickel process, Pyro = pyrometallurgical process, Hydro = hydrometallurgical process, BF = Blast furnace, ISF = Imperial smelting furnace, Elect = Electrolytic process. Indirect water consumption is that used in producing other raw materials used in various process stages, and also in generating the electricity that is required. (Reproduced by permission from T.E. Norgate, and R.R. Lovel, Sustainable water use in minerals and metal production, in *Water in Mining 2006*, Carleton, Australia: Australasian Institute of Mining and Metallurgy, Pub.10/2006, pp. 331–339, 2006.)

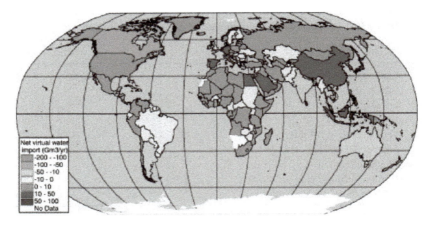

Figure 20.5

National virtual water balances over the period 1955–1999. (Adapted from A.Y. Hoekstra, and P.Q. Hung, Globalisation of water resources: International virtual water flows in relation to crop trade, *Global Environmental Change, 15*, 45–56, 2005.)

The trade of virtual water in agricultural crops permits large populations to exist in water-poor regions such as the Middle East. Figure 20.5 illustrates the virtual water balances for countries around the world. For the period 1995–1999, the countries exporting the most virtual water (mostly in wheat, soybeans, and rice) were the United States, Canada, Thailand, and Argentina. The largest importers were Japan, The Netherlands, Republic of Korea, and China. Not included in the analysis is the virtual water contained in industrial products as a consequence of manufacturing.

Hoekstra and Chapagain go on to estimate that industrial products have virtual water contents that range from 10 to 100 litres per $US of cost. Much research is needed to establish the degree of validity of this estimate.

20.4   THE WATER FOOTPRINT

Just as a country-level analysis can be made for energy or for other resources, a country-level water cycle can be generated. An example for Australia for the 2004–2005 fiscal year is given in Figure 20.6. In the detailed information required to generate this perspective, one can find the proportions of water consumption by the mining industry (1.5 percent) and manufacturing industries (2.5 percent), among other statistics.

A country-level water cycle can be enhanced by computing the "water footprint": the water used within the country and the virtual water supplied by the import of various commercial products, given by

$$\text{WFP} = \text{AWU} + \text{IWU} + \text{DWU} + \text{VW} \qquad (20.2)$$

where WFP = water footprint, AWU = agricultural water use, IWU = industrial water use, DWU = domestic water use, and VW = virtual water in international

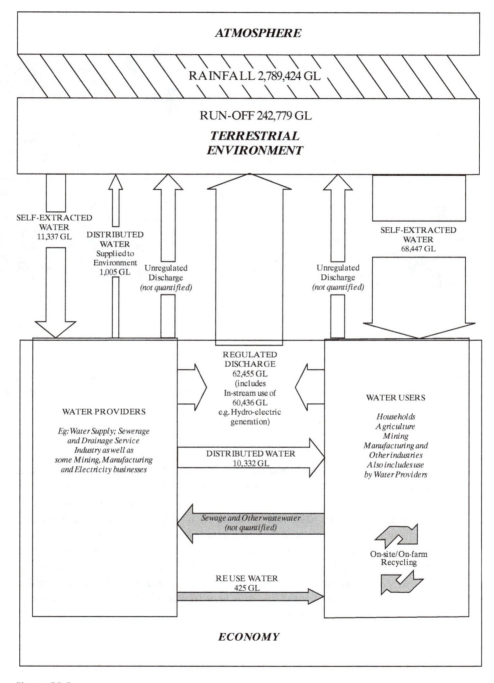

Figure 20.6

Water supply and use in the Australian economy, for the year beginning July 1, 2004. Source: *Water Account, Australia, 2004–2005*, Report 4610.0, Australian Bureau of Statistics, 2006. The units are Gigaliters (GL = 10^9 l) http://www.abs.gov.au/AUSSTATS/abs@.nsf/Lookup/4610.0Main+Features 1 2004-05?OpenDocument#, accessed December 26, 2007.

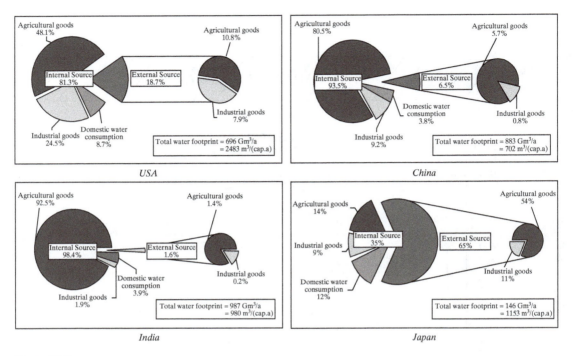

Figure 20.7

The average water footprints of four countries for the period 1997–2001. (Reproduced by permission from A.Y. Hoekstra, and A.K. Chapagain, Water footprints of nations: Water use by people as a function of their consumption pattern, *Water Resources Management, 21*, 35–48, 2007.)

trade (negative for export, positive for import). Figure 20.7 illustrates the results for four countries, both as quantitative totals and by component. The differences are striking. In the more developed countries, the water that is associated with high levels of industrial products plays a much larger role than in a developing country like China. India imports very little virtual water, while imports account for two-thirds of the Japanese water footprint. On an absolute magnitude basis, the per capita water footprint of the United States is three and one-half times that of China.

The global water footprint diagram is shown in Figure 20.8. Industrial use accounts for only 9 percent of the total; two-thirds of that use is domestic, one-third imported. On a per capita basis, the total footprint is 1240 m^3/yr.

20.5 WATER QUALITY

Water quantity is not the only aspect of water that needs to be considered. Also of major importance is the quality of the water—its acidity, its dissolved oxygen content, its impurity levels, and so on. Does it contain harmful organics? Does it contain reactive chemicals? Do its impurities require extensive treatment before use?

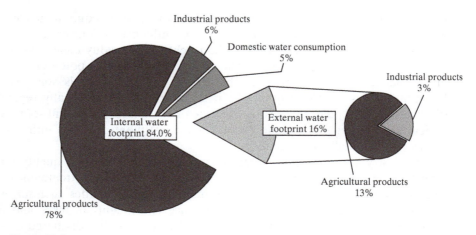

Figure 20.8

The average global water footprint for the period 1997–2001. (Reproduced by permission from A.Y. Hoekstra, and A.K. Chapagain, Water footprints of nations: Water use by people as a function of their consumption pattern, *Water Resources Management, 21*, 35–48, 2007.)

Consider the common sequence of Figure 20.9. Industry withdraws water of quality Q_1 from a convenient river (or a well, the arguments are similar). The quality of the water may be such that pretreatment is required. Following use, industry treats the water again, to quality Q_2, and discharges it. At the customer location further downstream, water of quality Q_3 is provided by the local government, and discarded water is

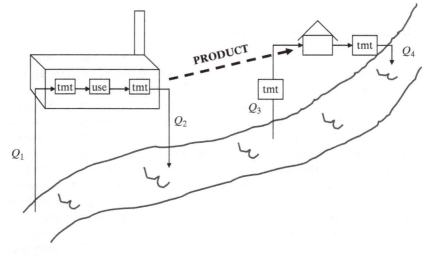

Figure 20.9

Sequential water treatment, industrial and domestic. The Q values refer to different water quality conditions, each of which may require or has already required treatment to meet a desired standard.

later treated to produce quality Q_4. In general, the quality requirements are different at each stage, and different actors at each stage influence or determine the quality.

In most cases, regulations limit the discharge of readily measured constituents such as acids, salts, and nutrients. Much less oversight exists for potentially problematic and bioactive constituents that may be present at very low levels: solvents, additives, lubricants, heavy metals, drug metabolites, and so forth. Additionally, regulations are often stricter for "point sources" such as factories than for "nonpoint sources" such as agriculture, meaning that some problematic constituents, such as nutrients in river basins, are essentially unregulated.

Industry has three quite different concerns related to water quality. One is that industry's own water cleanliness requirements may necessitate treatment before use. The second is the discharges from its manufacturing facilities themselves; though regulations set minimum standards, even trace constituents in lakes and rivers are increasingly linked to industrial discharges. The third is the discharge of industry's products—pesticides, detergents, and a wide variety of pharmaceutical and personal care products—by its customers. Many of these constituents are bioactive by design, and so are of particular concern from an ecosystem impact perspective.

The first line of industrial actions related to water quality is the obvious: discharge as little as possible and render benign what you do discharge. This often requires technologically advanced water treatment. The second activity is even more difficult—to design products that will cause little harm if discarded by the customer into municipal water supplies or into the environment. This challenge requires the use of the green chemistry techniques discussed in Chapter 8. It places a premium on reasonably prompt degradability, as well as ensuring that any transformation products are relatively benign. This two-pronged requirement raises industrial concerns related to water quality to a much higher level than has historically been the case.

20.6 INDUSTRIAL ECOLOGY AND WATER FUTURES

Because many actions of industry involve the use of water, water budgets, and responses to what the budgets reveal, are inherently part of industrial ecology. For products that use water during their service lifetimes—dishwashers, for example—there is a particular premium on designs that minimize water consumption during product use.

Water intensity differs greatly with industrial sector and with life stage. It is often the case that the initial extraction of resources is quite water intensive because of the material separation and purification involved. As a consequence, the nonagricultural sectors that have the highest rates of water use are petroleum and coal processing, primary metals, chemicals, and paper. Paradoxically, the benefits of modern technology often require very pure materials, and manufacturing those high purity materials often requires lots of water. Another aspect of industrial water assessment is that the extraction and processing stage for materials often occurs far from the fabrication of the materials into products, perhaps on another continent. The water used in extraction and processing constitutes a hidden flow essentially invisible to those using the material, but no less present for all that. In general, the water needed to provide materials through recycling is substantially less than that needed if virgin materials are used, so the use of recycled materials is desirable from both a material and a water perspective.

FURTHER READING

Gleick, P.H., Water use, *Annual Review of Environment and Resources, 28*, 275–314, 2003.

Morikawa, M., J. Morrison, and P. Gleick, *Corporate Reporting on Water: A Review of Eleven Global Industries*, Oakland, CA: Pacific Institute, 2007.

Pennisi, E., The blue revolution, drop by drop, gene by gene, *Science, 319*, 171–173, 2008.

Schüller, M., A. Estrada, and S. Bringezu, Mapping environmental performance of international raw material production flows: A comparative case study for the copper industry of Chile and Germany, *Minerals & Energy, 23* (1), 29–45, 2008.

Schwarzenback, R.P., B.I. Escher, K. Fenner, T.B. Hofstetter, C.A. Johnson, U. von Gunten, and B. Wehrli, The challenge of micropollutants in aquatic systems, *Science, 313*, 1072–1077, 2006.

Shiklomanov, I.A., Appraisal and assessment of world water resources, *Water International, 25*, 11–32, 2000.

U.S. Geological Survey, *The Quality of Our Nation's Waters: Nutrients and Pesticides*, Circular 1225, Reston, VA, 82 pp., 1999.

Vossolo, S., and P. Döll, Global-scale gridded estimates of thermoelectric power and manufacturing water use, *Water Resources Research, 41*, doi:10.1029/2004WR003360, 2005.

Younger, P.L., The water footprint of mining operations in space and time—A new paradigm for sustainability assessments? in *Water in Mining 2006*, Carleton, AU: Australasian Institute of Mining and Metallurgy, Pub.10/2006, pp. 13–21, 2006.

EXERCISES

20.1 Assume that the water requirements for chromium are represented by equation (20.1). How much water is needed to process one metric ton of chromite ore if the ore grade is 5 percent? If the ore grade is 1 percent?

20.2 You currently drive 10,000 km/year in a gasoline-powered car. If you switch to an all-electric car, how much additional water will need to be withdrawn in order to provide your annual needs, and how much of that will be consumed?

20.3 Redraw Figure 20.2 a as a Sankey diagram (e.g., similar to Figure 19.3). If you wish to make water use reductions, what recommendations might you make, and why?

20.4 All products carry with them embedded energy, water, and solid "hidden flows." Design a label for a new product of your choice that displays its "virtual resource" loads. How could the average customer use your label to make a more informed choice of products?

CHAPTER 21

Urban Industrial Ecology

21.1   THE CITY AS ORGANISM

Cities have often been compared to organisms. They are born; they grow; and they pulsate. As with other organisms, they utilize resources. The resource spectrum of a city is spectacular: apples from New Zealand, electronics from Japan, automobiles from Germany, leather from Argentina, lobsters from the United States. Cities also dissipate resources after use, sometimes to the local air and water, often to landfills, generally close to home. Cities are accomplished attracters and poor dispersers.

If a city can be regarded as an industrial ecology organism, what might we learn about cities from traditional methods of biology? What does a biologist learn from studying, for example, the metabolism of a squirrel? It turns out that a number of useful facts emerge—the amount and types of nutrients the animal requires, how the nutrients are utilized, what proportion is stored, how rapidly nutrients pass through the body, what fraction is devoted to growth, and how all of these change with time. The metabolic cycle may be pictured as shown in Figure 21.1a. The ingested material I, shown at the left of the diagram) has three fates. Some is not utilized at all (NU), as with the bones of small mammals eaten by owls. The remainder, the assimilated portion, consists of two parts: that devoted to respiration R (thus providing energy for the organism), and that devoted to the maintenance and growth of the organism's parts. The change in mass of the organism (its growth) is given by

$$\Delta M = I - E - R - NU \tag{21.1}$$

Analogously, the urban ecologist can, at least in principle, use the information in an urban metabolic analysis to gain perspective about steel beams, gasoline, water, food, and so on.

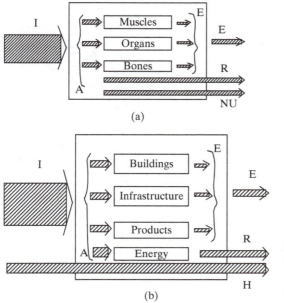

(a)

(b)

Figure 21.1

Metabolic flows of organisms (a) and of urban areas (b) I = ingestion, A = assimilation, E = egestion, R = respiration, NU = not used, H = hidden flows.

As with the squirrel, we wish to know the uses and fates of these "urban nutrients"—their rates of flow, their degree of spatial concentration, their modes of loss, their potential for reuse. Figure 21.1b is the urban version of Figure 21.1a. As before, there is ingestion (of crushed stone, diesel fuel, coffee makers, etc.) and egestion (of garbage, discarded cellular telephones, etc.). The assimilated material, as with the organismal analog, is of two parts. One is the energy sources—coal, fuel oil, natural gas—that provide energy to the urban system and generate respiratory products such as carbon dioxide. The second consists of the materials and products that enable our modern technological world to function. The urban organism changes mass with time according to

$$\Delta M = I - E - R \tag{21.2}$$

The difference between this formula and that for natural organisms is that there is no "not used" portion, at least if everything occurs as planned. A second difference is the arrow termed "hidden flows" (H); these represent the mobilization of waste rock and soil and the use of energy in the generation of the urban nutrients. These flows occur away from the city and so are hidden from it, but are properly chargeable to it.

A biological organism's metabolic flow consists largely of organic matter (which is often expressed in energy terms) and water. The flows connected to a city are much more complex. They include five components at minimum: nutrients, materials, water, energy, and dissipative flows, but these categories can be disaggregated considerably for various purposes. Nutrients can be subdivided into many food types, for example, and materials into a diversity of product groupings. Combining this broad scope with the challenge of assembling accurate and complete data renders an urban metabolic study a major challenge.

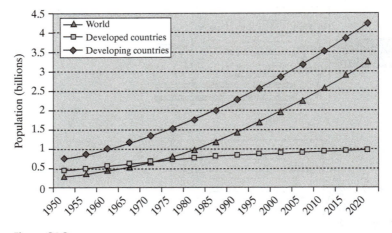

Figure 21.2

Urban population growth, 1950–2020. (Reproduced from United Nations, *World Urbanization Prospects: The 2001 Revision,* New York, 2001.)

Nonetheless, given that resources follow people, and people increasingly congregate in cities (Figure 21.2), the long-term sustainability of resource supplies, the ability to reuse them, and the environmental consequence of discarding them are all stories that will largely be told in cities.

21.2 URBAN METABOLIC FLOWS

A number of material flows are monitored at the national level, largely because taxes and import/export duties are collected at that level. No comparable information exists for cities, because there is little incentive for governments to collect it. As a result, urban metabolism information must be developed by determining typical information for households, businesses, and so forth, and then determining the number of each of those entities. This can be done for single nutrients or discards by evaluating all their flows into and from a given region.

The ability of a wealthy urban area to serve as an attractor for resources has been demonstrated by the metabolic analysis for London. In Figure 21.3 several of the metabolic parameters of that city are compared with those for the country and region in which it is located. London extracts almost nothing, of course, but imports more than twice as much per capita as the country or region, and it exports significantly more as well. Nonetheless, its urban character is reflected also by the fact that both its direct material input and direct material consumption are significantly lower.

A comprehensive urban metabolic study, representing both large and small flows into and from the urban region, has been described in some detail by Paul Brunner and his colleagues at the Swiss Federal Institute of Technology. They identified five sectors for investigation—agriculture, manufacturing, the private service sector, public services, and private households—and 12 materials—lead, phosphorus, water,

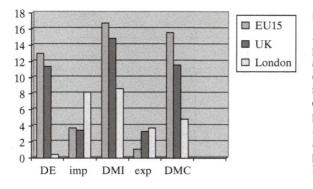

Figure 21.3

A comparison of metabolic parameters (megagrams per capita in 2000) for London, the United Kingdom, and the EU-15. The abbreviations are DE = domestic extraction, DMI = domestic material input, imp = imports, exp = exports, DMC = domestic material consumption. (The data are from H. Weisz, et al., The physical economy of the European Union, *Ecological Economics, 58*, 66–698, 2006, and B. Bongardt, *Material Flow Accounting for London in 2000*, paper presented at ConAccount Workshop 2003, Wuppertal Institute, Germany.)

construction materials, scrap, and so forth. They then used market research and personal interviews with representatives of the sectors in a specific Swiss region to estimate typical flows of the materials. Such an approach cannot have very high accuracy, of course, but it demonstrated clearly that water was the dominant input in mass terms, air (for combustion) was next, and construction materials followed. The largest outputs were waste water, carbon dioxide, and export goods. A significant fraction of input flow (water and air excepted) was being added to in-use stock. For lead, for example, the largest input flow was in scrap metal, and the largest output flow was in filter dust from the metals industry. For phosphorus, the largest input flow was in fertilizer and the largest output was the flow of biosolids to the river. The authors identified the largest uncertainty as a lack of information about the material composition of intermediate products and consumer goods.

Urban metabolic studies can also focus on the detailed analysis of a single resource. Such a study, for nitrogen in the Central Arizona–Phoenix ecosystem for year 2000, is shown in Figure 21.4. Of the 98 Gg/yr of fixed nitrogen entering the ecosystem, 52 percent was deliberate human inputs (human and animal food, nitrogen-containing chemicals, alfalfa fixation, and commercial fertilizers) and 36 percent was inadvertent human inputs (NO_x from combustion processes). Two-thirds of the losses were products of denitrification. The cycle contains many other details, all of them useful for aspects of science and policy.

21.3 URBAN METABOLIC STOCKS

Many measurements or estimates in science and technology are derivative; that is, they measure something related to what one wishes to determine rather than measuring the actual entity of interest. Perhaps the most famous examples of derivative measurements are those related to the structure of the atom—quarks and gluons are well understood because of analyses of the signatures of various types of collision fragments in atom accelerators, not because the entities have been weighed on a balance.

From the perspective of the atomic derivative measurements, we can address our desire to know the stocks of urban nutrients, especially materials of interest like copper or lead that exist in long-lived uses like building wiring and batteries. If we had good information on inputs and outputs, we could take the difference to compute stocks

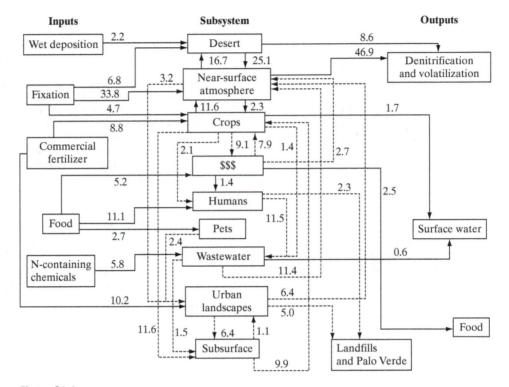

Figure 21.4

The nitrogen cycle for the Central Arizona–Phoenix urban region. The units are Gg N/yr. Internal transfers are shown as dashed lines, ecosystem inputs and outputs as solid lines. (Reprinted with permission from L.A. Baker, D. Hope, Y. Xu, J. Edmonds, and L. Lauver, Nitrogen balance for the Central Arizona–Phoenix (CAP) ecosystem, *Ecosystems, 4*, 582–602, 2001.)

over time. (This is the top-down method of Chapter 17; the analogous problem for the atmosphere is computing CO_2 concentrations by measuring or estimating the CO_2 emissions to the atmosphere and the rate at which CO_2 is removed by such processes as the growth of vegetation.) When input and output information for a long period of time is assembled, the difference (the standing stock) can be calculated. Only rarely do we have time series of such information for cities, however, either long or short. In addition, we would like to know more than just the amounts—we would like to know with some certainty where the material is located within the city. Such knowledge serves two purposes—it could make it easier to collect and reuse the material, and (if the material is environmentally toxic) it could help to anticipate environmental problems should some of the material be dissipated.

To determine the magnitude of stocks, we turn to inventories of products in which the resources occur, that is, to the bottom-up method of Chapter 17. For copper, for example, we determine the number of residences, automobiles, power distribution systems, and so forth in an urban area, often from GIS maps. We then multiply those

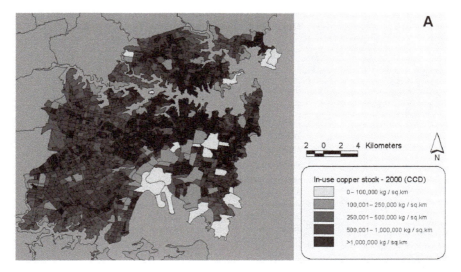

Figure 21.5

The spatial distribution of in-use copper in census collection districts in Inner Sydney, Australia, in 2000. (Reprinted with permission from D. van Beers, and T.E. Graedel, Spatial characterization of multi-level in-use copper and zinc stocks in Australia, *Journal of Cleaner Production, 15*, 849–861, 2007.)

reservoir numbers by typical copper concentrations within them to arrive at an estimate of both total copper stock and its location. (This is analogous to determining atmospheric CO_2 stock by direct measurement, as is done commonly.) If desired, spatial techniques can be used to distribute the stock determinations over the urban area. An example is shown in Figure 21.5, which is a study of copper in Inner Sydney, Australia. Large spatial variations are seen, reflecting the location of businesses, infrastructure, residential areas, and so forth. The result enables government officials to plan more wisely for the eventual recovery of those resources by knowing where they are located, when it is likely to be discarded, and how much can be expected to be available.

21.4   URBAN METABOLIC HISTORIES

An urban metabolic history monitors the physical metabolism of a city as it undergoes changes in population and state of development over time. We have only three extant examples of such studies, each for two example years, for Hong Kong, Sydney, and Toronto. On an absolute and on a per capita basis, inputs and outputs in Hong Kong and Sydney increased markedly over the 20 or so year intervals. (That for Sydney is shown in Table 21.1.) Toronto has much higher per capita input and output rates than do the other two cities, but those rates changed little over the 12-year interval of the Toronto determinations. Differences in methodology and level of detail make it very difficult to do more than give general impressions about this information, but even what little can be done is fascinating and demonstrates the great promise of urban metabolic histories for informing specialists and policy makers alike.

TABLE 21.1   Per Capita Material Flows in Sydney, Australia, in 1970 and 1990.

	1970	1990
Inputs		
Energy (GJ)	88.6	114.2
Food (Mg)	0.23	0.22
Water (Mg)	144	180
Outputs		
Solid waste (Mg)	0.59	0.77
Sewage (Mg)	108	128
Air waste (Mg)	7.6	9.3

Source: P.W.G. Newman, Sustainability and cities: Extending the metabolism model, *Landscape and Urban Planning, 44*, 209–226, 1999.

Case Study: Increasing Consumption and its Consequences

The propensity toward increasing consumption and its consequences can be appreciated from a study performed over nearly 30 years in Hong Kong. In the work, the urban metabolism was measured in 1971 and again in 1997, concentrating on resource consumption and waste generation. During that period, the population of the city rose from 3.94 million to 6.62 million. Equally dramatically, but perhaps somewhat less visibly, the per capita consumption of food, water, and materials increased by 29 percent, 40 percent, and 150 percent, respectively over that period, and total air emissions, CO_2 outputs, municipal solid waste, and sewage increased substantially, as shown in the table below. Major investments in infrastructure were required in order to sustain these increased flows without major environmental and human health consequences (Table CS-1).

TABLE CS-1   Selected Material Flows in Hong Kong, 1971 and 1997 (kg cap^{-1} yr^{-1})

Material flow	1971	1997
Food	570	680
Fossil fuel	1,000	2,000
Construction materials	1,000	3,800
Freshwater	99,010	137,980
Solid waste	782	2,086
Sewage	73,115	102,310
CO_2 emissions	2,285	4,776

Source: K. Warren-Rhodes, and A. Koenig, Escalating trends in the urban metabolism of Hong Kong, 1971–1997, *Ambio, 30*, 429–438, 2001.

21.5   URBAN MINING

A key concept in the discussion of resource reuse is that materials are used where people are located, and this is increasingly in cities. This means that materials in use present a potentially attractive opportunity for eventual resource recovery. Currently, the recycling of materials is largely opportunistic and short-term. However, the increasing urbanization of the global population is concentrating the materials in use, or, we might say, enriching the grade of the "urban ore bodies." This situation implies that a more knowledgeable and structured approach to resource recovery and reuse may be increasingly rewarding.

The specter of the exhaustion of the resources on which we have come to count has fostered an intellectual debate for decades. However, the looming scarcity of common materials seems increasingly possible. If it is difficult to predict the epoch in which exhaustion (or at least scarcity) of virgin geological resources will occur, it is undeniable that the stock is shrinking. Concomitantly, as cities grow and as their inhab-itants become more affluent, the in-use resource stock in cities keeps rising. Nonethe-less, cities are unlikely to be mines for everything. Silver and copper are surely concentrated in cities, for example, but cadmium's locations are more mixed—NiCd batteries are largely in cities, but cadmium-plated landing gear for aircraft resides at military and civilian airports, for example. For a few materials, such as uranium and plutonium, uses and locations in nuclear power plants and in national defense arsenals (as well as the inevitable radioactive decay) render recyclability moot and "urban ore" in cities for these materials nonexistent.

In-use stock is discarded only when it no longer has utility for the owner, and this occurs after quite different lengths of service, depending on the product. For example, the principal uses of copper have service lifetimes varying from about 8 years to about 50. This fact introduces a variable time delay into the system, meaning that material often becomes available for urban mining a decade or more after its initial entry into the system. This delay must be considered in any efforts to quantify the amount of mate-rial likely to be available for reuse and the forms in which it will be discarded.

Urban mining, or what might be termed "spatially oriented recycling," has a number of specific benefits for society. These include

- Mitigating problems of resource depletion by providing an alternative source
- Avoiding (for recycled materials) the environmental impacts of mining
- Saving energy
- Saving water

What are the practical requirements for implementing the extensive mining of urban ore? There are at least five:

- An accurate determination of in-use magnitudes or stocks of urban materials, good estimates of the discard flow rates as a function of time, and specification of the physical and chemical forms of the discards
- An accurate determination of the spatial distributions of the urban stocks

- The design of buildings, equipment, and products of all kinds so as to facilitate disassembly and reuse
- A system of efficient collection and sorting of discards
- The provision of incentives, monetary or otherwise, for material reuse

Can urban mines replace those of nature? This cannot happen so long as the total usage of materials is increasing, because even if 100 percent of discarded material is captured for reuse, this material, reentering the system after some years of service, can at best provide only a fraction of the total requirements. This situation can approach sustainability only as the total annual requirements for materials approach constancy.

Lastly, if we want to mine the cities, should we mine *all* cities? We do not know whether stock in cities scales linearly with population, how stock differs as a function of city wealth or location, or whether large cities tend to have more urban ore per capita or less. One could imagine unique situations as well; for example, that large commercial cities might be rich in copper, gallium, and other materials as a result of a large concentration of computers and other industrial electronics. A local manufacturing industry might result in a high density of steel. A community centered on agriculture might have lower densities of some of these technologically advanced materials. There is clearly a research challenge in characterizing these diverse urban stocks.

To have urban mines, we need urban miners and urban mining companies. Such entities exist now, but they are smaller than they might be, and less efficient. It is clear, however, that there is significant potential for urban mining as a contribution toward sustainability. It will take initiative, effort, and political will to make extensive urban mining a reality.

21.6   POTENTIAL BENEFITS OF URBAN METABOLIC STUDIES

We understand plants and animals in part because their metabolic information allows us to understand the functioning of those organisms. The same potential holds true for studies of urban physical metabolisms. The lure of such investigations is immediately apparent: allowing researchers and policy makers alike to examine what (and how much) enters a city, how long it is retained, and when and how it is discarded. This information has potential for informing discussions of resource availability, recycling, energy use, and environmental impact. It could enable the comparison of one city's performance with that of another, or with itself over time.

A comprehensive characterization of urban metabolism includes the need to measure not only the combined flows of all forms of materials, but the flows of *each* form. This is crucial because for recycling and reuse purposes the viability of the stock of a material generally depends on the efficiency with which it can be recovered. A material discarded in pure form, such as copper pipe, is reused readily. The same amount of material in complex manufactured products—cellular telephones, for example—may be much more likely to be consigned to the landfill. Impacts to the environment are also highly dependent on chemical form—a chromium-plated bumper is environmentally benign, but the Cr^{6+} ion is not.

As with any organism, a city's metabolism is characterized by the quantities of flows of resources, their distribution throughout the organism, and the dynamic nature of the flows. The sustainability of a city cannot be evaluated unless those factors are quantified. If they are, they define the urban industrial ecology of the city and permit decisions to be made on an informed basis.

FURTHER READING

Bessey, K.M., Structure and dynamics in an urban landscape: Toward a multiscale view, *Ecosystems, 5*, 360–375, 2002.

Grimm, N.B., et al., Global change and the ecology of cities, *Science, 319*, 756–760, 2008.

Jacobs, J., *The Economy of Cities*, New York: Random House, 1970.

Kaye, J.P., P.M. Groffman, N.B. Grimm, L.A. Baker, and R.V. Pouyat, A distinct urban biogeochemistry? *Trends in Ecology and Evolution, 21*, 192–199, 2006.

Petaki, D.E., et al., Urban ecosystems and the North American carbon cycle, *Global Change Biology, 12*, 2092–2102, 2006.

Shen, L., S. Cheng, A.J. Gunson, and H. Wan, Urbanization, sustainability, and the utilization of energy and mineral resources in China, *Cities, 22*, 287–302, 2005.

Simon, D., Urban environments: Issues on the peri-urban fringe, *Annual Review of Environment and Resources, 33*, 167–285, 2008.

Wolman, A., The metabolism of cities, *Scientific American, 213* (3), 179–190, 1965.

EXERCISES

21.1 Devise a diagram along the lines of Figure 21.4 for the contemporary metabolism of copper in a technologically advanced city (copper is used primarily in wiring, plumbing, and power distribution). Identify any assumptions you needed to make.

21.2 Repeat Exercise 21.1 for a poor but rapidly developing city. If you were mayor, does this diagram suggest any policy initiatives?

21.3 Is the movement of people from rural to urban locales positive or negative for the environment? Why?

21.4 Some cities around the world refer to themselves as "sustainable cities." If you were mayor of one of them, how could you justify this appellation?

CHAPTER 22

Modeling in Industrial Ecology

22.1 WHAT IS AN INDUSTRIAL ECOLOGY MODEL?

Once a system of interest has been defined, the logical next step is to want to determine how the system works. In the case of the automotive system, for example, how is automotive use related to the characteristics of infrastructure or environment or the culture within which it must operate? The intellectual construct that attempts to describe those interrelationships is termed a "model."

A model is a representation or description designed to show the structure or operation of an object or a system. In its initial formulation, a model is *conceptual* and speaks to the structural aspects of the definition. Such a construction seeks to represent the components of the object or system of interest, and the potential for interaction among the components. The food web representations, ecological and industrial, of Chapter 16 are examples of conceptual models.

If we wish to understand not only the structure of an object or system but also its operation, we must comprehend and represent not only the *potential* for interaction, but also the *magnitude* of the interactions among the system components. The model is then no longer conceptual, but *mathematical*, and is almost always realized on the computer.

We emphasized in the discussion of systems in Chapter 15 that industrial ecology in its fullest realization deals with holarchic systems that connect technology with human actions. Because subsequent interactions with the environment are always part of the perspective, we expand the industrial holarchic systems of Figure 7.1 to incorporate the environment, as shown on the left panel of Figure 22.1a. We can add a bit of structure to our thinking with the diagram in Figure 22.1b. Here the technological activities of interest form the horizontal portion of the diagram; the human actions that motivate the

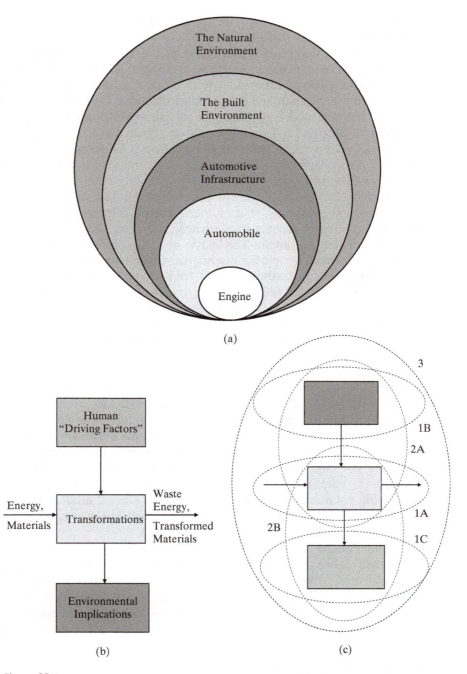

Figure 22.1

(a) Adding the environment to form an industrial ecology holarchy; (b) The basic components of industrial ecology models; (c) Classes of models in industrial ecology (Class 2C, combining human driving factors and the environment, is not shown).

technological activities constitute the top box; and the environmental responses to these activities constitute the bottom box.

It turns out that many models, even quite detailed ones, address only part of this framework. Those that attempt to represent only a single entity, even if they acknowledge that connections with others exist, we term "Class 1" models. In industrial ecology, Class 1 models focus on materials and energy transformations; these are the Class 1A models of Figure 22.1c. These models are predictive and purposeful, from the perspective of the classifications of Chapter 15. A model focused on human decision making related to technology, but not addressing the technology itself from a modeling perspective, is a Class 1B model. Similarly, an environmental model recognizing but not directly addressing technology is a Class 1C model.

Figure 22.1b provides one box for each of the three components—technological, human, and environmental—but of course each is a holarchy itself. The technological holarchy—piece part, product, infrastructure, and so on—is, in principle, perfectly predictive, and definitely purposeful. The human holarchy—individual, family, community, and so on—is purposeful as well, but not predictive. The environmental holarchy—organism, community ecosystem, landscape ecosystem, and so on—is neither purposeful nor predictive. True emergent behavior is not possible in the technological holarchy, but is expected in both of the others, and in combinations of the technological holarchy with either of the others. In all cases, these are "open" systems—those in which energy and materials can be exchanged across the system boundaries (recall the diagram of the Types I, II, and III ecosystems in Chapter 4).

Models increase in complexity and in interest when two of the components of Figure 22.1b are incorporated to a significant degree. We term these constructs "multidisciplinary industrial ecology models" (MIEMs). They might comprise technological and human components (Class 2A models) or technological and environmental components (Class 2B models). (We also acknowledge the existence and utility of Class 2C models, treating both human and environmental components, but these are not industrial ecology models because they do not address technology). Because the human and environmental components are not predictive, they are self-organizing holarchic open (SOHO) entities, so Class 2 models could be designated MIEM/SOHO. They can also be considered *collaborative* models, as they are not organized around a centrally managed purpose.

The ultimate in industrial ecology models is the incorporation of all three components, each realized in some detail. These are designated Class 3 and are *comprehensive industrial ecology models* (CIEMs). As with Class 2 models, they can be termed CIEM/SOHO entities, and they are collaborative. Class 3 models address the central, overriding question in industrial ecology: "How does individual human choice manifest itself in the interactions that involve technology, the environment, and sustainability?"

22.2 BUILDING THE CONCEPTUAL MODEL

22.2.1 The Class 1 Industrial Ecology Model

Given that we understand what a model is, how does one get built? Ideally, a model is realized in a sequential process, initiated by defining the question we wish the model to answer. It is unrealistic, of course, to expect a model to help understand the totality of structure and operation of any but the simplest systems, and holarchies are not simple. In addition, models require data to guide them and enable their functioning, and data are very often limited. Identifying an important and interesting question that will justify and guide the construction of the model is therefore a challenging activity, a balance between expansive desires and realistic limitations.

Let us illustrate model building by choosing to ask, "What fraction of lead in discarded computers in Country A was recycled in year 2000 within the country?" (Computers have historically used tin–lead solder to fasten electronic components.) This seemingly straightforward question requires a model because, typically, only some of the information is available. Answering the question requires only two pieces of information: how much lead was discarded in computers in year 2000, and how much of it entered the recycling stream. In practice, we may instead know how many computers were discarded but not their lead content, and we may know how much lead was recycled in the country but not how much of it came from computers. (It may help to refer to Figure 17.3 in this regard.) Nonetheless, the answer to this question could be of interest for any of a number of reasons: (1) a lead recycler in another country wants to evaluate a business opportunity; (2) an environmental organization wants to estimate damage from lead loss; (3) computer manufacturers want to understand their potential liability, and so on. We thus have identified a question whose answer requires a model, the first step in the sequence shown in Figure 22.2.

The second step in model building is that of "scale." (This term and others are defined in Table 22.1), which is where the builder determines what sort of information is needed to respond to the question. In our example, four scales are required: the number of discarded computers, the lead content of the computers, the number of computers recycled, and the total lead recycling flow.

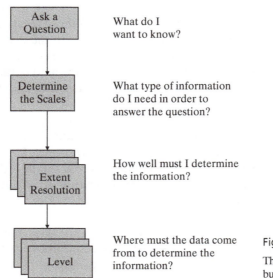

Figure 22.2

The sequence of decision making in building a conceptual model.

TABLE 22.1   Definitions of Terms Related to Model Building and Data Acquisition

Term	Definition
Scales	The spatial, temporal, quantitative, or analytical dimensions used to measure and study an object or process of interest.
Extent	The size of a scale dimension.
Resolution	The precision used in measurement.
Levels	The units of analysis that are located in the same position on a scale. Many conceptual scales contain levels that are ordered hierarchically, but not all levels are necessarily linked to one another in a hierarchical system.

Source: Adapted from C.C. Gibson, E. Ostrom, and T.K. Ahn, The concept of scale and the human dimensions of global change: A survey, *Ecological Economics, 32,* 217–239, 2000.

Next, we consider the *extent* of the information. In all cases it is for year 2000 and, if available, for the country as a whole. The number of computers discarded and recycled is straightforward as to extent, as is the total recycling flow. However, computers are made by different manufactures and have differing lead contents. It is unrealistic to determine the elemental composition of every computer, so a decision on extent is required—perhaps to take the average lead content of computers made by the top four manufacturers from 1993 to 1997 (the approximate manufacturing period for computers discarded in 2000).

The *resolution* relates to how accurately we need to know the information we seek. This decision often has major implications for the time required to gather and verify the information. In the present case, we may choose to be satisfied with lead content figures and computer counts with estimated uncertainties of 20 percent.

A further consideration is of the *level* from which information is needed. In our example, the entire country is the *desired* level because that corresponds precisely to the question. If data are only available at province or city level, however, or if it is anticipated that the original question may be expanded to those levels in the future, a multilevel approach to data acquisition may be required. With the decision regarding level having been made, our conceptual model is complete. The analytical model is then constructed by defining the mathematical relationships among the data entities (e.g., discarded lead equals number of discarded computers multiplied by the average lead content).

Two types of variables are present within models. One is the *exogenous* variables, whose values are specified by the model builder. These might be, for example, the rate of virgin resource use, or the cost of a resource. The second type is *endogenous* variables, whose values are calculated by the model—the amount of steel in recycled automobiles, for example. The specification of exogenous variables is necessary to run the model; the values calculated for endogenous variables are the result.

Questions often have a dynamic component, such as "Has the recycled lead function changed over time?" If so, all the data discussed above will have a time dimension, a very significant addition of effort and complexity.

22.2.2 The Class 2 Industrial Ecology Model

The model described above is a Class 1 model, because it directly addresses only the technological aspects of lead use. Suppose, however, that the question that was asked was a broader one, such as "What are the environmental impacts of the human use of lead in Country A in 2000?" The expansion here is twofold. First, we are asked to consider all the ways in which humans use lead, not just in computers, because we imagine that Country A still used leaded gasoline and lead paint in 2000 (an expansion of our Class 1 model), and we asked a detailed environmental question as well (which means that a Class 2 model is required). Our hypothetical model is thus linked to issues of risk assessment, as discussed in Chapter 6.

After a bit of thought we tighten the question we have been asked, so as to make the model tractable and decide to address lead toxicity in humans and animals, rather than all possible lead impacts. We then consider scale and immediately realize that people and animals are not distributed uniformly in Country A, nor are lead emissions, so the model must have two spatial dimensions. The scales for which spatially resolved information will be needed are (1) human population, (2) animal population, (3) lead emissions. We also need nonspatial information on two other scales: lead sensitivity of humans and lead sensitivity of animals. And, of course, we need to retain the scales from the computer-centered Class 1 model we described earlier.

We will not belabor this example further, except to say that there will be questions of extent (what is the minimal lead concentration to be considered), of resolution (what spatial resolution is required, and are data available at that resolution), of exposure (are dose-response data available), and perhaps of level (can lead from gasoline be considered on anything but a regional basis and can animal toxicity be considered on anything but a patch basis).

22.2.3 The Class 3 Industrial Ecology Model

It might be the case that the question posed to the model builder is broader still: "How do human choices produce the environmental impacts of lead in Country A in 2000?" Here the question retains the features that were addressed by the Class 1 and Class 2 models, but the addition of the human element elevates the challenge to that of a Class 3 model. We must now address an additional set of scale dimensions: the human choice of how much to drive, how much lead paint to use, and whether to buy a computer with a (leaded) cathode ray tube display or a (lead-free) liquid crystal display, for example. Extent issues may emerge as well (sampling only a portion of the population), as do those of resolution (state or neighborhood) and level (educational level of the responders).

With the entry of human behavior into the picture (as was also the case with the Class 2A and 2C models), the mathematical model must add stochastic aspects to its previously analytic framework. This is generally done with "agent-based models," in which each decision maker (the agent) is given a certain probability of making a decision that has implications for the question the model is seeking to answer. This approach recognizes that different agents aim for different goals and so do not necessarily make the same decisions, yet interact directly by sharing information and services and indirectly by sharing the resources of the system. The model creates an assembly of agents; a protocol for agent interactions; and connections between those interactions, their associated technologies, and the environment. Properly designing such a model requires detailed knowledge of technological holarchies, human dynamics, and environmental processes, and is supremely interdisciplinary.

22.3 RUNNING AND EVALUATING INDUSTRIAL ECOLOGY MODELS

22.3.1 Implementing the Model

Models in industrial ecology typically take one of three forms: system dynamics, linear programming, or agent-based modeling. In all cases, the system of interest is defined by the mathematical relationships linking the exogenous and endogenous variables. In the system dynamics case, the goal is to solve a set of equations of the form

$$e_i = F_i(x_1, x_2, \ldots, x_n) \tag{22.1}$$

where no limits are placed on the form of F. In most cases in industrial ecology, however, the equations are linear. [Substance flow analysis models are an example, as in the stocks and flows related to the reservoir in Figure 17.1b, where $B = (A + B) - (C + D)$]

The goal in linear programming, a technique often employed in economic models, is to maximize a linear function whose variables are required to satisfy a system of linear constraints, that is,

$$z = F(x_1, x_2, \ldots, x_n) \tag{22.2}$$

Where

$$a_{i,1}x_1 + a_{i,2}x_2 \ldots + a_{i,n}x_n \leq b_i \tag{22.3}$$

and z is the quantity to be maximized.

In the case of agent-based models, the equations of the form of Equation 22.1 represent the sums of decisions made by individual agents, who might be potential customers for a specific product, managers deciding whether to build a new facility, or automobile owners deciding when to discard their vehicle and replace it. The mathematical functions are thus the probabilities that certain decisions will be made. In such systems, the lack of deterministic structure allows for the emergence of unpredicted behavior (as discussed for complex systems in Chapter 15).

Computer code can be written in order to implement models of all types, but convenient software packages are available for all but the most complex problems. These include STELLA, VENSIM, and SIMULINK for systems analysis models, SIMPLEX for linear programming models, and ANYLOGIC and SWARM for agent-based models.

22.3.2 Model Validation

How do we know whether model results are valid? The typical output of a model consists of computed values for a number of dependent variables. Ideally, these results are compared with actual data for conditions similar to those simulated by the model. In our example, one might investigate whether the actual flows of lead scrap are consistent with the model results. If so, an initial measure of confidence is provided. If the model fails to reproduce the data then the model is faulty in some way, and studying those discrepancies often provides clues to increased understanding. Finally, the model can be used to make a prediction (e.g., lead recycling in 2010), and the prediction evaluated by actual measurement. When a large number of such steps have been carried out, more or less successfully, the model is said to have been *validated*. The model results can then be regarded as likely to be good representations of the actual situation.

A model that is validated is one that does not contain known or detectable flaws and is internally consistent. That is not to say that the model results necessarily reflect the behavior of the real world. The results depend in part on the quality and quantity of the inputs, the accuracy with which processes are represented, and the extensiveness of model testing. A model might generate results that are reasonable and useful on one spatial or temporal scale, but not at another. Depending on the purpose of the model, such performance may be perfectly satisfactory.

Naomi Oreskes of Dartmouth College and her colleagues have written on the uses to which models can legitimately be put, even if they cannot be regarded as "perfectly correct." They say

> Models can corroborate a hypothesis by offering evidence to strengthen what may be already partly established through other means. Models can elucidate discrepancies in other models. Models can be also be used for sensitivity analysis—for exploring "what if" questions—thereby illuminating which aspects of the system are most in need of further study, and where more empirical data are most needed. Thus, the primary value of models is heuristic: Models are representations, useful for guiding further study but not susceptible to proof.

22.4   EXAMPLES OF INDUSTRIAL ECOLOGY MODELS

Class 2 models explicitly address transformations of energy and/or materials (the center box on Figure 22.1b) plus either the human driving factors or the environmental implications. This can be done in either of two ways. If the models for each of the two areas of focus are *linked*, they are run separately but the output of one serves as input to the other. If the models are *coupled*, they are run simultaneously, with continuous feedback between them. Coupled models are generally preferred to linked models, as they are more accurate representations of reality, but tend to be considerably more difficult to construct, run, and interpret.

In Chapter 17 we showed a characterization of the life cycle of nickel, derived for 52 countries worldwide. This work actually incorporated a Class 1A model, because data for some of the flows were not available. The results demonstrated a new perspective on global nickel trade, achieved by comparing nickel trade flows as derived from the model for different life stages. Figure 22.3 plots the net import flows of nickel in final products as a function of the sum of those of primary and intermediate nickel goods. The size of the dots reflects the summarized net import flows of primary, intermediate, and final nickel. It becomes apparent that most

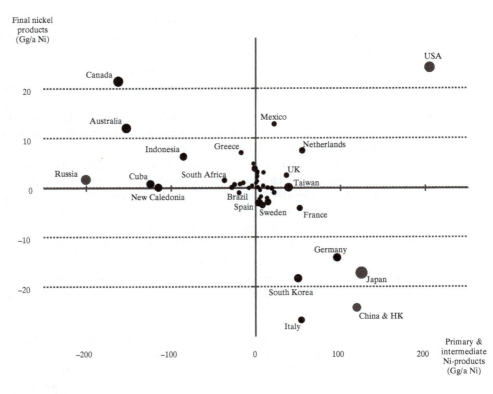

Figure 22.3

A scatterplot of traded nickel products, showing the net import of nickel in final goods as a function of the net import of primary plus intermediate nickel goods. The size of the bubbles reflects the sum of the three net imports (primary, intermediate, final). (Reproduced with permission from B. Reck, et al., The anthropogenic nickel cycle: Insights into use, trade, and recycling, *Environmental Science & Technology, 42*, 3394–3400, 2008.)

countries either export their primary and intermediate nickel products and import final goods (upper left quadrant; e.g., Canada, Australia, Russia) or vice versa (lower right quadrant; e.g., Japan, China). Only a few countries import at all levels, with the United States standing out as being the largest importer at both levels shown.

Class 2A models in industrial ecology treat both the transformations of materials and/or energy and the human driving factors that influence them. An example is the study of regional bio-based management in Canton Zurich, Switzerland. The transformation model was designed to explicate the flow of biowaste through composting or anaerobic digestion to final users. In parallel, the flows of money that accompanied the flows of material were computed. Finally, a questionnaire-based interview process measured the current use of fresh and mature compost and the potential for that rate of use to increase.

The biowaste flow model results for 2002 are shown in Figure 22.4a. Nearly equal amounts of the material was composted and anaerobically digested. Farmers were by far the predominant consumer group, receiving roughly equivalent amounts of compost and digested matter. More compost was lost to decay than was reused.

The money flow model results in Figure 22.4b present quite a different story. Large sums were paid by the bio-waste generators to the processors, but subsequent consumers (professional gardeners, hobbyists) either paid small sums or were paid by the processors (farmers) for receiving the processed output. The subsequent interview process suggested that some of the customers could be willing to receive additional processed bio-waste provided they were properly informed and that the financial implications were favorable.

Class 2B modeling in industrial ecology is exemplified by the Dutch combination of FLUX, a model for the flows and accumulations of metals in The Netherlands, and DYNABOX, an environmental multimedia model. An example of some

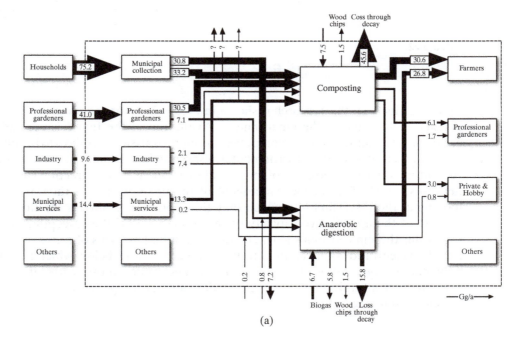

(a)

Figure 22.4

Results of a quasi-stationary flow model of biowaste and biowaste transformation products in Canton Zurich, Switzerland, 2002; (b) The money flow model corresponding to the system of Figure 22.4a; (Reproduced with permission from D.J. Lang, et al., Material and money flows as a means for industry analysis of recycling schemes: A case study of regional bio-waste management, *Resources, Conservation, and Recycling, 49*, 159–190, 2006.)

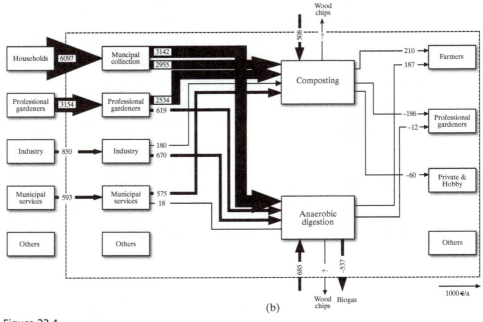

Figure 22.4

of the results from this research is shown in Figure 22.5, for cadmium in the anthropogenic system and in the environment. At the top of the figure is the cadmium cycle for The Netherlands. It computes cadmium emissions to the environment as shown, together with the environmental reservoirs into which the cadmium flows. At bottom left, the disposition of impact flows is shown, a large fraction (net) entering in-use stock. The second component of the model classifies the emissions by looking at the reservoirs into which they flow and then evaluates the flows in the context of the assimilative capacity of the reservoirs and the susceptibility of the organisms therein (bottom right). The risks are stated in terms of ratios of the computed concentrations of cadmium to those concentrations below which no effect is anticipated, hence ratios above unity are of concern and those below are not. The flows to agricultural lands are of particular interest for policy, as are the human toxicity potentials.

22.5 THE STATUS OF INDUSTRIAL ECOLOGY MODELS

Industrial ecology models, nonexistent a decade ago, are now becoming more common and are beginning to provide results that are useful for decision makers. This is particularly true of the Class 1A analytic models arising from material flow analysis. Thus, model building, at least to a modest degree, has become an inherent part of material flow analysis. The Class 1A models are still far removed from the complexities of state-of-the-art Class 1B and 1C models, but are developing rapidly.

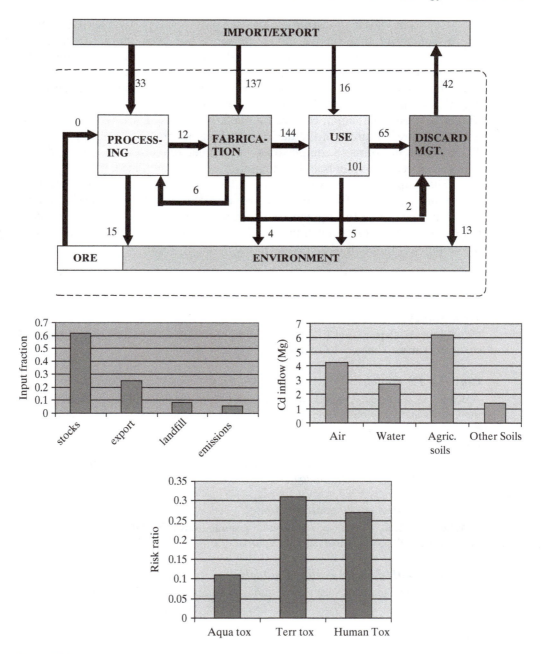

Figure 22.5

Results of a Class 2B industrial ecology model for cadmium in The Netherlands in 1990. The cadmium cycle derived by the FLUX SFA model is at the top, and its computed emissions (bottom left) and their fates (bottom center) serve as inputs to the DYNABOX ecosystems model, which computes several risk ratios (see text). The mass balances in the cycle are not perfect because of data limitations and rounding. Source: E. van der Voet, J.B. Guinée, and H. A. Udo de Haas, *Heavy Metals: A Problem Solved?* Dordrecht: Kluwer Academic Publishers, 2000.

Class 2 models in industrial ecology are rare. This is partly due to the challenge of interdisciplinarity that such a model presents, but even more to the problem of mismatches of scale, extent, and resolution. It is perfectly feasible (and appropriate for many purposes) to carry out a technological model with a focus on an entire country. That approach is seldom suitable for a technology-environment model, however—an aquatic system behaves much differently than does a desert or a grassland. The same challenge exists for human-centered models, which recognize that national averages are poor representations of the decision making of urbanites, or the poor, or the old. The consequence of these complexities is that much remains to be done to devise Class 2 industrial ecology models and to demonstrate their usefulness.

There appear to be, as yet, no true Class 3 industrial ecology models. This situation is not surprising for a relatively young field. However, increasingly important questions related to the human transformations of materials and energy are being asked, and models offer much potential for responding to them. We anticipate rapid development of models of all classes in industrial ecology.

FURTHER READING

Terminology:

C.C. Gibson, E. Ostrom, and T.K. Ahn, The concept of scale and the human dimensions of global change: A survey, *Ecological Economics, 32*, 217–239, 2000.

Class 1A IE Model:

A. Kapur, G. Keoleian, A. Kendall, and S. Kesler, Dynamic modeling of in-use cement stocks in the United States, *Journal of Industrial Ecology, 12*, 539–556, 2008.

Class 1A IE Model:

B. Reck, et al., The anthropogenic nickel cycle: Insights into use, trade, and recycling, *Environmental Science & Technology, 42*, 3394–3400, 2008.

Class 1B (nonindustrial ecology) model:

G.A. Schmidt, et al., Present-day atmospheric simulations using GISS Model E: Comparison to the in situ, satellite, and reanalysis data, *Journal of Climate, 19*, 153–192, 2006.

Class 1C (nonindustrial ecology) model:

J. Delgado, Emergence of social conventions in complex networks, *Artificial Intelligence, 141*, 171–185, 2002.

Class 2A IE model:

D.J. Lang, et al., Material and money flows as a means for industry analysis of recycling schemes: A case study of regional bio-waste management, *Resources, Conservation, and Recycling, 49*, 159–190, 2006.

Class 2B IE model:

J.B. Guinée, et al., Evaluation of risks of metal flows and accumulation in economy and environment, *Ecological Economics, 30*, 47–65, 1999.

Class 2C (nonindustrial ecology) model:

A. Voinov, et al., Patuxent landscape model: Integrated ecological economic modeling of a watershed, *Environmental Modeling and Software, 14*, 473–491, 1999.

System dynamics model methodology:

M. Ruth, and B. Hannon, *Modeling Dynamic Economic Systems*, New York: Springer, 1997.

Linear programming model methodology:

D. Gale, Linear programming and the simplex method, *Notices of the AMS, 54*, 364–369, 2007.

Linear programming model methodology:

J.W. Chinneck, *Practical Optimization: A Gentle Introduction*, http://www.sce.carlton.ca/faculty/chinneck/po.html, 2001, accessed December 12, 2007.

Agent-based model methodology:

C.J.E. Castle, and A.T. Crooks, *Principles and Concepts of Agent-Based Modelling for Developing Geospatial Simulations*, Working Paper 110, Centre for Advanced Spatial Analysis, University College London, 2006, http://www.casa.ucl.ac.uk/working_papers/paper110.pdf, accessed January 3, 2008.

Agent-based model methodology:

F. Bousquet, and C. LePage, Multi-agent simulations and ecosystem management: A review, *Ecological Modeling, 176*, 313–332, 2004.

Agent-based modeling in manufacturing:

W. Shen, Q. Hao, H.J. Yoon, and D.H. Norrie, Applications of agent-based systems in intelligent manufacturing: An updated review, *Advanced Engineering Informatics, 20*, 415–431, 2006.

Model structure and philosophy:

N. Oreskes, K. Shrader-Frechette, and K. Belitz, Verification, validation, and confirmation of numerical models in the earth sciences, *Science, 263*, 641–646, 1994.

Model structure and philosophy:

R. Costanza, L. Wainger, C. Folke, and K.-G. Mäler, Modeling complex ecological economic systems: Toward an evolutionary, dynamic understanding of people and nature, *BioScience, 43*, 545–554, 1993.

MODELING SOFTWARE WEBSITES

Stella	http://www.iseesystems.com
Vensim	http://www.vensim.com
Simulink	http://www.mathworks.com/products/simulink
Anylogic	http://www.xjtek.com
Swarm	http://swarm.org/wiki/ABM_Resources

Note: These sites were current as of January 3, 2009. This listing is for the convenience of the reader and does not imply endorsement of the software.

EXERCISES

22.1 What industrial ecology model class is each of the following? Why?
- Characterizing the Japanese annual cycle of tin.
- Describing the chemistry of the Antarctic ozone hole.
- Understanding human demand for water in Arizona, the way in which water is provided, and the environmental implications of doing so.
- Quantifying the environmental effects of zinc ore extraction and refining.

22.2 A friend poses the following question: "How have the transport choices of Chinese citizens influenced China's need for high-strength steel over the past decade?" Devise a conceptual model to answer this question. What is the model's class? Discuss issues of scale, extent, resolution, and level. Diagram your model.

22.3 Develop a detailed conceptual model for the operation of a toaster. What question do you wish to answer with your model? What is the model's class? Discuss issues of scale, extent, resolution, and level. Diagram your model.

PART V   Thinking Ahead

C H A P T E R   2 3

Industrial Ecology Scenarios

23.1   WHAT IS AN INDUSTRIAL ECOLOGY SCENARIO?

Scenarios are stories about possible futures. They are not predictions, but mental explorations. They are mechanisms for exploring complex systems when we feel challenged to predict their future behavior but nonetheless want to understand them better and perhaps prepare potential responses in case things go wrong. A good scenario, like a good story, is compelling in its possibilities—it engages the listener in thought, and perhaps eventually in preparation. In industrial ecology, scenarios provide the inspiration for the development of corporate and governmental policy involving technology, society, and the environment.

Scenarios have long been used by corporations, military organizations, and many others. They are usually presented in a conceptual form in those arenas. For example, here is an abstracted version of a scenario by Allen Hammond in his book *Which World? Scenarios for the 21st Century* (Washington, DC: Island Press, 1998), written from the perspective of the year 2050:

> Oil is no longer much of a geopolitical factor today; even the vast Saudi reserves could not keep up with Asia's huge energy needs, and earlier in the century prices surged as chronic shortages developed, eventually forcing a global shift to natural gas and, increasingly, to renewable forms of energy such as solar power, wind power, and biofuels. Renewable sources of energy now account for a third of global energy supplies.
>
> Despite that shift, environmental conditions are worse than they were half a century ago. The climate is distinctly warmer and more variable and is expected to become more so. In some of the small island countries now at risk of submersion by

raising seas, preparations for evacuation are under way. Climate isn't the only concern. Despite urgent attempts now being made to preserve threatened ecosystems, for many it is too late. Living coral reefs survive in only a few remote atolls.

Crop failures are common, reflecting a shortage of wild cultivars from which to breed plants resistant to new diseases and capable of coping with changed climatic conditions. Air and water pollution is an urgent, universal concern throughout newly industrialized Asia and Latin America.

People have adjusted. Electric and other new-generation automobiles now predominate in most urban areas because pollution taxes and high oil prices have made gasoline and diesel vehicles prohibitively expensive.

Most conceptual scenarios are forward-looking rather then fictionally retrospective, and need not be as sweeping as that outlined above. They might be as simple as "Suppose the price of one of the principal materials used to make our products doubles in price—what are our options?" In practice, scenarios that are most useful are fairly detailed, and every feature is examined carefully to make sure it is potentially realistic. In many cases a scenario is entirely conceptual—it consists of a "story line" such as the one reproduced above and serves as a stimulus to think in detail about the story line's implications. Perhaps, more often, the features of the scenario are expressed in mathematical form so that the consequences of the story line can be explored quantitatively. A common approach is to construct a mathematical model of the present or the past, in order to get an appreciation of the details of the system under study, and then develop the scenario(s) from that framework.

Scenarios can be divided in another way as well. Some treat evolutionary situations such as the gradual change in climate and ecosystems resulting from the emissions of greenhouse gases. Others are directed to disruptive events, such as the consequences of a national disaster. Table 23.1 illustrates these attributes of scenarios.

23.2   BUILDING THE SCENARIO

In developing a scenario it is useful to take advantage of the typology shown in Figure 23.1. The first task is to determine the project goal, that is, to define the purpose of the scenario to be constructed. In the world of science, the goal is one of *exploration*, to investigate how a holarchic system might change over time. In the

TABLE 23.1   An Analytic Framework for Scenarios, with Examples

	Evolutionary behavior	Disruptive behavior
Conceptual scenarios	Business as usual[1]	Arab oil shock[2]
Mathematical scenarios	Climate change[3]	Heart attack[4]

[1] Moyer, K., Scenario planning at British Airways—A case study, *Long Range Planning*, 29, 172–181, 1996.
[2] Schwartz, P., *The Art of the Long View*, New York: Doubleday, 1991.
[3] Intergovernmental Panel on Climate Change, *Emissions Scenarios*, Cambridge, UK: Cambridge University Press, 2000.
[4] F.F.-T. Ch'en, et al., Modeling myocardial ischaemia and reperfusion, *Progress in Biophysics & Molecular Biology, 69*, 515–538, 1998.

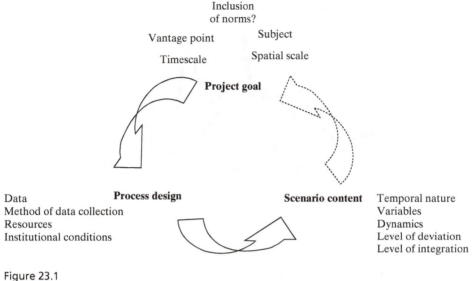

Figure 23.1

The scenario typology of van Notten, et al., showing the three overarching themes and some of their characteristics. (Reproduced with permission from P.W.F. van Notten, J. Rotmans, M.B.R. van Asselt, and D.S. Rothmans, An updated scenario typology, *Futures, 35,* 423–443, 2003.)

world of business, the goal is likely to be *decision support*, to provide enhanced perspective on a potential situation so that corporate actions have a higher probability of soundness. In the world of policy, the goal is likely to be a hybrid of these two, as policy makers seek options that respond to changing conditions while meeting political constraints and practical limitations. Industrial ecology scenarios might have any of these aims.

Process design is the second of the scenario themes. Its features follow from the project goals and from the availability of information. The task here is to determine what information is needed in order to position the scenario on a solid footing—the current price of oil, for example, or the response of a molecule of ozone to an industrial chemical. If social or cultural features are to be included, methods for developing that information must be characterized as well.

The final theme is scenario content—the timescale, the exact nature of the variables, the level of integration, and so forth. This stage is where a decision would be made as to how many scenario variations are desired and how multidisciplinary the scenarios will be. This last attribute establishes the industrial ecology *scenario class*, which is identical with the class structure of industrial ecology modeling (Figure 22.1).

23.3   EXAMPLES OF INDUSTRIAL ECOLOGY SCENARIOS

Because good scenarios generally rest on a foundation of reliable information, industrial ecologists did not really consider scenario building until they had generated such a foundation by extensive field work, data mining, and so forth. That job began to be accomplished in a few relevant topics by about the year 2000; as a consequence, scenarios in

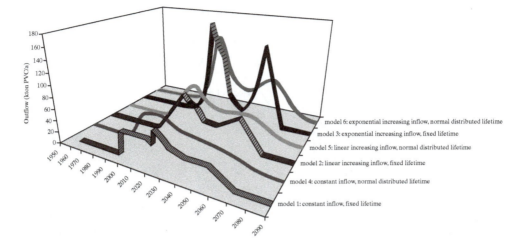

Figure 23.2

The evolution over time of discards of PVC pipe, flooring, and cables in Sweden under different scenario assumptions. (Reproduced with permission from R. Kleijn, R. Huele, and E. van der Voet, Dynamic substance flow analysis: The delaying mechanism of stocks, with the case of PVC in Sweden, *Ecological Economics, 32*, 221–254, 2000.)

industrial ecology from that time forward were increasingly being formulated and the results published. We present here a few examples of industrial ecology scenarios and the significance of their results.

A six-scenario study of the in-use stocks of poly (vinyl chloride) (PVC) products is shown in Figure 23.2; this is a Class 1A (industrial ecology) scenario that deals with technology in some detail, but not with society nor the environment. The study addressed the concern that PVC-containing products contain a variety of potentially hazardous additives. In this work, performed with the country of Sweden in mind, the scenarios differed in the rate of inflow of new products and the duration and form of the in-service lifetimes. The computations generate quite complicated outflows, partly because the assumed lifetimes of PVC-containing flooring and cables are much less than that of PVC pipe. Municipal governments or recyclers face quite different challenges in responding to the amounts of material being discarded and the temporal variations that occur, depending on which of the scenarios is closest to the situation that actually emerges.

The second example, also a Class 1A industrial ecology scenario, is illustrated in Figure 23.3. This is a study of the iron demand of China over the twenty-first century, where the demand is stimulated by (mostly) the construction of iron-containing steel buildings, bridges, and other infrastructure. The results are based on anticipated per capita steel needs; they show a strong peak in about 2025, by which time China's population has mostly stabilized and most of the needed construction to provide for its requirements has been accomplished—at least under this scenario.

Iron is made in two quite different ways—in blast furnaces, which use mostly iron ore, and in electric arc furnaces, which use mostly scrap iron. (The latter are more modern and efficient.) The scenario predicts the flow of scrap iron from Chinese discards, based on

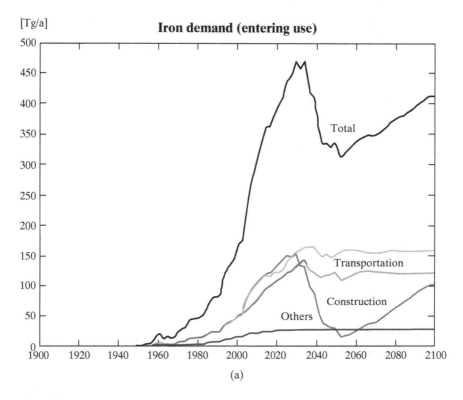

Figure 23.3

Iron demand in China, modeled from 1980 to 2000 and in a "current trends" development scenario from 2000 to 2100; (b) The generation of iron scrap in China modeled from 1980 to 2000 and in a "current trends" development scenario from 2000 to 2100. (Courtesy of D. Müller.)

the flow into use (Figure 23.3a) and the lifetimes of those uses. The results are shown in Figure 23.3b. They demonstrate that not until about 2050 will enough scrap iron be available to meet the demand. Until that time, under this scenario, the Chinese government would have to employ mostly blast furnaces in order to supply the iron that will be needed.

A third example is for a more holarchic set of scenarios, Class 2A, done to examine the effect of individual decision making upon the development of energy networks. The setting is in a rural part of South Africa, which has a number of sugar mills. The managers can invest in technology to burn bagasse (a sugar mill waste product) to produce energy; they can sell bagasse to a rather distant power station for the same purpose; or (if enough of them agree) they could induce the formation of an independent power producer. Unlike the Class 1A scenario examples above, this set of scenarios makes use of agent-based modeling, with the behavior of the managers under different financial and social conditions producing different emergent behavior.

A sampling of results is shown in Figure 23.4 for four scenarios. All the scenarios included the same financial incentives; they differed by whether the agent-based decision probabilities involved three-year contracts with the distant power station,

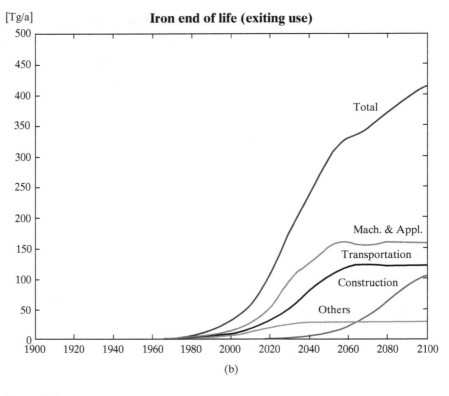

Figure 23.3

Continued

whether managerial decisions influenced local electrification (labeled MCDM on the figure), and whether loyalty and trust among the managers (labeled S.E.) was taken into account. The differences among the results are dramatic and provide significant food for thought for sugar mill managers, electrical grid planners, and politicians at various levels.

Class 2B scenarios in industrial ecology are exemplified by the Dutch FLUX and DYNABOX models of Chapter 22 when used in the scenario mode. Some of the results from this research are shown in Figure 23.5 for metals in the anthropogenic system and in the environment. At the top of the figure are the calculated emissions of two of the metals into various environmental reservoirs at the eventual steady state. The second component of the model classifies these emissions by looking at the reservoirs into which they flow and then evaluates the flows in the context of the assimilative capacity of the reservoirs and the susceptibility of the organisms therein (bottom). The risks are stated in terms of ratios of the computed concentrations of metals to those concentrations below which no effect is anticipated, hence ratios above unity are of concern and those below are not. It is only under the two stringent control scenarios that the ratios decrease below the level thought necessary to avoid environmental damage.

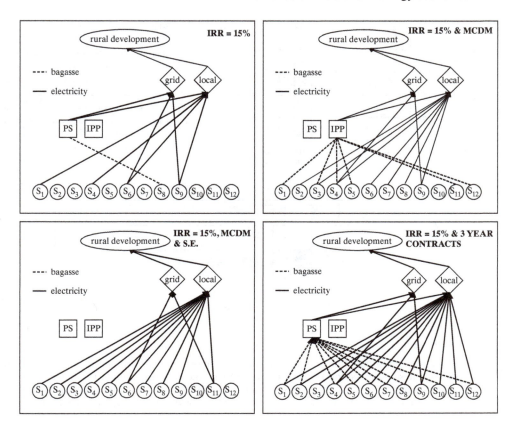

Figure 23.4

The effects of agent behavior on rural electrification network configuration scenarios in KwaZulu Natal, South Africa. The diagrams illustrate the decision of sugar mill managers to sell bagasse for power generation or to generate power themselves, as described in more detail in the text. The results are for year 22 of the scenarios. PS = existing power station, IPP = independent power producer, S_x = sugar mills, IRR = internal rate of return, MCDM = multicriteria decision making, S.E. = social equity. (Reproduced with permission from J. Beck, R. Kempener, B. Cohen, and J. Petrie, A complex systems approach to planning, optimization and decision making for energy networks, *Energy Policy, 36*, 2795–2805, 2008.)

23.4 THE STATUS OF INDUSTRIAL ECOLOGY SCENARIOS

Scenarios offer the opportunity to perform experiments on paper or with a computer, rather than in real life. They have the potential to provide very much improved perspective on issues of importance and to give a framework for the consequences of impulsive and disruptive events of various kinds. One can readily imagine industrial ecology scenarios addressing such topics as

- The future of metal use given decision choices on individual products by potential customers
- Changes in energy use

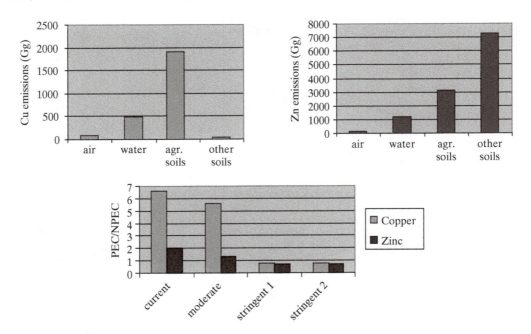

Figure 23.5

Steady state emissions of copper (top left) and zinc (top right) to the environment in The Netherlands under present rates of use; (bottom) Steady state terrestrial ecotoxicity risk ratios in The Netherlands under three enhanced regulatory scenarios, compared with those for today's regulatory structure. Source: E. van der Voet, J.B. Guinée, and H. A. Udo de Haas, *Heavy Metals: A Problem Solved?* Dordrecht: Kluwer Academic Publishers, 2000.

- The development of ecoindustrial parks
- Industrial relationships over time given a changing water cycle
- And many others

Despite this promise, industrial ecology scenarios present substantial challenges in scope, data availability, and implementation. This is especially true with Class 2 and Class 3 scenarios. Nonetheless, the Class 1A and 2A examples illustrated here are evidence that the development of industrial ecology scenarios has begun and will continue to be an important part of the science in the future.

FURTHER READING

Scenario philosophy and implementation:

Kickert, R.N., G. Tonella, A. Simonov, and S.V. Krupa, Predictive modeling of effects under global change, *Environmental Pollution, 100*, 87–132, 1999.

Schwartz, P., *The Art of the Long View*, New York: Doubleday, 1996.

Van Notten, P.W.F., J. Rotmens, M.B.A. van Asselt, and D.S. Rothman, An updated scenario typology, *Futures, 35*, 423–443, 2003.

Class 1A IE Scenario:

Kleijn, R., R. Huele, and E. van der Voet, Dynamic substance flow analysis: The delaying mechanism of stocks, with the case of PVC in Sweden, *Ecological Economics, 32*, 221–254, 2000.

Class 1A IE Scenario:

Davidsdottir, B., and M. Ruth, Pulp nonfiction: Regionalized dynamic model of the U.S. pulp and paper industry, *Journal of Industrial Ecology, 9* (3), 191–211, 2005.

Class 1C (nonindustrial ecology) evolutionary scenario:

Kripalani, R.H., et al., South Asian summer monsoon precipitation variability: Coupled climate model simulations and projections under IPCC AR4, *Theoretical and Applied Climatology, 90*, 133–159, 2007.

Class 1C (nonindustrial ecology) disruptive scenario:

Ch'en, F.F.-T., et al., Modeling myocardial ischaemia and reperfusion, *Progress in Biophysics & Molecular Biology, 69*, 515–538, 1998.

Class 2A IE Scenario:

Beck, J., R. Kempener, B. Cohen, and J. Petrie, A complex systems approach to planning, optimization and decision making for energy networks, *Energy Policy, 36*, 2795–2805, 2008.

Class 2B IE Scenario:

van der Voet, E., J.B. Guinée, and H. A. Udo de Haas, *Heavy Metals: A Problem Solved?* Dordrecht: Kluwer Academic Publishers, 2000.

EXERCISES

23.1 The automobile technology system is illustrated in Figure 22.1a. Construct a similar diagram for the pharmaceutical technology system. Describe a technological change that would noticeably perturb the system. Repeat for a cultural change and for a government policy change.

23.2 Develop three scenarios for the next three decades for the geographical region of which you are a part. (You may choose the scale, but no larger than national.) What are the characteristics of your scenarios? What governmental policies might be suggested, and when should they be considered?

23.3 Repeat Exercise 23.2, but for a multinational manufacturing firm instead of a government.

CHAPTER 24

The Status of Resources

24.1 INTRODUCTION

It is a trivial statement to say that modern technology could not exist without renewable and nonrenewable resources. Nonetheless, the availability of these resources has seldom been a matter of much concern, especially for a product designer who has never been taught to regard materials as anything but commodities to be employed as necessary or convenient. The result is nuclear reactors dependent on the availability of zirconium, catalytic converters dependent on the availability of rhodium, hybrid vehicle batteries dependent on the availability of lanthanum, cellular telephones dependent on the availability of tantalum, and a vast panoply of products whose manufacture and use assume the availability of energy and water in abundance. Are these unwise design decisions? They may prove to be, because many suggest that resources in general, or at least some specific resources, may become quite scarce in the years to come.

However, there are strong voices in opposition to the idea of any constraints on the availability of resources. These "cornucopians" are of the opinion that supplies of resources are ample. If supplies should somehow fall short anyway, the cornucopians predict that substitutes will rapidly be developed to fill the need.

Is one of these positions true and the other false? Or, like so many contentious issues, is the correct answer somewhere in between? Or, is the question incomplete, so that resource availability should be considered in a different framework?

24.2   MINERAL RESOURCES SCARCITY

The mineral resources used for millennia have come from rich deposits located near Earth's surface (the crust), and thus potentially mineable. These deposits largely stem from geologic processes that occurred hundreds of millions of years ago. The most important are hydrothermal (heated water) and magma (molten rock) processes, both involving fluids that dissolve elements and eventually deposit them in crystal-lized form. The driving force for these processes is plate tectonics, which is the move-ment of lithospheric plate about Earth's surface as a result of the planet's internal energy dissipation. It is an occasional fortunate accident when such a deposit is rich enough to be useful to us, and when it occurs in an accessible location. Such deposits are occasionally discovered but most of the likely locations on Earth have now been explored. It is unrealistic to anticipate that major new ore deposits lie hidden and waiting to be found.

Mineral deposits are, of course, of varying degrees of richness. Economic consid-erations generally require that the richest deposits are mines first, the next-richest second, and so on. As the use of poorer deposits is contemplated, it is useful to know the distribution of occurrence of resources of interest. Each element is present to some degree in each kind of rock, and the occurrence spectrum of at least the common chemical elements (e.g., silicon, iron, calcium) appear to display lognormal distribu-tions in common rocks, as suggested in Figure 24.1a. The lognormal distribution has not been well established, however, especially for trace elements, and there is some weight of evidence that the abundance spectrum of most technologically desirable minerals forms a bimodal ore grade distribution as shown in Figure 24.1b. A "mineralogical barrier" is indicated on the figure; this is the ore grade above which atoms are present in sufficient concentration to form distinct mineral deposits. Below that concentration, the individual metal atoms substitute for other atoms in minerals of the most abundant elements in common rocks. The quantity of the resource residing in deposits above the mineralogical barrier (assuming Figure 24.1b is correct) is termed the "aggregate resource," and is estimated by geological techniques.

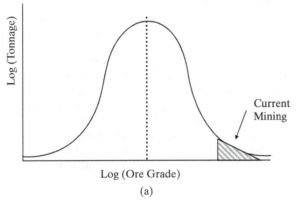

Current Mining

Log (Ore Grade)

(a)

Figure 24.1

(a) Lognormal distribution of a technologically useful material in Earth's crust. (b) A bimodal dis-tribution of a technologically useful material in Earth's crust. In each diagram, the area represented by current mining is shaded and the position of the mineralogical barrier indicated. (Reproduced with permission from R.B. Gordon, T.C. Koopmans, W.D. Nordhaus, and B.J. Skinner, *Toward a New Iron Age?* Cambridge, MA: Harvard University Press, 1987.)

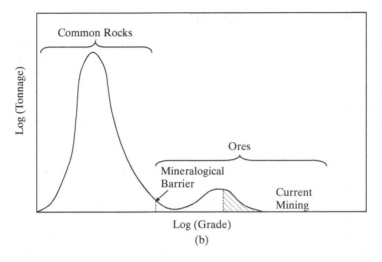

Log (Grade)
(b)

Figure 24.1

Continued

When the availability of a resource is thought to be in doubt, it is sometimes regarded as *critical.* The U.S. National Research Council (NRC) views criticality as a function of two variables: *importance in use* and *availability.* To quote from the report:

> Importance in use embodies the idea that some minerals or materials are more important in use than others. Substitution is the key concept here. For example if substitution of one mineral for another in a product is easy technically, or relatively inexpensive, one can say that its importance is low. In this case, the cost or impact of a restriction in the supply of the mineral would be low. On the other hand, if substitution is difficult technically or is very costly, then the importance of the mineral is high, as would be the cost or impact of a restriction in its supply. . . .
>
> Availability is the second dimension of criticality. Fundamentally, society obtains all minerals through a process of mining and mineral processing (primary supply). Later, however, in the course of fabrication and manufacturing and ultimately after products reach the end of their useful lives, society can obtain mineral products through the processing of scrap material (secondary supply). Availability reflects a number of medium- to long-term considerations: geologic (does the mineral exist?), technical (do we know how to extract and process it?), environmental (can we extract and process it with a level of environmental damage that society considers acceptable and with effects on local communities and regions that society considers appropriate?), societal and technological (can we collect and recycle discards to minimize virgin material requirements?), political (how do policies affect availability both positively and negatively?), and economic (can we produce a mineral or mineral product at costs consumers are willing and able to pay?).

This two-variable concept of criticality can be effectively communicated by a graphical representation in which the vertical axis reflects importance in use and the

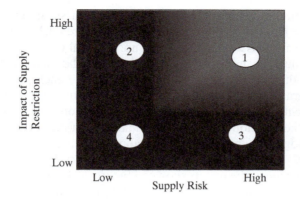

Figure 24.2

The criticality diagram.

horizontal axis is a measure of availability, as is pictured in Figure 24.2. The features of the diagram can be readily explained and understood. As shown in Figure 24.3a, a material whose uses are important and where supply restrictions would have high impact would be located toward the top of the diagram. Similarly, a material with high supply risk would be located toward the right of the diagram (Figure 24.3b). Consequently, criticality becomes not an absolute state, but one of degree. Figure 24.3c shows that criticality increases to the degree that a material's location on the diagram is toward the upper right. The criticality concept thus identifies a "region of danger"

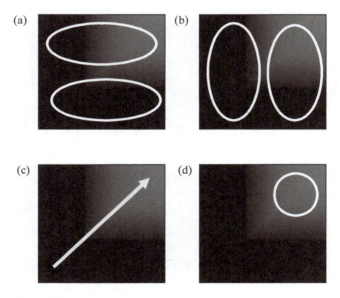

Figure 24.3

Features of the criticality diagram: (a) Regions of high (upper oval) and low (lower oval) supply restriction impact; (b) Regions of high (right oval) and low (left oval) supply risk; (c) The "arrow of increasing criticality;" (d) The "region of danger."

(Figure 24.3d) for materials high in importance and with substantial risk of supply constraints. Hence, a material located in region 1 of Figure 24.2 is more critical than one located in either regions 2 or 3, and much more critical than a material located in region 4. For a product, an industry, or a country (and the analyses for each might be quite different), the goal should be to rigorously locate each possible material of interest on this diagram, and then to attempt to use substitution, exploration, or other methods to avoid being dependent on materials in the region of danger.

Determining the location of a particular material on the criticality matrix is an exercise in expert judgment, and we will not belabor the details here. (Those interested can review the NRC report, 2007.) However, a preliminary NRC effort to carry out criticality analysis produced the results shown in Figure 24.4. Of the metals surveyed, a number fall within the region of danger—rhodium, platinum, manganese, niobium, indium, and the rare earths. Copper is considered not critical, not because the impacts of supply restriction would be minimal, but because supply risk is judged to be low. A number of other elements fall between these extremes.

Potential resource scarcity is sometimes argued to be indicated by the "depletion time τ_D," given by

$$\tau_D = \frac{\text{Reserve base}}{\text{Annual use}} \tag{24.1}$$

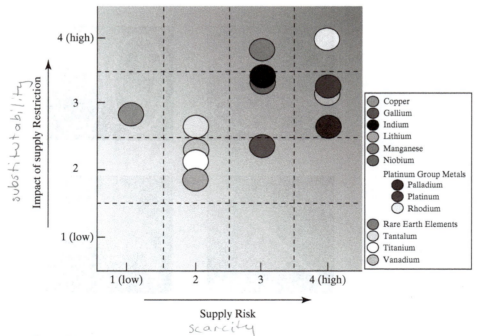

Figure 24.4

The criticality matrix for 12 elements and the rare earth elements group. (Courtesy of the U.S. National Research Council, 2007.)

TABLE 24.1   Classes of Abundance Relative to Rate of Use

Abundant ($t_D > 100$ yr)	Al, B, C, Ca, coal, Cr, Fe, Ir, K, Li, Mg, Na, Nb, Os, Pt, rare earths, Ru, Si, Ti, V, Yt
Common ($t_D = 50$–100 yr)	Co, Hf, natural gas, Ni, P, Pd, Rh, Sb, Ta, W, Zr
Constrained ($t_D = 25$–50 yr)	Ba, Bi, Cd, Cs, Cu, Mn, Mo, oil, Se, Sn, Sr, U
Rare ($t_D < 25$ yr)	Ag, Au, Hg, In, Pb, S, Th, Zn

This list is limited to major and selected trace elements and energy sources. (Based on S. Kesler, *Mineral Resources, Economics, and the Environment,* New York: Macmillan, 1994.)

(The reserve base is that part of an identified resource that meets specified minimum physical or chemical criteria related to current mining and production practices.) A long depletion time suggests less concern than does a short one. As seen in Table 24.1, however, rare earths, niobium, and platinum all have $\tau_D > 100$ yr. Clearly τ_D does not take into account such factors as nongeological supply risk, nor the importance of use.

There is no simple answer to the question of whether any mineral resources should be regarded as critical or whether current use is unsustainable. Part of the challenge is that technology is not static, nor are the materials that technology uses. As this is written, there are supply concerns for indium, currently a necessary constituent of flat-panel displays, and for rhodium, the only known catalyst capable of efficiently transforming oxides of nitrogen in vehicles exhaust systems. Tomorrow may bring other issues with other materials. The fact that it is easy to cite examples of potential trouble, however, supports that the sustainable supply of any and all resources should not be assumed.

Ultimately, of course, the issue here is the balance between supply and demand. In principle, demand for mineral resources can be met at least in part by recycling obsolete products. It is clear that much more effort in the recycling area is warranted. Nonetheless, for at least some materials, in at least some countries and corporations, the availability of certain specific materials will continue to be a concern. Additional analysis of resource supply and demand is needed to engender more informed understanding of the status of a wide spectrum of materials.

24.3   CUMULATIVE SUPPLY CURVES

A helpful perspective on the relationships of total supply of nonrenewable materials to their price over all time is provided by cumulative supply curves, developed by John Tilton of the Colorado School of Mines and Brian Skinner of Yale University. There are three scenarios of interest. That of Figure 24.4a assumes that moderate increases in demand will result over time in gradual increases in price. Figures 24.4b and 24.4c, however, represent situations in which high-grade ore deposits are exhausted and much lower-grade deposits are utilized.

Several groups of factors relate to the cumulative supply curves. The first group is geological and determines the slope of the curve. As described earlier in this chapter and illustrated in Figure 24.1, if ore grade distributions are lognormal, they will lead to a cumulative supply curve of the form of Figure 24.5a. If they are bimodal, the result is

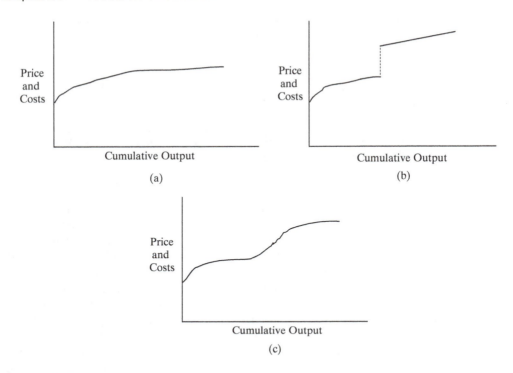

Figure 24.5

Cumulative supply curves for nonrenewable resources. (a) Slowly rising slope due to gradual increase in costs. (b) Discontinuity in slope due to jump in costs. (c) Sharply rising slope due to rapid increase in costs. (Adapted from Tilton, J.E., and B.J. Skinner, The meaning of resources, in *Resources and World Development*, D.J. McLaren and B.J. Skinner, Eds., New York: John Wiley, pp. 13–27, 1987.)

likely to be a cumulative supply curve of the form of Figure 24.5b or Figure 24.5c. There is, at present, insufficient data to determine which alternative is correct.

The second group of factors is related to demand and determines how rapidly the world moves along the supply curve. The factors here include global population, per capita income, and intensity of material use (the consumption of a mineral commodity per unit of income). As discussed in Chapter 1, the first two factors are clearly and strongly increasing. The intensity of use, the third factor, appears to be trending downward for most resources as a consequence of new technologies and cultural preferences for "upscale" products. A fourth factor in the demand equation for minerals (but not for energy resources) is the availability of recycled material as a substitute for virgin material. While a helpful supplement to extracted stock, recycled material cannot be expected to play a dominant role in material supply in a time of rapidly increasing consumption.

The third group of factors has the potential to cause the cumulative supply curve to shift its position. The dominant factor in this group has historically been new technology, which shifts the curve downward. The second factor, changes in input cost as a result of labor, capital, or energy transitions, could shift the curve either up or down. These possibilities have historically proven difficult or impossible to predict reliably.

Case Study: Sustainability of Copper

One of the few elements that has been studied sufficiently for ultimate supply and demand to be estimated is copper. Various researchers have put the quantity of the resource (the possible ultimate supply) between 1.6 and 2.2 PgCu. For ultimate demand, one possible estimation method is to relate demand to the average per capita copper stock, which enables such services as electricity distribution, plumbing, heating, and air conditioning. This figure for North America in 2000 was 170 kg. Suppose that by the mid-twenty-first century the average world citizen (there will be about nine billion of them) will have a quality of life similar to that of the year 2000 North American and could thus require a similar level of copper stock. This implies a total stock in use of about 1.5 PgCu. Both the supply and demand numbers are very rough estimates, of course, and the amount of copper needed might be reduced due to (1) more efficient product design, (2) efficient copper recycling, or (3) substitution of other materials for copper. Nonetheless, the fact that the supply and demand estimates are essentially identical suggests that our knowledge of these factors for copper needs to be improved, and copper supply and demand should be followed carefully through the next few decades in order to ensure the adequacy of the resource for human uses.

Source: S. Spatari, et al., Twentieth century copper stocks and flows in North America: A dynamic analysis, *Ecological Economics, 54*, 37–51, 2005.

24.4 ENERGY RESOURCES

Fossil fuels, by far the most common source of energy, are traded and used around the globe. The resulting energy provides a variety of service—heating, illumination, transportation, and so forth. The amount used by individuals varies with their wealth, their geographical location, their culture, and their personal habits. Differences in these factors can result in very different rates of energy use. This situation is captured by examining the relationship between per capita energy use and the human development index (a composite indicator developed by the United Nation Development Program to show countries relative well-being in social as well as economic terms). Figure 24.6 shows that about 1 metric ton of oil equivalent (about 42 GJ) is required to maintain an HDI level of 0.8 or above. Countries that use much more energy per capita show only very modest increases in HDI.

In Chapter 19, we pointed out that industrial ecology was closely related to the use of energy. Although energy is generated in a variety of ways, fossil fuels dominate industry's contemporary energy budget, as well as that of humanity as a whole. As seen in Table 24.2, oil is the largest current energy source, accounting for about a third of the world's primary energy generation in recent years. Nuclear power and hydropower account for a few percent each. Of the final energy amount, only about 15 percent is used industrially. As a consequence, sectors other than industry will be strong competitors for whatever energy is available.

A vigorous debate for many years has centered on the (ill-posed) question, "Will we run out of oil?" As with other commodities, if the price becomes high enough, purchases will be made only by those who are willing to pay the price. Nonetheless,

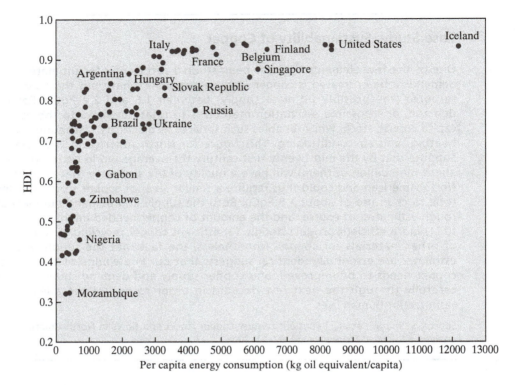

Figure 24.6

The relationship between the human development index (HDI) and per capita energy use in 1999–2000. (Reproduced from *World Energy Assessment Overview 2004*, United Nations Development Programme, http://www.undp.org/energy/weaover2004.htm, accessed January 2, 2008.)

TABLE 24.2   Global Energy Supply and Use in 2004, in Gtoe/yr

	Coal	Oil	Gas	Nuclear	Hydro	Other
Energy supply	2.75	3.96	2.32	0.71	0.23	0.06
Energy use						
Electricity	1.63	0.25	0.49	0.70	0.23	0.06
Heat	0.27	0.05	0.39	0.01	–	–
Industry	0.50	0.32	0.46	–	–	–
Transport	0.01	1.86	0.07	–	–	–
Residential	0.07	0.23	0.40	–	–	–
Commercial	0.02	0.12	0.15	–	–	–
Agricultural	0.01	0.11	0.06	–	–	–
Other	0.21	0.41	0.11	–	–	–
Nonenergy use	0.03	0.60	0.19	–	–	–

Source: International Energy Association, *IEA Energy Statistics*, http://www.iea.org/statist, accessed December 31, 2007. Gtoe = 10^9 metric tons oil equivalent.

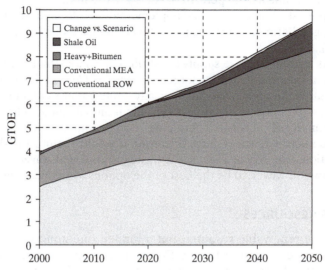

World Oil Production form Conventional and Unconventional Resources: Reference/USGS

Figure 24.7

World oil production from conventional and unconventional sources. (Reproduced with permission from D.L. Greene, J.L. Hopson, and J. Li, Have we run out of oil yet? Oil peaking analysis from an optimist's perspective, *Energy Policy, 34,* 515–531, 2006.)

because oil (and coal and natural gas) is the result of high-temperature, high-pressure processing of buried biomass over millions of years, there could well be ultimate supply limitations. In addition, because fossil fuel combination results in the greenhouse gas carbon dioxide, regulatory restrictions on use could be imposed.

Different research groups have constructed scenarios of future oil production. These are subject to many assumptions about technology, accuracy of reservoir estimates, and so forth. Specialists differ, but a typical result is that shown in Figure 24.7. It suggests that world oil production from conventional sources is likely to peak around 2020. To the extent that the unconventional bitumen and shale oil are employed (they require much more energy to extract and process and generate more carbon dioxide per unit of energy generated), they may provide significant additional amounts.

The outlook for natural gas is no more optimistic. Its future is likely to be similar to that of oil, but perhaps slightly later in time. Coal, the third fossil fuel, is abundant, with a depletion time greater than 200 years. Strip-mining for coal, a common practice, is problematic, however, as are the environmental challenges of coal's combustion. Coal will doubtless continue to be a major component in energy supply, but will need to be developed with care.

Uranium, the fuel for nuclear power, has a depletion time similar to that for oil. Nuclear reactors do not have wide public acceptance in many parts of the world, however, and long-term storage of spent fuel is a continuing concern. As a result, nuclear power use is now beginning to expand after many years of stasis. As a result, the criticality of uranium will need to be carefully addressed in coming years.

Other sources of energy—biomass, hydro, geothermal, solar, wind—appear unlikely to provide as much as 25 percent of energy needs in the next few decades even under the most optimistic scenarios. Broad implementation is limited by supplies (biomass), nature (hydro, geothermal, wind), technology and nature (solar), and environmental concerns (hydro).

As a result of these supply constraints, it is likely that the mid-twenty-first century will see a major shift in the sources of energy. Coal gasification may be a significant contributor, although it is technologically limited at present and the economics are problematic. A potential alternative, especially if global climate change accelerates the shift away from fossil fuels and a "bridging" energy source is needed, is a new generation of fail-safe nuclear reactors. Such a transition would require the development of reliable international approaches to nuclear waste disposal, and considerable public education. It may prove, however, to be the best of the options available within a few decades.

24.5 WATER RESOURCES

Water is a renewable resource, but a limited one nonetheless. It has been said that "water is the oil of the twenty-first century," and there is every prospect that the availability of water will increasingly limit industrial processes that utilize it. Unlike wood or other renewable resources, the total average quantity of water available to humans is fixed. What is not fixed, however, is any constancy in rate of supply, since droughts, floods, and average water flows are all relatively common experiences. Climate change promises to increase these uncertainties in the coming decades.

Water is essential in many parts of industry for cooling, as a solvent, as a transport medium, and sometimes as a constituent of finished products. Actual consumption is generally a small fraction of intake—most cooling water is returned to the source, for example. Overall, depending on the country and its level of industry, between 5 and 20 percent of global consumptive use is industry related. A much larger amount, 70–85 percent globally, is employed by agriculture. Absent major gains in the efficiency of agricultural water use, the growing global population will demand additional water in order to increase the food supply. The remaining consumptive water uses, perhaps 5–15 percent, are for domestic purposes, again a use unlikely to diminish in coming years. Industrial acquisitions to water thus need to surmount the growing demand for food and personal needs.

Water stress and scarcity for all uses varies widely with location, as shown in Figure 24.8. Under various development scenarios, the number of people living under high water stress is anticipated to be between three and five billion (i.e., one-third to half the global population) by 2050. This picture indicates the challenge anticipated for the next few decades. It will demand a very high level of water in efficiency, recycling, and reuse if human demands and water supplies are to approach equivalency.

24.6 SUMMARY

As with so many other aspects of industrial ecology, the sustainability of resources is a complex systems problem. One cannot look at the aggregate resources of copper as the metric for copper's sustainability, because as the grade of copper ore being mined decreases with time, the implications for energy and water become increasingly

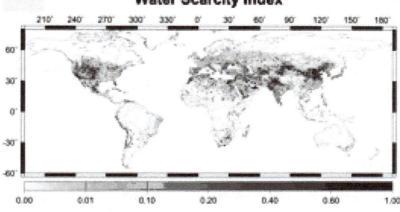

Figure 24.8

Global distribution of the water scarcity index. (Reproduced with permission from T. Oki, and S. Kanae, Global hydrological cycles and world water resources, *Science, 313,* 1068–1072, 2006.)

relevant (recall the demonstrations in Chapters 19 and 20 showing that energy and water use in mining and processing depend dramatically on ore grade). Recall as well that technological uses of energy and water are small fractions of total use, and must compete with energy used for heating and transport and water for agriculture and personal hygiene. Recall that Earth's people all wish to live comfortable lives, with the associated demands for resources of all types.

So, are we running out of resources? This question has been explored from the standpoint of absolute amounts of materials, the form in which they are available to us, their geographical distribution, the energy and water required for extraction and processing, and the environmental impacts that limit availability. We conclude that each one of these considerations has the potential to limit supplies of certain materials, though at different temporal and spatial scales. Materials whose cumulative supply curves follow predictable supply-cost patterns, as in Figure 24.5a, are unlikely to suffer from significant scarcity for at least several decades, although cost increases may over time change the spectrum of uses. For materials whose cumulative supply curves contain discontinuities (Figure 24.5b) or costs that rise rapidly with small increases in supply (Figure 24.5c), however, virgin material scarcity is a real possibility. For energy resources, gradual transitions to a higher proportion of renewable sources are indicated. In the case of water, many geographical locations appear likely to be increasingly constrained.

It is very unlikely that any resources will completely disappear, because if they become very scarce the price will become very high and use will decline. However, there appears to be a real potential over the next several decades for a number of widely used resources to become scarce enough that their increased price will force a new spectrum of conservation, substitution, and more thoughtful use.

FURTHER READING

Mineral Resources:

Gordon, R.B., T.C. Koopmans, W.D. Nordhaus, and B.J. Skinner, *Toward a New Iron Age?* Cambridge, MA: Harvard University Press, 1987.

Kesler, S.E., *Mineral Resources, Economics, and the Environment*, New York: Macmillan, 1994.

Tilton, J.E., *Depletion and the Long-Run Availability of Mineral Commodities*, Washington, DC: Resources for the Future, 2001.

U.S. National Research Council, *Minerals, Critical Minerals, and the U.S. Economy*, Washington, DC: National Academy Press, 2007.

Energy:

BP, Inc., *BP Statistical Review of World Energy 2008*, London, available at http://www.bp.com/productlanding.do?categoryId=6929&contentId=7044622, accessed December 22, 2008.

Greene, D.L., J.L. Hopson, and J. Li, Have we run out of oil yet? Oil peaking analysis from an optimist's perspective, *Energy Policy, 34*, 515–531, 2006.

Nakićenović, N., A. Grübler, and A. McDonald, *Global Energy Perspectives*, Cambridge, UK: Cambridge University Press, 1998.

Witze, A., That's oil, folks, *Nature, 445*, 14–17, 2007.

Water:

Gleick, P.H., *The World's Water, 2000–2001: The Biennial Report on Freshwater Resources*, Washington, DC: Island Press, 2000.

Islam, M.S., T. Oki, S. Kanae, N. Hanasaki, Y. Agata, and K. Yoshimura, A grid-based assessment of global water scarcity including virtual water trading, *Water Resources Management, 21*, 19–33, 2007.

Seckler, D., R. Barker, and U. Amarasinghe, Water scarcity in the twenty-first century, *Water Resources Development, 15*, 29–42, 1999.

EXERCISES

24.1 Manganese is widely used as an additive in steel making and in other applications. Investigate its supply risk and use information on its major uses to estimate the impact of supply restriction. Do you think manganese is properly located on Figure 24.4 (an evaluate known to be only a rough estimate)? Why or why not?

24.2 If the average human development index for African countries increased to 0.85, what does this imply for energy supply in Africa? (Refer to Figure 24.6.)

24.3 Using Figure 19.11, estimate the energy required to extract and process copper from ore of grade 3.0 percent. Repeat for ore of grade 0.3 percent, the minimum ore grade now being actively mined.

24.4 Lead is a material widely used in automotive and stationary batteries. On the basis of whatever information you can locate on lead, evaluate its availability from the standpoint of abundance, co-occurrence, and geographical occurrence. What do you predict for lead as an industrial material in the next few decades?

CHAPTER 25

Industrial Ecology and Sustainable Engineering in Developing Economies

25.1 THE THREE GROUPINGS

Economic activity, and thereby industrial ecology, is being restructured as a result of globalization. Earlier industrialization patterns arose in a particular geographic location and then spread toward rim economies. The United Kingdom was the core of the Industrial Revolution, for example, as the United States was for mass production and consumption. Recent patterns are more globalized—Europe, the U.S., and Asian economies all play important roles in ICT, for example. The firms that embody technological competence are increasingly global and multicultural, as illustrated by IBM's significant R&D capability and production and management functions in India, China, Europe, the United States, and elsewhere. Thus, the industrial pattern is shifting away from a "core to rim" geographic model to one that minimizes the nation-state in preference to a multinational matrix model that transcends political boundaries.

Nonetheless, a simple division of nation-states can still provide valuable structure. For many purposes a useful schematic breakdown is (1) highly developed countries (HDCs); (2) rapidly developing countries (RDCs); and (3) slowly developing countries (SDCs). Highly developed countries tend to have stable financial and governance institutions, high GDP per capita, the capability of mobilizing large amounts of financial and human capital internally, and indigenous research and development competency. Examples would include most of the Organization for Economic Co-operation and Development (OECD) countries, such as Australia, Canada, the United States, European Union members, and Japan, as well as Singapore, a successful city-state. Close behind these countries is a second group that is

TABLE 25.1   Representative Data for the "BRIC" Countries and the United States for 2004

Metric	Brazil	Russia	India	China	USA
Population (millions)	181	142	1,081	1,313	297
GDP ($Billion)	604	581	691	1,932	11,712
Global rank in purchasing power(GDP PPP)	9	10	4	2	1
TVs per 100 people	87.5	75.2	No data	No data	99.6
Cars per 1000 people	130	142	6	10	468

Note: PPP, purchasing power parity, meaning that the GDP is adjusted for the average price of consumer goods in each economy.

Source: The Economist, *Pocket World in Figures: 2007*, London: Profile Books, 2007.

rapidly developing. A notable subset of this category is the so-called BRIC countries: Brazil, Russia, India, and China. Despite their obvious differences, from an industrial ecology perspective they share several salient characteristics (Table 25.1). First, they have large populations and are rapidly moving toward levels of economic development that will support high levels of demand for mass consumer goods, including automobiles, housing, and higher-quality food consumption. This transition demands attention because of the pent up demand for energy and resources implied by this growth. For example, as Table 25.1 indicates, while the United States had some 468 cars per 1000 people, Russia had only 142, Brazil 130, and India and China 6 and 10 respectively. Even more notably, this gap in material consumption extends across all goods, as the similar figures for television ownership indicate. The fact that there are a number of other countries with BRIC characteristics — for example, Thailand, Malaysia, Viet Nam, Chile, Peru, Mexico, and Argentina — indicates that the challenges of rapid growth are not limited to a small group of nations, but constitute a global consideration.

Finally, there are countries that are developing quite slowly and have little immediate prospect of accelerating their growth. Many of these are in Africa, especially sub-Saharan Africa, and they pose unique challenges of governance and economic development. They are challenged not just by a lack of financial resources, stable and transparent governments, viable legal and regulatory structures, and educational infrastructures, but by concomitant stresses such as AIDS and other more local diseases such as malaria which, because of the poverty of much of the population, do not attract the research and drug development of so-called "rich world diseases." Additionally, many African countries exhibit high population growth rates: of the fastest growing populations in the world, nine of the top ten are in Africa.

From a sustainability and industrial ecology perspective, therefore, the three groups of nations offer very different analytical challenges: the HDC economies with their current consumption patterns; the RDC economies, especially the BRIC countries, with their already large and accelerating demand for resources; and the SDCs, where it is difficult to see a path forward given the implications of globalization and rapid growth of competing economies. A comparison of some of the significant characteristics of these groups is given in Table 25.2.

TABLE 25.2   Characteristics of Nation-State Development Groups

	HDC	RDC	SDC
Population	Moderate	High	High
Waste management infrastructure	Good	Inadequate	Virtually nonexistent
Energy supply	Excellent	Marginal	Poor
Water supply	Mostly adequate	Adequate to marginal	Often inadequate
Technological level	High	Intermediate	Generally low
Rate of technological change	High	Very high	Generally low
Unemployment	Low	Moderate	High
Education level	High	Moderate	Low
Legal system	Extensive and effective	Generally adequate to marginal	Often ineffective
Transportation network	Extensive	Adequate to marginal	Inadequate

25.2   RDC/SDC DYNAMICS AND PERSPECTIVES

That lesser-developed countries (LDCs) differ from more-developed countries in many ways is an obvious statement. For the present purposes, however, the important issues relate to the implications of those differences for industrial ecology and sustainable engineering. In this section, we explore some of the properties of LDCs (we have grouped RDCs and SDCs together) from that perspective.

LDCs have large and growing populations.   The population of HDCs are nearly stable and likely to remain so. In contrast, those of most of the LDCs are rapidly increasing. If we recall the IPAT equation from Chapter 1, population will be a major driver of environmental performance for the LDCs in the first half of this century, at least.

LDCs have a different technological starting point.   It is instinctive to regard LDCs as located at an earlier point on a hypothetical development curve, with HDCs further "up the curve." This idea is inaccurate. Whereas HDC development largely proceeded within country boundaries, today's global economic system means that no country is isolated, and new initiatives both draw from and are in competition with similar initiatives around the world. Technologies, once grown indigenously, are readily available as aides to development.

LDCs have large rates of change.   Industrial development in the HDCs took between one and two centuries. In marked contrast, LDCs today can undergo major transformation in a decade or two, depending on a number of factors (infrastructure, intellectual and capital resources, institutional stability, and the like). Figure 25.1, the copper cycle in China for three years from 1994 to 2004, is not atypical. It shows that the flow of copper use and the quantity of that flow added to in-use stocks increased by a factor of more than four over the decade, aided by an increase of almost a factor of six in copper imports. Growth in cellular telephone use, personal vehicle registrations, and many other signals (both positive and negative) of development are similarly spectacular.

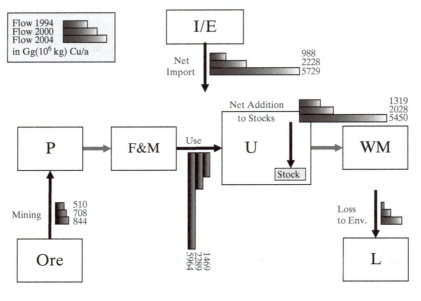

Figure 25.1

Flows of copper in China: 1994, 2000, and 2005. (Courtesy of T. Wang, Yale University.)

LDCs are rich in resources. A large fraction of Earth's accessible resources are in LDCs, as shown for a selection of metals in Figure 25.2. The major proportion of the materials extracted from these repositories is quickly exported. Monetary trade statistics tend to obscure this situation, because HDCs typically import low value, high mass goods and export high value, low mass goods, as shown in Figure 25.3.

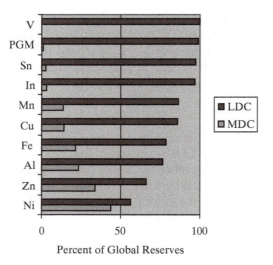

Figure 25.2

The percentages of global reserves of a number of industrial metals that are located in developing countries. PGM = platinum group metals. Source: U.S. Geological Survey, *Mineral Commodity Summaries*, 2008.

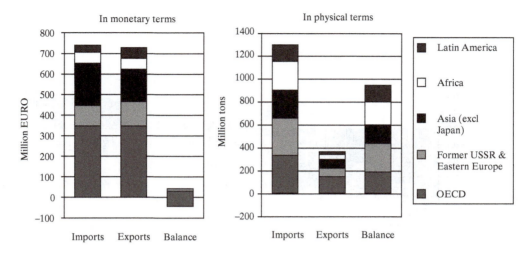

Figure 25.3

The external trade balance of the EU-15, expressed in monetary terms (left) and physical terms (right). (Reprinted by permission from S. Giljum, and N. Eisenmenger, North–South trade and the distribution of environmental goods and burdens: A biophysical perspective, *Journal of Environment and Development, 13,* 73–100, 2004.)

LDCs have low domestic rates of resource use. LDC economies are marked by very limited rates of use of many domestic resources. They also have low rates of imports, as exemplified by the iron and chromium statistics shown in Figure 25.4. As a consequence, an LDC undergoing rapid development has an opportunity to choose a development path not highly constrained by embedded technology or infrastructure.

LDCs are major factors in pollution, locally and globally. It is a commonplace that most pollution happens early in the life cycle—mining, logging, initial resource processing. As a result, a large fraction of Earth's anthropogenic pollution occurs in LDCs. Countries that import processed resources from LDCs have, in effect, outsourced the pollution attributable to those imports. Figure 25.5 illustrates this situation for six pollutants, using the metric of environmental emissions from imports minus environmental emissions from exports. For Western Europe over a period of nearly 20 years, substantial "virtual pollutants" were outsourced each year for each of the pollutants.

The pollution is not necessarily local in impact. It is estimated that CO_2 emissions from China now exceed those of the United States, and the LDC/HDC balance of CO_2 emissions will continue to shift thereafter (Figure 25.6). Much of this emission is a consequence of the outsourcing of the early stages of resource acquisition.

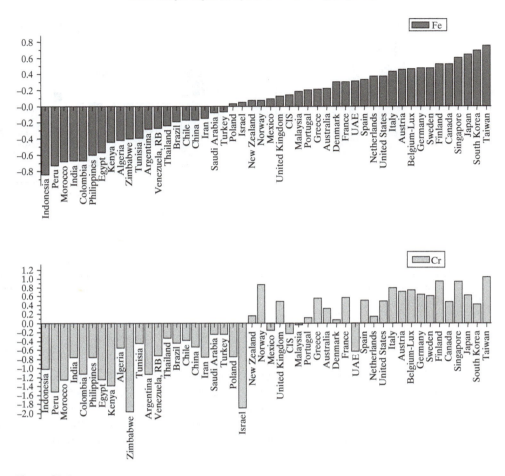

Figure 25.4

Iron and chromium use per capita for a selection of 49 countries. Source: Stocks and Flows Project, Yale University, 2008.

LDCs are becoming the world's manufacturers. Spurred on by low labor costs and high worker commitment, manufacturing is shifting to LDCs, even for very high technology products. As Figure 25.7 demonstrates, for example, the fabrication of semiconductors has shifted substantially to Asia and the LDCs in a very few years.

LDCs have leapfrogging potential. Because of the diffusion of technology and the wide availability of information, LDCs do not have to repeat the development patterns of the HDCs. A frequently cited example is that cellular technology enables communications without the requirement for extensive wired infrastructure. Another is China's public commitment to a circular economy.

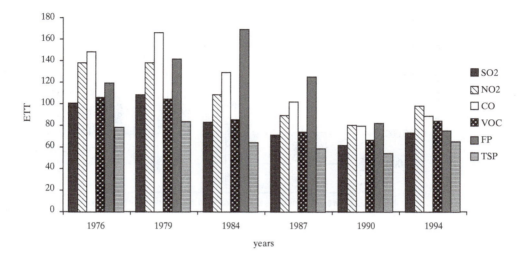

Figure 25.5

The "environmental terms of trade" analysis for Western Europe from the developing countries. (Reproduced with permission from R. Muradian, et al., Embodied pollution in trade: Estimating the "environmental load displacement" of industrialized countries, *Ecological Economics, 41*, 51–67, 2002.)

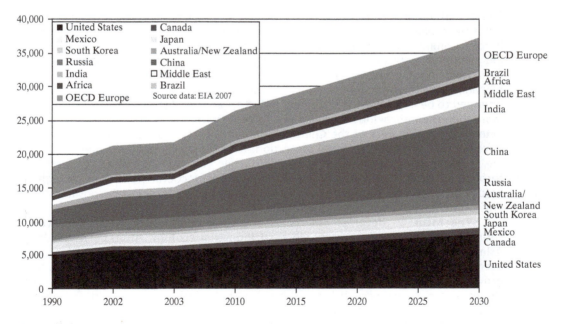

Figure 25.6

Country and region emissions of carbon dioxide, historic and projected. (Reproduced with permission from http://rainforests.mongabay.com/09-carbon_emissions.htm.)

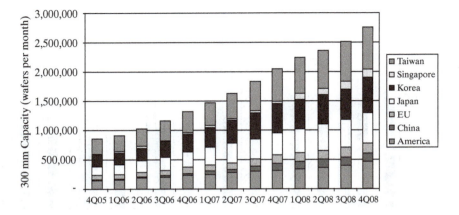

Figure 25.7

The fabrication capacity for 300 mm semiconductor wafers, 2005–2008. Source: http://
wps2a.semi.org/cms/groups/public/documents/marketinfo/export/P042367~2~P033599/
75886-4.jpg.

Case Study: The "Circular Economy" in China

The idea of the circular economy (CE) is in some sense the application of the fundamental industrial ecology metaphor to an entire economy. As China's National Development and Reform Commission, in charge of cleaner production policy, puts it, "The accepted working definition may be interlinked manufacturing and service businesses seeking the enhancement of economy and environmental performance through collaboration in managing environmental and resource issues." But, especially as the circular economy is intended to be implemented in Chinese industrial and legal structures, it is necessarily far more than theoretical industrial ecology. To begin with, the goal of CE, as framed by the 16th National Congress of the Chinese Communist Party in 2002, is primarily economic development, not environmental protection. The primary performance metric, for example, is quadrupling of China's GDP by 2020, while enhancing social equity (particularly through creation of new jobs) and environmental protection. The focus is on high technology products and good economic returns, while managing resource consumption and environmental pollution to the degree feasible.

The focus on development marks one clear demarcation between developed country and developing country views of industrial ecology. Another, particularly in Asia, is the continued challenge of extremely rapid growth. China, for example, has averaged economic growth of over 8.5 percent per year for the past 24 years, a rate of growth that creates enormous social, structural, industrial, and political challenges. Under such circumstances, it is not surprising that relative priorities differ from those usually found in industrial ecology literature, which tends to reflect Western economic contexts. Moreover, the conceptualization of the CE reflects Chinese culture and morals, not those of the West. Thus, the five-dimensional framing of CE reflects the traditional Chinese metaphysics of matter, organized around fire, soil, wood, water, and metal, and its reinterpretation in the CE includes particular attention to the urban–rural tension, which is powerful in countries such as China and India, but scarcely recognized in Europe or the United States where the demographic transition to an urbanized, service- and manufacturing-oriented, society has already occurred.

The CE concept and implementation are in their infancy; the first CE regulations were promulgated in Guiyang, the capital of Guizhou Province, by the local people's congress in 2004. The first national law, adopted in 2008, does not just reaffirm the role of cleaner production in China's environmental regulatory system (the Law on Promoting Clean Production was adopted in 2003), but requires product and packaging takeback and establishes energy efficiency requirements and recycling programs for materials and for larger products such as automobiles and electronics. Additionally, it creates financial incentives, including preferential tax treatment and investment regimes and encourages development of appropriate government purchasing policies. Skeptics have questioned whether many of the operational aspects of the law, such as product takeback, can be implemented in light of the almost complete absence of necessary infrastructure and noted that enforcement provisions are somewhat feeble.

It is far too early to know whether the CE model will work in China, and if so how well. But it is apparent that it represents a reinterpretation of environmentally focused industrial ecology to fit the characteristics of a non-Western culture in a very different phase of its life cycle. Moreover, even at this stage many Chinese are seeking information on more efficient technologies and practices and reaching out for the industrial ecology methods and tools that can enable them to evaluate their industrial systems. Thus, the CE experiment in China offers practitioners of industrial ecology and sustainable engineering both a unique learning opportunity and a chance to contribute significantly to the evolution of a socially, economically, and environmentally preferable Chinese industrial system.

25.3 INDUSTRIAL ECOLOGY AND SUSTAINABLE ENGINEERING PRACTICE IN LDCs

As noted earlier, developing countries are carrying out large and increasing shares of global manufacturing. This provides an imperative for the practice of green engineering, especially because such an approach is easier to "lock in" when facilities are being designed and constructed than after they have been in operation. Pollution prevention focused on air and water emissions is central to achieving good performance goals.

The situation is perhaps more challenging and more important with respect to recycling and remanufacturing activities. The widely known problems engendered by low technology recovery of metals from electronic discards shipped to Asia by HDCs serve as a model for how not to approach end of life. The low-technology methods harm workers and the local environment, while recovering only modest fractions of the valuable metals contained in the products. This situation needs policy attention by HDCs, as well as the application of sustainable engineering by the LDCs.

For most customer products, LDCs and HDCs are served by similar or identical corporations, generally multinationals. A distinction is that some products depend on reliable local infrastructures such as energy and water. To the extent that those requirements can be minimized or avoided, locally designed products can sometimes be both more suitable and more satisfying.

Unlike these customer products, buildings in LDCs are constructed in place and generally make use of local materials. A life cycle assessment, at least a rudimentary one, is often helpful in making material selections. Rubble foundations, timber and bamboo wall and window construction, and tile floors are examples of LCA-preferred materials in many LDC areas.

Approaches to infrastructure can also be very different in LDCs. There are many examples of HDC water treatment systems that were inoperable soon after being built because local expertise in operations and maintenance was not available. To avoid these problems, contemporary approaches are more likely to advocate point-of-use treatment in which flocculant/disinfectant packets are used to treat water on a household by household basis. Similarly, ventilated pit latrines provide an alternative to resource-intensive sewage systems.

The rapid advances in distributed energy generation cited in Chapter 11 also mesh nicely with the energy needs of LDCs. Solar cells, microturbines, and other state-of-the-art equipment are increasingly inexpensive and enable energy provisioning to occur more and more at the local level, thus minimizing or avoiding extensive energy distribution infrastructures.

Cellular telephony is also ideally adapted to LDCs. It requires little in the way of distribution systems, minimizing both the use of materials and the capital costs of a traditional HDC installation.

25.4 THOUGHTS ON DEVELOPMENT IN LDCs

The LDCs need to choose their own development paths, rather than having them externally imposed. Nonetheless, a few issues seem desirable for LDCs in general, and perhaps for HDCs as well.

- Educate on, and practice, industrial ecology and sustainable engineering as early in the development sequence as possible. Many things tangible and intangible are difficult to remove once in place.
- Take advantage of modern technology in industrial processes and in the services provided to individuals. Importing technology enhances productivity and lowers environmental impact.
- Encourage more sustainable development: energy minimization, abundant public transport, avoidance of "sprawl," encouragement of recycling and reuse. Improve on the often poor performance by the HDCs in these areas.
- Promote moderate lifestyles. Data tend to indicate that happiness is not directly related to consumerism, but that it is related to a moderately high quality of life, including both social quality (e.g., social liberalization and democracy) and environmental quality.

We sketch in Figure 25.8 the evolution of development in the three country groupings. The HDCs are already at or near the penultimate state. The RDCs are on their way, but can still choose their target level. The SDCs are earlier on the journey and have the spectrum of opportunities before them. In particular, they are developing from a much more advanced technological base than did the MDCs; they have the

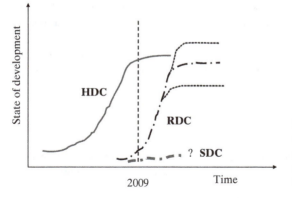

Figure 25.8

The evolution of states of development in highly developed countries (HDCs), rapidly developing countries (RDCs) (for which optional paths are suggested), and slowly developing countries (SDCs).

opportunity to profit from observing both salutary and problematic behavior in MDCs; and without major amounts of infrastructure or buildings in place, they have the opportunity to "do it right the first time." It is not too expansive to suggest that the future of the world as we know it may rest upon how well they take advantage of these opportunities.

FURTHER READINGS

Abeysundara, U.G.Y., S. Babel, and S. Gheewala, A matrix in life-cycle perspective for selecting sustainable materials for buildings in Sri Lanka, *Building and Environment*, doi: 10.1016/j.buildenv.2008.07.005, 2008.

Bai, X., et al., Enabling sustainability transitions in Asia: The importance of vertical and horizontal linkages, *Technological Forecasting and Social Change*, doi: 10.1016/j.techfore.2008.03.022, 2008.

Erkman, S., and R. Ramaswamy, *Applied Industrial Ecology—A New Strategy for Planning Sustainable Societies*, Bangalore, India: AICRA Publishers, 2003.

Fischer-Kowalski, M., and C. Amann, Beyond IPAT and Kuznets curves: Globalization as a vital factor in analyzing the environmental impact of socio-economic metabolism, *Population and Environment*, 23, 7–47, 2001.

Freeman, C., and F. Louca, 2001, *As Time Goes By: From the Industrial Revolutions to the Information Revolution* (Oxford: Oxford University Press).

Galloway, J.N., et al., The environmental reach of Asia, *Annual Review of Environment and Resources*, 33, 461–481, 2008.

Krausmann, F., M. Fischer-Kowalski, H. Schandl, and N. Eisenmenger, The global sociometabolic transition: Past and present metabolic profiles and their future trajectories, *Journal of Industrial Ecology*, 12, 637–656, 2008.

Landes, D. S., 1998, *The Wealth and Poverty of Nations* (New York: W. W. Norton and Company).

Mani, M., and D. Wheeler, In search of pollution havens? Dirty industry in the world economy, 1960–1995, *Journal of Environment and Development*, 7, 215–231, 1998.

Montgomery, M.A., and M. Elimelech, Water and sanitation in developing countries: Including health in the equation, *Environmental Science & Technology*, 41, 17–24, 2007.

Muradian, R., and J. Matrinez-Alier, Trade and the environment: from a 'Southern' perspective, *Ecological Economics*, 36, 281–297, 2001.

Resource Optimization Initiative (ROI), 2008, www.roi-online.org, accessed January 2008. (This is the website of a nonprofit organization dedicated to bringing industrial ecology solutions to businesses and governments in India.)

Streicher-Porte, M., et al., Key drivers of the e-waste recycling system: Assessing and modelling e-waste processing in the informal sector in Delhi, *Environmental Impact Assessment Review, 25*, 472–491, 2005.

EXERCISES

25.1   As they grow, how might LDCs develop their infrastructure (transportation, energy, water, waste treatment) in ways different from the traditional HDC model? What gains could be made by doing so?

25.2   What policy initiatives are suggested by the "virtual pollution" of Figure 25.5?

25.3   You work for an environmental NGO on international issues involving water. Your current assignment is to write a report on an area in East Africa where a long drought has placed significant stress on traditional herding practices, and a low level conflict between Islamic and nationalistic groups is strengthening. What policy areas would you consider in your report, and why?

25.4   You represent a BRIC nation in international sustainability negotiations. Highly developed countries are taking the position that, regardless of your state of development, you should contribute equally to economically costly global mitigation strategies. What is your response, and why?

25.5   You are the prime minister of a developing country and are interested in being remembered for contributing to the rise of your nation to global importance. There has just been a major discovery by a large multinational of an important ore body within your borders, which promises significant contributions to your GDP over the next two decades.

    (a)   What steps would you take to ensure that the revenues from exploitation of the resource are used to best effect for your nation?

    (b)   Your major market is a developed country with high environmental standards. It demands that you apply its standards to all aspects of your operation, including production and smelting of ore, and all transportation and infrastructure activities. This is impossible for you in the short run. What is your response?

CHAPTER 26

Industrial Ecology and Sustainability in the Corporation

A major goal of this book is to enable scientists, engineers, and managers in the private sector to gain perspective, develop targets, and learn to use tools that relate to the environment and sustainability. This chapter brings this information together from a corporate perspective. The first two sections deal with *what to do*, first from the manufacturing standpoint, then from the service sector. The next two sections relate to *why firms should want to embrace industrial ecology and sustainability* and address the strategic and economic benefits that such actions can bring about. The final section discusses *how to do it*—implementing industrial ecology and sustainability within the corporation.

26.1   THE MANUFACTURING SECTOR, INDUSTRIAL ECOLOGY, AND SUSTAINABILITY

The preceding 25 chapters, in some way, relate to the inherent relationships that link the manufacturing sector to issues of the environment and sustainability. These issues can be encapsulated in five themes:

> *Evolving perspectives.*   Industrial ecology thinking adds new vistas to the thinking of the firm: the interactions of technology with society (Chapter 7), the concept of factories and service facilities as industrial organisms (Chapter 4), and the relevance and importance of sustainability (Chapter 2).
>
> *Evolving tools.*   Industrial ecology tools are becoming increasingly diverse and robust: metabolic analysis (Chapter 5), sustainable engineering (Chapter 8), design for environment and sustainability (Chapters 10–11), life cycle assessment (Chapters 12–14), and material flow analysis (Chapter 17).

Resource limitations.   For various resources and in various locales, limitations are beginning to be felt and must be addressed by corporate product design and overall management, as discussed in Chapter 19 (energy), Chapter 20 (water), and Chapter 24 (materials).

Changing markets.   The traditional markets for industrial products, long directed to industrial customers or wealthy suburbanites, are evolving toward the less-developed countries (Chapter 25) and especially the large urban regions within those countries (Chapter 21).

Links to larger systems.   Like it or not, industry increasingly realizes that it is part of a number of holarchic systems (Chapter 15), as discussed from the perspective of industrial symbiosis (Chapter 16), national material accounts (Chapter 18), models (Chapter 22), and scenarios (Chapter 23).

These topics form the basis of a corporate manager's what-to-do list of issues related to the environment and sustainability.

26.2   THE SERVICE SECTOR, INDUSTRIAL ECOLOGY, AND SUSTAINABILITY

Traditional environmental policies, and thus many of the sustainability initiatives that flow from them, have been based on the implicit assumption that manufacturing is the principal source of environmental stress. This perception reflects the visibility of environmental impacts arising from manufacturing, their associated human health impacts, and the often straightforward technological solutions. However, the assumption that regulating and managing manufacturing emissions will create an environmentally and economically sustainable world is simply wrong. Sustainability issues arise from the confluence of population numbers, technological and economic development, and demographic change that characterizes humanity at the beginning of the twenty-first century, and an important part of that evolutionary story is the shift in many economies to services (Figure 26.1). In most developed economies, the service sector encompasses some 65–85 percent of economic activity. This proportion differs by country—the Japanese economy is more manufacturing oriented than the U.S. economy, for example, and China's economic development is heavily oriented toward manufacturing.

A service is any commercial activity where the predominant characteristic is not the production of an artifact, or, as one wag put it, anything you buy that can't be dropped on your foot. Manufacturing an automobile is not a service activity, but leasing the automobile is, because what is being transferred in the transaction is not the title and ownership of the product, but its use. From the same perspective, if a pesticide manufacturer provides a chemical formulation to a farmer, it is selling a manufactured product rather than offering a service. If the pesticide manufacturer instead offers the farmer a comprehensive pest control system that includes applying pesticides, giving advice on planting rotations and crop selection, encouraging changes in mulching and plowing practices, employing biological pest controls, and performing real time monitoring of field conditions with implanted sensors and satellite systems—a so-called integrated pest management (IPM) package—it is servicizing—that is, satisfying a

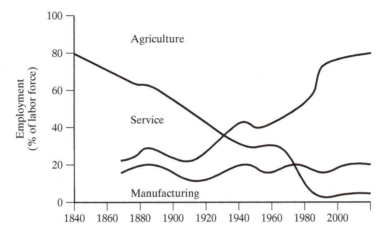

Figure 26.1

Sector employment in the United States from 1840 to 2008. Notice that the shift is not primarily from manufacturing, where employment as a percent of the labor force has remained relatively constant, to service sectors, but rather reflects the long-term mechanization of agriculture and the concomitant release of people for employment in services. Similar patterns characterize other developed countries.

need by providing a service rather than providing products that meet the need. Between these two extremes lie numerous business models that combine traditional pesticide manufacturing and IPM.

A historical example of the differences between product and service firms is provided by the now-outdated model of national telephone monopolies, under which most of the companies supplied telephones and other products on a lease basis. Thus, the customer purchased the use of a telephone, not the telephone itself; should the telephone become nonoperational it was simply exchanged for another. The telephones themselves were refurbished and recycled, sometimes for decades. This approach changed both the functionality the customer perceived, and the design of the product underlying the service, because a product that is designed for a service environment will usually be different from one designed for production and sale. To note just one obvious difference, where a product is designed for recycling and reuse, it will usually be more robust than one which will only be used for one life cycle, then disposed off. Especially where product take-back is part of the service environment, costs that may have been previously external to the manufacturer (and designer) of the product are now internalized to that firm, and over time may affect the way the product is designed. Rather than including just (inexpensive) initial cost, for example, the firm's cost function must expand to include such elements as reverse logistics systems (to get the product back), and the cost of disassembling, refurbishing, or recycling the product, its components, or its materials. The result in most cases of this evolution of product from standalone to service platform is greater environmental efficiency.

Traditionally, industrial ecology has focused on the artifacts, material choices, and energy consumption of information technologies. It is increasingly important to consider the services that such technologies support, however, because of their high cultural and

TABLE 26.1   Product and Process Characteristics of Typical Commercial and Industrial
Service Facilities

Facility	Product	Process
Alpha services		
Dry cleaner	Clean clothing	Solvent cleaning
Hair salon	Hair maintenance	Chemical and physical treatments
Hospital	Health maintenance	Medical care
Beta services		
Appliance repair	Reconditioned appliance	Part and function maintenance
Grounds care	Property maintenance	Moving, fertilizing, etc.
Package delivery	Transport of packages	Pickup, movement, delivery
Gamma services		
Bank	Financial services	Electronic transactions
Burglar alarms	Building monitoring	Electronic communication

economic impact. A life cycle assessment of a personal computer, for example, will identify some salient characteristics of its operation—energy consumption, for example—but if that computer is part of a network that enables widespread teleworking, the social and environmental benefits of that service may well dwarf the impacts embedded in the machine itself. This is a relatively underdeveloped and challenging domain for industrial ecology and sustainable engineering, and we expect considerable development in this area going forward.

It is useful to divide the components of the service sector into three generic types, because they operate differently and because their environmental implications differ as well (Table 26.1). In one common provisioning of services, termed "alpha services," a service is provided in a fixed location and the customer travels to the service. An example is the dry cleaner, who receives clothing, treats it by one or more chemical and/or physical processes, and returns it. The product involved (the clothing) is owned by the customer, transferred temporarily to the service provider, and then recovered. A second type of service is termed "beta service," and its distinguishing characteristic is that the provider performs the service at a customer location. Examples might include appliance repair in the customer's home, or the commercial provision and servicing of leased photocopy machines. In the latter case, the material is owned by the service provider, transferred temporarily to the user, and then recovered.

The advent of modern technology has enabled a new class of services to arise in recent years. These "gamma services" are provided without either the customer traveling to the service or the service traveling to the customer. Rather, the service is provided by electronic means, such as bank transfers by telephone, or software upgrades delivered over the Internet rather than on disks sent by regular mail. In its more extensive implementations, gamma services permit entire careers to be accomplished with a minimum of physical movement, through computers, telephones, the Internet, and associated software and hardware.

Evaluating the environment and sustainability implications of services is a complex process. From an environmental perspective, all services operate on technology

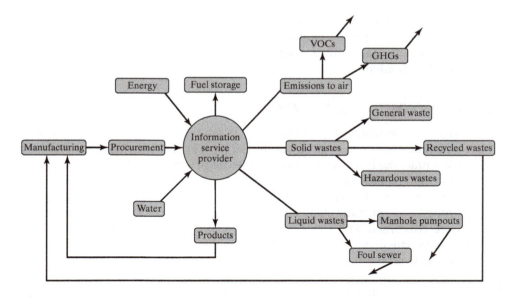

Figure 26.2

The information networks operated by telecommunications companies provide a framework for many environmentally preferable services, such as teleworking, but they have their own environmental impacts associated with their operation. (Adapted from *Annual Report,* London: British Telecom, 1997.)

platforms: education services require school buildings; e-retailing requires transportation networks; and teleworking requires information infrastructures. In fact, services tend to be slightly more capital intensive per employee than manufacturing. For example, consider the provisioning of Internet services. Not only is a physical network, with all its operational aspects, required (Figure 26.2), but the components attached to that network, such as computers, have environmental impacts associated with their manufacture, use, and end-of-life stages as well.

The policies of service firms can have significant sustainability implications. As regards the environmental dimensions of sustainability, service firms can fulfill new and important roles. Especially as firms such as WalMart become dominant in their sectors, they gain the capability to leverage suppliers to improve the environmental efficiency of the supply chain. They are also in a position to educate consumers to make informed choices, thus facilitating environmentally preferable resource consumption and product use. Different kinds of service firms can help customers substitute services for material and energy use: an example is the support of telework and virtual office arrangements by communications providers.

Service firms share many attributes with manufacturing firms yet have a somewhat different suite of opportunities and liabilities. Products are involved in both cases, but a manufacturing firm designs and makes a product and then (usually) hands over responsibility for it to the service producer. The service firm buys not only the equipment but also the "virtual impacts" that relate to the actual impacts of the manufacturing chain and the responsibility for product end of life. From a social standpoint, the service firm is generally in a more strategic position to influence personal or corporate choices. Overall, it is very clear that service firms are vital components of improved environment and sustainability performance, not uninterested bystanders.

26.3 ENVIRONMENT AND SUSTAINABILITY AS STRATEGIC

The environmental program of a firm is seldom equivalent to what we would term a true industrial ecology program. In the past, environmental issues for the firm arose from practices involving emissions or waste streams, and the governmental response—remediation and compliance—was driven by regulation. Moreover, the focus of environmental activity within and outside the firm was primarily on the manufacturing process and generally characterized by end-of-pipe controls on emissions. Regulations were concerned with single media, individual substances, specific sites, or particular process emissions. There was little recognition (or incentive to recognize) that impacts arising from the firm's activities were fundamentally linked with regional and global natural, technological, and economic systems. Industrial ecology, however, requires that recognition and encourages a fundamental integration of these issues at the level of the firm.

The principal incentives to action also shift in this process. Although well-crafted regulation remains important, market demand, which tends to be far more important in the cultural model of businesspeople, becomes increasingly powerful. Sophisticated customers that drive much commercial procurement, such as governments in Japan, Europe, and the United States, increasingly demand environmentally preferable products and services, even though defining such offerings is often difficult. Ecolabeling schemes, such as the U.S. Energy Star for energy-efficient electronic products or the German Blue Angel ecolabeling scheme, impose product design and operation requirements, a far cry from end-of-pipe controls on manufacturing facilities.

It is important to recognize the difference in kind, not just degree, in this transition. The Blue Angel requirements for personal computers, for example, include modular design of computer systems, customer-replaceable subassemblies, avoidance of bonding between incompatible materials such as plastics and metals, and post-consumer product take-back. Implementing take-back requirements alone requires development of a reverse logistics system to get the products back to a central location where they can be disassembled. This establishes a new relationship with suppliers as they become responsible for the appropriate initial design of their components or subassemblies, it develops a new corporate capability (disassembly and management of end-of-life products and material streams), and it influences the business planning process. It is obvious that none of these requirements has anything to do with traditional environmental approaches but rather involve the strategic activities that are the core of any manufacturing company. Failure to effectively perform these functions does not, as with traditional environmental regulation, simply expose the firm to liability. Rather, failure to perform in this new mode of operation affects the firm's ability to market its product, and the cost structure of each product. Failure to design products, logistics systems, and business and marketing plans for post-consumer take-back, efficient disassembly, and recycling does not simply raise a firm's overhead. Instead, it prices a firm's products out of the market, regardless of how efficiently the product can be made initially. Environmental capabilities thus change from being a way to control liabilities to becoming a potential source of sustainable competitive advantage.

When environmental issues are regarded as overhead, they are only peripheral concerns of corporate management. As they become more strategic, however, environmental considerations become one of the routine objectives and constraints that the firm must manage, whether in product design, manufacturing, or business planning. For example, the environmental specialists in an electronics manufacturing firm may identify the use of lead solder in printed wiring board connections as a concern, but the complexity and technological constraints of bonding technology may make substitution of a less toxic alternative for all solder uses impractical, at least for a time. This is because the environmental considerations are part of the design equation, but not the only part; economic, competitive, technological, and other considerations are also balanced in the process. The evolution of environmental issues from overhead to strategic may appear to devalue environmental issues because they no longer stand apart. However, including them in the strategic decisions of the firm provides a far higher level of environmental efficiency over the longer term.

This shift in environment from overhead to strategic for the firm entails substantial organizational change. The environment, health, and safety (EHS) group in most firms has traditionally had almost nothing to do with any of the strategic or substantive decision-making processes; in organizational terms, it has weak or nonexistent coupling to other functional units within the company. Once the firm begins to understand the implications of the shift, however, EHS couplings to corporate operations are greatly strengthened. Complying with Blue Angel requirements clearly is a challenge to R&D organizations, business planning and product management operations, and other operational elements; it cannot be done by any single organization—and, in particular, it cannot be done by the EHS group alone.

26.4   THE CORPORATE ECONOMIC BENEFITS OF ENVIRONMENT AND SUSTAINABILITY

Corporations are in business to make money for their stockholders. This goal is in conflict with purchasing equipment to minimize pollution, the traditional interaction of managers with environmental issues. As a result, many managers regard anything related to the environment and sustainability as detrimental to financial performance. However, the opposite is often, perhaps even, generally, true. Ab Stevels of Philips Electronics has listed five ways (slightly revised by us) in which considering the environment and sustainability can be financially beneficial:

1. *Making money by Design for Environment and Sustainability (DfES).* A primary goal of DfES is to use less of everything—materials, energy, water—and to emit less—discards, by-products, and so on. All of these actions are financially beneficial.

2. *Making money by using the supply chain.* Suppliers provide what they are asked to provide, but enlisting them in green design efforts brings benefits to both supplier and purchaser, such as lower weight components, elimination of problematic chemicals, and environmentally superior packaging.

3. *Making money by green marketing.* It is a truism that few customers will pay extra for green products, but green attributes such as energy minimization during product use or added durability can tip the scales in the direction of a purchase. Note the transition here from simply "green products," which include only environmental concerns, to "sustainable products." The latter addresses not only environmental considerations but, as in the case of energy efficiency, also enhances social benefit (e.g., by providing the customer with an energy efficient, and therefore preferable, product).

4. *Making money by increasing quality.* A design review focused on environment and sustainability can often minimize complexity and thereby improve product yield, with fewer rejects at either supplier or manufacturer level.

5. *Making money by paradigm shifts.* Paradigm shifts are changes in mindset—realizing that a design or an operation need not be done the way it always has been done in the past. DfES activities sometimes initiate these shifts, as in using folded cardboard instead of Styrofoam in packaging (more compact, easier to recycle) or marketing services instead of products (enhances take-back, retains customer contact).

The bottom line for actions related to the environment and sustainability is that they need to make economic sense. When these actions are seen to be economically beneficial, they are much more easily integrated into corporate functions, top to bottom.

Case Study: Green Chemistry at Glaxosmithkline

The initial perception of many firms is that "greener" products and processes are inherently more expensive. That perception is refuted regularly, nowhere more than in the green chemistry emphasis of GlaxoSmithKline, a large pharmaceutical manufacturer. David Constable, who heads the green chemistry program, points out that the synthesis of drugs typically takes six or more steps, each with solvents and often with by-products. Traditional organic solvents constitute about 90 percent of overall reaction mass, and their manufacture and recovery represents 70–80 percent of all life cycle impacts. The situation can be much improved by carefully choosing the synthesis route and solvent. As seen in illustration CS-1 below, different synthesis routes generate vastly different accounts of waste material. Furthermore, because solvent recovery and waste disposal are minimized, large monetary savings are common.

26.5   IMPLEMENTING INDUSTRIAL ECOLOGY IN THE CORPORATION

The kinds of organizational restructuring required to implement the principles of industrial ecology and sustainability in existing firms involve changes in both the internal context—within the firm's existing divisions and organizational entities—and externally, as the firm seeks to achieve new relationships with customers and suppliers. While it is true that developing more socially, environmentally, and economically efficient technologies is an important support for this evolution, it is also true in many firms that the most difficult barriers are cultural, not technological.

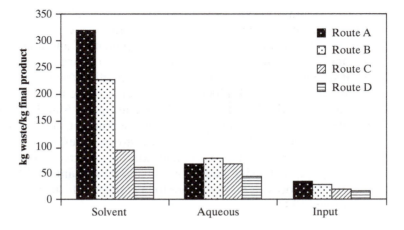

Figure CS-1

Solvent waste, aqueous waste, and input waste generated by different synthesis routes for the same compound (Courtesy of D.J. Constable)

Among the obvious challenges is determining where responsibility for various components of an industrial ecology program rests. In some cases, such as traditional environmental compliance, ownership is relatively clear; in other cases, such as implementation of global worker policies, from pay to safety issues, ownership may be fragmented. Implementing industrial ecology and sustainability in the firm involves the types of activities and programs usually associated with any culture change in complex organizations:

1. Individual champions who ideally come from several appropriate organizations within the firm and are willing to take the associated risks must be identified and supported. For example, green accounting issues would be the province of a financial organization champion, while DfES programs might be championed by someone from the product research and development group.

2. Barriers to change, especially those which arise from corporate culture and informal patterns of behavior, must be identified and reduced. Support from the highest levels of the corporation, combined with extensive training programs, is essential in this regard.

3. The least threatening method of introducing new techniques, tools, and systems, such as design teams that include specialists from many disciplines, should be identified and used. The best changes are often those that are never recognized by those that implement them.

4. Strong rationales for new activities, defined in terms of the target audience's interests and cultural models, must be developed. For example, if a company manufactures personal computers the Blue Angel requirements can be used with the design team as examples of customer demand patterns; for a service firm, the potential to create a market for services that replace environmentally problematic products can be a driver for behavior change.

5. Managers and firms are more likely to be comfortable with practices and procedures that are widely accepted, even if there are some economic penalties involved. For example, an important component of an industrial ecology implementation program at many firms is reliance on a formal environmental management system (EMS). EMSs range from the relatively simple to the quite complex, as illustrated by ISO 14000, a set of environmental systems requirements developed by the International Organization for Standards (ISO) through a negotiating process involving industrial, government, and public stakeholders. In many cases, internally developed EMSs are more effective, more efficient, less bureaucratic, and less expensive than external systems, especially where certification requires extensive audits by outside consultants. Still, because ISO is a recognized process, it may be easier to implement a 14000 process than to develop an internal EMS.

Especially for those beginning their corporate career, it is important to remember that most firms are open to environment and sustainability initiatives, but those initiatives must align with the economic interests and internal culture of the firm if they are to be successful. Policies that encourage such alignment and reward such initiatives on the part of firms, reinforce internal sustainability champions and reward appropriate institutional innovation and are thus important components of any successful corporate governance portfolio.

Philips Electronics's Ab Stevels has discussed three alternative approaches that corporations can take toward issues of environment and sustainability (Table 26.2). One is the defensive approach classically adopted by most industries over the years.

TABLE 26.2   Alternative Corporate Approaches to Environment and Sustainability

Item	Defensive approach	Cost-oriented approach	Proactive approach
Driver	Legislation	Money saving	Customer satisfaction
Management	• Environmental declaration	• Policy	• Vision
	• Command and control	• Projects	• Integrated into the business
Main objectives	• Compliance	• Continuous development	• Transformational change
	• Stasis	• Evolutionary change	• Exceed competitors
Control	Afterward	Built-in	Up-front
Activities	• Substance reduction	• Material reduction	• Lower cost for user
	• Product take-back	• Disassembly time reduction	• Green designs
			• Durable products
Training	How to comply	How to reduce	How to integrate into business
Financial implications	Costly	Neutral	Beneficial
Benefits	Green and societal	Green and company	Green and customer

Source: Adapted from Stevels, A.L.N., *Adventures in EcoDesign of Electronic Products*, Delft, The Netherlands: Delft University of Technology, ISBN/EAN 978-90-5155-039-9, 2008.

Driven by legislation, it aims solely for compliance, is invariably costly, and is not industrial ecology. The second approach is cost oriented. It aims toward evolutionary change, saving money by minimizing resource use and involving suppliers. We might term this "industrial ecology light." The third approach is transformational and proactive, aimed at fully integrating industrial ecology and sustainability throughout the firm. It is obvious that this transition from approaches one to two and, we hope, to three, is what we wish for every firm.

FURTHER READING

Allenby, B.R., *Industrial Ecology: Policy Framework and Implementation*, Upper Saddle River, NJ: Prentice-Hall, 1999.

Castells, M., and P. Hall, *Technopoles of the World: The Making of 21st Century Industrial Complexes*, London: Routledge, 1994.

Esty, D.C., and A.S. Winston, *Green to Gold: How Smart Companies Use Environmental Strategy to Innovate, Create Value, and Build Competitive Advantage*, New Haven, CT: Yale University Press, 2006.

Graedel, T.E., The life cycle assessment of services, *Journal of Industrial Ecology, 1* (4), 57–70, 1998.

Pascale, R.T., Surfing the edge of chaos, *Sloan Management Review, 40* (3), 83–94, 1999.

Porter, M.E., Green and competitive, *Harvard Business Review, 73* (5), 120–134, 1995.

Reiskin, E.D., A.L. White, J.K. Johnson, and T.J. Votta, Servicizing the chemical supply chain, *Journal of Industrial Ecology, 3* (2&3), 19–31, 2000.

Stevels, A.L.N., Adventures in EcoDesign of Electronic Products, Delft, The Netherlands: Delft University of Technology, ISBN/EAN 978-90-5155-039-9, 2008.

U.S. Environmental Protection Agency, The Lean and Green Supply Chain, EPA Report 742-R-00-001, Washington, DC, 2000. Available at www.epa.gov/oppt/library/pubs/archive/acct-archive/pubs/lean.pdf.

EXERCISES

26.1 What advantages accrue to a firm when it regards issues of environment and sustainability as strategic, not as overhead?

26.2 Choose a service facility with which you are familiar, perhaps because you have been employed in helping to provide the service or because you use the service frequently. Classify the service as alpha, beta, or gamma and describe the service provided and the process by which it is provided.

26.3 A hospital can monitor the status of patients with heart irregularities either by having them visit the outpatient clinic once a week or by giving each patient an electronic monitor and periodically transferring data electronically from home to clinic. Assume the medical care is equivalent either way. Discuss the environmental advantages and disadvantages of the two options.

26.4 The president of your company, a business-to-consumer Internet firm that sells CDs and books all around the world, has just called you into her office and asked you to make the company "Triple Bottom Line (TBL) compatible."

    (a) What are the specific issues a company such as yours should consider in evaluating its TBL status?

(b) What kinds of industrial ecology research might you want to initiate to get a better idea of your TBL status?

(c) What types of additional information about your operations might you want to collect to evaluate the social dimensions of your firm's operations?

26.5 Your firm is investing in a poor developing country. You are going to manufacture widgets there, but are concerned that there is no infrastructure in the country to handle the wastes defined as hazardous by your home country, a highly developed country with high environmental standards. If you have to build the infrastructure yourself, it would make your investment in the developing country uneconomic, given the risk profile of the venture. You know the jobs that you will create and the technology transfer that would occur if your project goes forward are important to the developing country, but are concerned about the health impacts and potential liability issues. What would you do?

CHAPTER 27

Sustainable Engineering in Government and Society

27.1 ECOLOGICAL ENGINEERING

In earlier chapters, we have employed ecological concepts and analytic approaches as guides to activities that might be undertaken by industrial ecologists. Ecological engineering, as defined by William Mitsch, one of the specialty's founders, is "the design of sustainable ecosystems that integrate human society with its natural environment for the benefit of both." Ecological engineering is most often encountered in the design of water treatment systems, where the filtration and cleansing activities of plants and soil substitute for a conventional water treatment plant. The key feature of ecological engineering is that the engineer abandons the concept of total control and instead aims to enable functionality by the ecosystems themselves.

Three attributes separate ecological engineering from other engineering specialties:

Ecological engineering is based on self-organizing hierarchical open (SOHO) systems, as discussed in Chapter 15. The goal of the ecological engineer is to initiate and enable a suitable SOHO system by seeding appropriate species into the area being engineered and then allowing self-organization to occur.

Ecological engineering designs and enables systems that are primarily or entirely self-sustaining. These systems are usually solar based, requiring minimal or negligible energy from traditional, human-engineered sources such as fossil fuels or electricity. As a result, SOHO systems are generally less costly than conventional systems but require more land. Ecological engineering supports ecosystem conservation and development, thus providing benefits to nature as nature provides benefits to humans.

An example of ecological engineering is the rainwater harvesting system of Kroon Hall at Yale University (Figure 27.1). In this design, rainwater falling on the

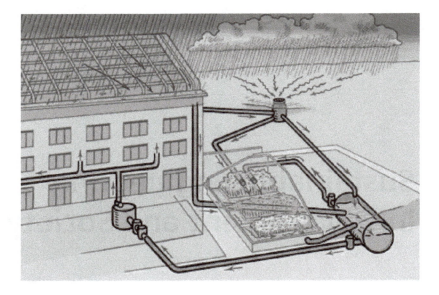

Figure 27.1

The rainwater harvesting system of Kroon Hall, Yale University. Water collected from the roof of the building and from the ground (tank on upper right) empties into the pond, where it is cleansed by aquatic plants. It is then sent to the large tank for storage and for subsequent use in toilets and for landscape irrigation. (Courtesy of Yale University.)

roof and grounds is directed to a pond containing aquatic plants that serve as biofilters. Once cleaned, the water is used for toilets in the building and for landscape irrigation. The system is expected to save some 2 million liters of potable water annually and to pay for itself in decreased potable water charges in 10 years.

Ecological engineering is also employed in the restoration of ecosystems that have been degraded or destroyed by traditional engineering approaches. However, ecological engineering neither includes the cleanup of contaminated "brownfield" sites so that they can be used by humans again, nor the dredging of rivers, lakes, and harbors to facilitate human use, because in these cases ecosystem restoration and function are not taken into account, at least not in any central way. It does, however, encompass activities such as the restoration of wetlands degraded or destroyed by human development projects. In such restorations, functionality useful to humans such as water purification or erosion control may result so long as ecosystem integrity is among the primary project goals.

27.2   EARTH SYSTEMS ENGINEERING AND MANAGEMENT

Most of industrial ecology, and indeed most of this book, is directed toward altering our technological society so that we balance social, environmental, and economic domains and reduce or eliminate impacts on the environment. This is essentially a proactive approach. Other engineering activities that some might term "industrial ecology" relate to trying to manage major Earth systems that are already being affected by human activity: We term this "Earth system engineering and management" (ESEM).

ESEM arises from the realization that because of changes in culture, technology, economic systems, human population levels, and economic activity, the dynamics of many fundamental natural systems are increasingly dominated by the activities of a single species—ours. This is not a new concept but is made vital by the increasing scale of human impact, and the increasing coupling between human activities—an increasingly complex "built environment"—and different natural systems.

ESEM is defined as "the engineering and management of Earth systems (including human systems) so as to provide desired human-related functionality in an ethical manner." Important elements of this definition include treating human and natural systems as coherent complexes, to be addressed from a unified perspective, as well as an understanding that requisite functionality includes not just the desired output of the technological system, but also respect for, and protection of, the relevant aspects of coupled natural systems. This can include things valued by humans, such as aesthetics, or ecosystem services such as flood control, as well as respecting biodiversity or the global water cycle as independent values in themselves.

27.3   REGIONAL SCALE ESEM: THE FLORIDA EVERGLADES

The degradation of the Florida Everglades occurred because South Florida is an area of significant agricultural activity, rapidly increasing population and economic activity, ecological importance, and contrasting local cultures of economic development and environmental protection. Human intervention has been ad hoc but substantial; nearly 1800 miles of canals have been built over the past 50 years, diverting some 1.7 billion gallons a day of water flow to service agriculture and people and to manage flooding. Equally important, water quality has changed significantly over the same period, and a number of invasive species are increasingly successful. The role of technology in shaping the Everglades is worth noting from an industrial ecology and sustainable engineering perspective, for the most important contributions were not necessarily the obvious ones, such as hydraulic engineering. In particular, a critical technology shaping Southern Florida was air conditioning for both homes and cars, without which highly uncomfortable hot and humid summers would discourage many potential immigrants.

As each component of the complicated human patterns surrounding the Everglades continued its growth, the Everglades ecosystem and the associated human communities that depend upon it became increasingly unsustainable. In response, the Comprehensive Everglades Restoration Plan (CERP), an effort spearheaded by the U.S. Army Corps of Engineers and the South Florida Water Management District, was developed in 1999 (Figure 27.2). Some idea of the magnitude of such an effort is indicated by CERP's scale: It includes more than 40 major projects and 68 project components, at a cost estimated at $10.9 billion in 2004 dollars, over a timeframe of three decades. Its goal is to restore water quality and flow in natural systems to functional levels, while continuing to support industrial, agricultural, settlement, and other human

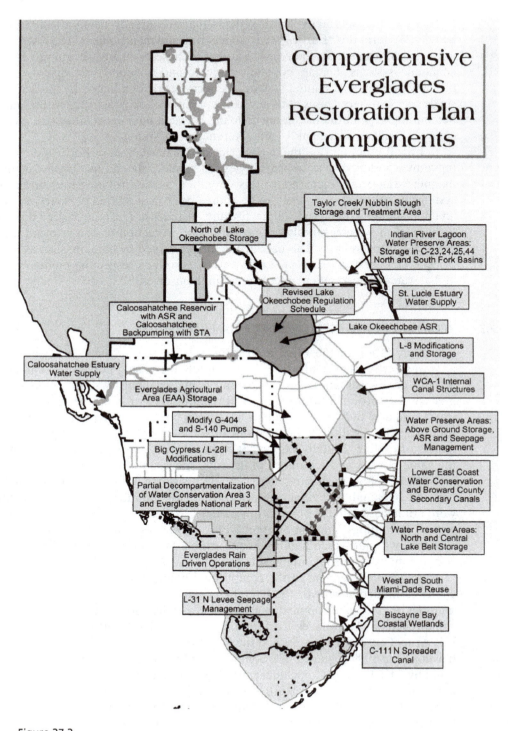

Figure 27.2

The 2001 restoration plan for the Florida Everglades. (Reproduced from www.evergladesplan.org)

activities, and avoiding flooding and other undesirable events. The hope is to maintain a desirable ecological community and human activity at politically acceptable levels, a challenge that will require implementation of over 200 separate projects. It remains to be seen to what extent this heroic goal can be realized.

27.4   GLOBAL SCALE ESEM: STRATOSPHERIC OZONE AND CFCS

In 1928, the American chemist Thomas Midgley developed the first of a chemical family known as chloroflurocarbons, or CFCs, as a substitute for ammonia, chloromethane, and sulfur dioxide—toxic materials that were then being used as refrigerants. Because CFCs were very stable molecules, they were both low in toxicity and not very volatile, and thus proved to be important industrial cleaning compounds and propellants as well as refrigerants. Unfortunately, their stability meant that upon release they migrated through the troposphere to the stratosphere where, upon absorbing high energy photons, they split off chlorine radicals that catalyzed the destruction of stratospheric ozone. Ozone in the troposphere is a component of smog, and undesirable, but stratospheric ozone performs the important function of absorbing high energy ultraviolet (UV) energy before it strikes Earth's surface. Because life at the Earth's surface evolved without significant high energy UV exposure, it was significantly affected by such radiation, which had the potential to kill algae and plankton and generate high levels of cataracts and skin cancer in humans.

The catalytic action of chlorine released from CFCs on stratospheric ozone was first described by Molina and Roland in 1973. Their work and that of others eventually led to the Montreal Protocol on Substances that Deplete the Ozone Layer, which focused on stratospheric ozone recovery to pre-CFC levels. This agreement and subsequent actions have come to be a prime example of a significant positive achievement of international environmental policy.

For the industrial ecologist, this episode has several important lessons. Perhaps most trenchantly, when CFCs were first developed they could have been poster molecules for "green engineering"—they were effective, nontoxic, had no obvious environmental impacts, substituted for much more problematic materials, and were easy and cheap to produce. It was only much later that the true impact of these materials on a major Earth system was recognized. It is important to strive for more benign and appropriate chemistry systems; it is just as important to remember the need for constant vigilance and recognize how little we understand regarding the complex systems we are inevitably affecting.

CFCs were not only a case of a relatively small human activity destabilizing a major Earth system. They were also a classic example of how human, natural, and industrial systems are now integrated at regional and global scales: industrial economics and human values (e.g., health of workers, safety of refrigerant systems), stratospheric physics and chemistry, radiation balances and living systems, international politics—all were coupled.

But perhaps the most important thing to note in this case study is that it is a successful example of ESEM. A problem was identified, a transparent and active political dialog resulted, agreement among the preponderance of stakeholders was accomplished before critical tipping points in the Earth system at issue were reached, and subsequent tracking of important metrics—in this case, the size of the annual ozone hole above Antarctica—helped ensure that the policy initiatives were having the desired effects on the systems at issue.

27.5   GLOBAL SCALE ESEM: COMBATING GLOBAL WARMING

Global warming is the phenomenon in which human-associated emissions of gases to the atmosphere result in increased trapping of outgoing infrared radiation. The principal anthropogenic "greenhouse" gas is carbon dioxide, though methane, CFCs, nitrous oxide, and other gases also contribute. Because greenhouse gases are directly related to the use of energy, and because energy is the enabler of modern technology and modern life and culture, an ESEM approach to climate is much more complicated and challenging than was the case with CFCs. A further complication is that the climate system is tightly coupled to a great many other parts of Earth systems functioning (Figure 27.3).

Because of the potential severity of global warming, a number of mitigative approaches have been proposed. We discuss them below.

27.5.1   Capturing Carbon Dioxide

Human emissions of carbon dioxide, the main anthropogenic greenhouse gas, are not just a phenomenon of the Industrial Revolution. In fact, initial perturbations to atmospheric CO_2 concentrations arose from the deforestation of Europe and North Africa between the tenth and fourteenth centuries. Much later, the development of the internal combustion engine and, as a result, the automotive industry, greatly accelerated emissions of carbon dioxide. As fossil fuel use increases in our modern world, atmospheric CO_2 concentrations continue to increase as well. A possible ESEM alternative is to capture carbon dioxide from stack gases, liquify it, and inject it into underground or undersea aquifers and geologic formations. There, it may remain almost indefinitely.

Strictly speaking, CO_2 sequestration from power plant stacks is not ESEM, but pollution control, just as is the capture of volatile organic gases before they can leave an industrial facility. In common usage, however, all proposals for dealing with global warming tend to be lumped into an ESEM framework. Regardless of how it is classified, if a fossil fuel power plant is designed to combust carbon-based feedstocks, transfer the combustion products directly to long life reservoirs and produce energy in the form of electricity or hydrogen, the perspective on power generation and the environment undergoes fundamental change. As Figure 27.4 illustrates, rather than being part of a significant environmental problem—a large emitter of greenhouse gases—the power plant becomes part of the solution—a factor in the control of greenhouse gas atmospheric concentrations.

27.5.2   Sequestering Carbon in Vegetation

CO_2 sequestration, as described above, is an approach that aims to prevent CO_2 from being emitted into the atmosphere. Once it is there, however, another potential ESEM approach is to remove a portion of it. One technique that has been fairly widely

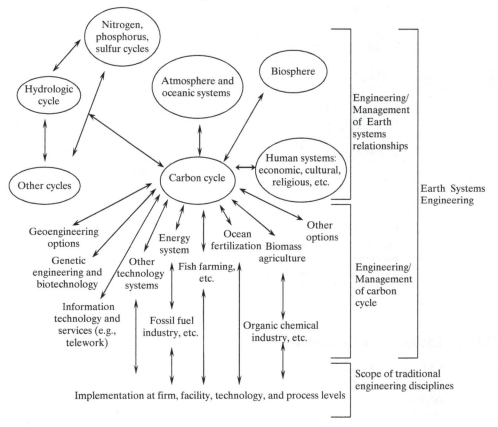

Earth Systems Engineering and Management:
Climate Change–Carbon Cycle Schematic

Figure 27.3

A high-level ESEM mapping of some of the systems implicit in the climate change management challenge. The diagram illustrates the complicated relationships that make climate change a much more difficult issue to address than stratospheric ozone depletion by CFC emissions.

embraced has been the planting of fast-growing trees, since atmospheric CO_2 is the building block for the cellulose from which trees are made. It is clear that reforestation of previously forested areas will indeed store CO_2 for at least a period of time, though the absorption rate slows as the trees age. Although the degree of long-term gain is imperfectly understood, tree planting has many beneficial aspects besides carbon storage, and may be increasingly adopted. Increased biomass proposals raise other issues, however: Can significant increases in biomass production be done in such a way as to avoid destabilizing the global nitrogen cycle? How critical are genetically engineered forms of biomass (e.g., trees designed to fix their own nitrogen) to the implementation of this plan? Once again we are challenged to think of any action from a very broad systems perspective.

Figure 27.4

The Sleipner natural gas/condensate complex in the North Sea off Norway. The natural gas in this region typically contains about 9 percent CO_2, which must be reduced to less than 2.5 percent before it can be sold. The larger platform comprises the extraction and condensation equipment, while the smaller platform to the left houses the CO_2 extraction and sequestration equipment. (Photo courtesy of Statoil.)

27.5.3 Sequestering Carbon in Marine Organisms

CO_2 is a building block for phytoplankton, the tiny marine organisms that carry out nearly half of the photosynthesis on Earth. The reproduction and growth of those organisms is limited in most parts of the oceans by the availability of nutrients, particularly iron. It was thus proposed in the early 1990s that if the oceans were fertilized with iron, the resulting growth of organisms would remove considerable CO_2 from the atmosphere.

Spreading fertilizer on the ocean surface on a regular and widespread basis is an enormously ambitious project, but a few tests have been made to assess the feasibility of the idea. In the most extensive of these, a seeding experiment in the Southern Ocean south of Tasmania, increased phytoplankton growth was stimulated and maintained for over a month. It was not clear that the CO_2 that was incorporated was then transferred to the deep ocean, as would be required for the approach to be effective. In addition, there is concern for unintended side effects such as deoxygenating the deep ocean and disrupting the structure of marine food webs. Given the current scientific uncertainty, it remains unclear whether, and at what scale, this approach should be employed.

27.5.4 Scattering Solar Radiation with Sulfur Particles

Another potential ESEM approach to the mitigation of global warming avoids dealing with CO_2, but rather with preventing incoming solar radiation from reaching the planet's surface. This idea was proposed some years ago by Russian climatologist

Mikhail Budyko, who envisioned injecting some 35 Gg of sulfur dioxide annually, about 25 percent of the amount presently released by fossil fuel burning, directly into the stratosphere. He calculated that such an amount, once converted there to sulfate aerosol particles, should significantly enhance the backscattering of solar radiation to space. The success of the technique depends on an accurate assessment of stratospheric sulfur dioxide to sulfate transition rates at all latitudes and seasons, and it is uncertain whether this information can be precisely derived. Still more challenging, however, are the logistical difficulties of delivering many gigagrams of gases or particles to an altitude near the limit of modern aircraft by fleets of thousands of planes. A possible alternative is to load sulfur particles into ballistic shells and shoot them into the stratosphere using the guns of the world's large naval vessels (several thousand rounds per day, day after day, year after year). By either method, the cost would be in the tens of billions of U.S. dollars annually, and the potential environmental impacts appear highly problematic.

27.5.5   Reflecting Solar Radiation with Mirrors in Space

An alternative to injecting scattering particles into Earth's upper atmosphere is to send mirrors or other reflective devices by space satellite to the Lagrangian L1 point (a point along the Earth–Sun line where no net forces act on a small object), so that the objects might remain at that location and reflect radiation away from Earth indefinitely (Figure 27.5). Unfortunately for the concept, gravitational displacement forces from planets and other celestial objects are to be expected at the L1 point, so the objects would require an active positioning system in order to remain in place. Active positioning systems imply such things as mechanical components and compressed gases, and preliminary assessments of this idea suggest that spacecraft stabilization over long periods of time may be impractical. In addition, of course, the enterprise would be extremely expensive, not to mention beset with political intricacies, and the size and reflectivity of the devices would have to be determined very precisely in order not to heat or cool Earth more than desired.

Figure 27.5

Conceptual diagram for solar reflectors placed at Lagrangian point L1 of the Earth–Sun system to decrease the amount of solar radiation incident on Earth. There are five Lagrangian points, three of which are shown (the other two are at the points of equilateral triangles from the Earth–Sun line). Objects located at these points are stable, or nearly so, because gravitational acceleration and centrifugal acceleration are exactly in balance. L1 is approximately 1 percent of the distance from the Earth to the Sun.

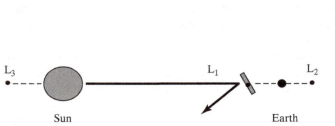

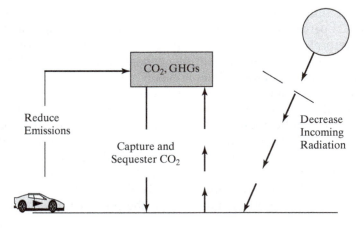

Figure 27.6

The system of potential ESEM approaches to combat global warming.

27.5.6 Global Warming ESEM

As we have seen, there are many ways in which technologies, economic practices, and cultures can be used to modify human impacts on the carbon cycle and climate systems. Figure 27.6 illustrates how these options might be gathered together into a portfolio of options to enable management of the carbon cycle and climate. It is important to recognize that these options address symptoms rather than root causes, and thus do not adequately address other implications of global climate change. In particular, none of these geoengineering approaches would reduce the increasing acidification of the ocean as it continues to absorb CO_2 and create carbonic acid, thereby affecting marine organisms that use calcium carbonate for their shells. This is evidence of the high degree of interconnections among fundamental natural and human cycles, many of which are not addressed by targeted approaches. Imperfectly understood couplings and complexities raise caution flags for most or all geoengineering ideas.

27.6   THE PRINCIPLES OF ESEM

As the examples in the previous section suggest, the institutional, cultural, and knowledge capabilities to carry on ESEM do not yet exist. The ethical structure necessary to support integrated ESEM activities is lacking as well. However, Earth systems that are so severely degraded that they are or threaten to become nonfunctional may nonetheless require that we take some action. In these cases, one can draw on experience to date with complex systems engineering projects and related fields such as adaptive management to generate a basic set of ESEM principles that, although still illustrative, create an operational foundation. These principles can be sorted into three categories: theory, governance, and design and engineering.

27.6.1 Theoretical Principles of ESEM

The theoretical underpinnings of ESEM reflect the complexity of the systems involved and our current levels of ignorance. The overriding dictum in ESEM is thus cautionary.

- Only intervene when necessary, and then only to the extent required, because minimal interventions reduce the probability and potential scale of unanticipated and undesirable system responses.
- ESEM projects and programs are not just scientific and technical in nature, but unavoidably have powerful economic, political, cultural, ethical, and religious dimensions. Social engineering—efforts to change cultures, values, or existing behavior—and technical engineering both need to be integrated in ESEM projects, but they draw on different disciplines and knowledge domains and involve different issues and world views. An ESEM approach should integrate all these factors in order to arrive at a satisfactory outcome.
- ESEM requires a focus on the characteristics and dynamics of the relevant systems as systems, rather than as sums of individual components. The components will, of course, also have to be considered: ESEM augments, rather than replaces, traditional engineering activities.
- Boundaries around ESEM initiatives should reflect real-world linkages through time, rather than disciplinary simplicity.
- Major shifts in technologies and technological systems should be evaluated before, rather than after, implementation of policies and initiatives designed to encourage them. For example, it is apparent that corn-based ethanol as a biomass fuel leads to higher-cost food and, in some countries, political unrest. But more fundamentally, encouraging reliance on biomass plantations as a global climate change mitigation effort should not become policy until predictable implications such as further disruption of nitrogen, phosphorus, and hydrologic cycles are explored.

27.6.2 Governance Principles of ESEM

As discussed in Chapter 7, the global governance system is rapidly evolving and becoming more complex. These changes, especially when combined with the inherent complexity of human and natural systems, give rise to a second category of principles involving ESEM governance.

- ESEM initiatives by definition raise important scientific, technical, economic, political, ethical, theological, and cultural issues in the context of global polity. Given the need for consensus and long-term commitment, the only workable governance model is one that is inclusive, transparent, and accountable.
- ESEM governance models that deal with complex, unpredictable systems must accept high levels of uncertainty as inherent in the process. ESEM policy development and deployment must be understood as a continuing dialog with the relevant systems rather than a definitive endpoint and should thus emphasize flexibility. Moreover, the policy maker must be understood as part of an

evolving ESEM system, rather than an agent outside the system guiding or defining it.

- Because Earth systems are self-organizing and open and are thus capable of emergent behavior, continual learning at the personal and institutional level must be built into the process.
- There must be adequate resources available to support both the project, and the science and technology research and development that are necessary to ensure that the responses of the relevant systems are understood.

27.6.3   Design and Engineering Principles of ESEM

Finally, there is a set of principles that informs the design and engineering of ESEM systems:

- Know from the beginning what the desired and reasonably expected outcomes of any intervention are and establish quantitative metrics by which progress may be tracked.
- Unlike simple, well-known systems, the complex, information dense and unpredictable systems that are the subject of ESEM cannot be centrally or explicitly controlled. Rather than being outside the system, the Earth systems engineer will have to see herself or himself as an integral component of the system itself, closely coupled with its evolution and subject to many of its dynamics.
- Whenever possible, engineered changes should be incremental and reversible, rather than fundamental and irreversible. Any scale-up should allow for the fact that in complex systems, discontinuities and emergent characteristics are the rule, not the exception.
- An important goal in Earth systems engineering projects should be to support the evolution of resiliency, not just redundancy, in the system. Thus, inherently safe systems are to be preferred to engineered safe systems.

27.7   FACING THE ESEM QUESTION

While the global climate change discussions provide, perhaps, the most dramatic example of prospective ESEM, one might also cite the ongoing efforts to manage the Baltic Sea, managing regional forests to be sustainable, restricting exploitation of local and regional fisheries, understanding the dynamics of powerful emerging technology systems (nanotechnology, biotechnology, robotics, ICT, and cognitive science), and meeting continued challenges from invasive species. This last example is a product in large part of previous patterns of human migration and illustrates that ESEM is not something that humans should now begin to do, but that we have been overtly influencing natural systems for centuries. Similarly, it is not unreasonable to view global agricultural and energy systems, tightly linked as they now are by trade and commodity markets, as an ESEM process—and, obviously, another one that has been going on for centuries.

For the most part, the specialties of ecological engineering and ESEM emphasize broad systems thinking as contrasted with the narrower (but very useful) goal of "doing it right the first time." They thus enhance the technological society–environment interaction in potentially useful ways not incorporated into less expansive sustainable engineering approaches.

In Table 27.1, we collect characteristics of implemented or proposed ESEM projects. Some important lessons are revealed by the table. It is interesting to ask why some ESEM proposals are being implemented and others are not. The table shows that implementation is not related to the potential severity of the environmental challenge, nor to the spatial scale of the proposed ESEM activity, nor to the spatial scale of the impact. It is related, however, to the public visibility of the environmental challenge and (to a lesser extent) to the degree of scientific understanding. The implication is that if we can readily witness a problem and know how to attack it, ESEM implementation will occur. Conversely, with more complex but less visible perturbations such as climate change, it is likely to be difficult to engage the public in addressing them.

ESEM raises a fundamental issue: What level of ESEM is appropriate? ESEM by its nature is a means to an end which can only be defined in ethical terms. Simply put, the question "To what end are humans engineering, or *should* humans engineer, the planet Earth?" is a moral as well as a technical one. It is also not just hypothetical: Human institutions are implicitly answering that question every day by invoking ESEM activities on a variety of spatial and temporal scales.

ESEM will assume increased relevance as discussions move from a goal of environmental improvement to one of sustainability. The latter implies some sort of targets for technology–environment interactions, together with policies designed to meet those targets, monitoring to evaluate progress toward those targets, and periodic review to assess whether mid-course corrections are needed. If we as a society are serious about sustainable development, we will need a family of approaches to deal with the societal, economic, and environmental goals that sustainable development implies.

27.8 PROACTIVE INDUSTRIAL ECOLOGY

This chapter has discussed ways in which humans are restoring degraded systems, or pondering ways to do so. There are numerous situations where such activities are appropriate—brownfields in urban areas, the Florida Everglades, and the like. Nonetheless, this sort of approach clearly represents reactive thinking. It is contentious, time-consuming, and costly. In some cases, with global climate change, for example, it remains unclear whether amelioration can even be accomplished. Nonetheless, the recognition that human society is now practicing ESEM in various ways needs to call forth a global discussion: How well do we understand where we are headed? What collaborative approaches are suggested by that realization?

Industrial ecology and its implementation in sustainable engineering practices and methodologies take a quite different, proactive approach, seeking to "do it right the first time." At the product level, doing so may be complicated but the concept is straightforward. As the system in question grows—to urban area, to country, to the planet, and from single discipline domains to cut across social, engineered, and natural systems at all scales—complexity increases. Goals must be set, often from a sustainability perspective, and implementation of the goals then achieved. It is here that industrial ecology and sustainable engineering become central specialties, showing

TABLE 27.1 Characteristics of Implemented or Proposed ESEM Projects

ESEM Activity	Environmental challenge	Potential severity	Scale of impact	Understanding of challenge	Scale of activity	Public visibility	State of implementation
"Brownfield" redevelopment	Human toxicity	Moderate	20–100 m	Good	20–100 m	Very high	Extensive
Dredging the waters	Pollutants in sediment	Low	100–1,000 m	Good	100–1,000 m	Modest	Several projects
Wetlands restoration	Ecosystem degradation	High	100–10,000 km	Modest	100–10,000 km	High	Several projects
CO_2 sequestration	Global warming	Very high	Global	Modest	1–10 km	None	None
Tree planting	Global warming	Very high	Global	Modest	1–10 km	None	Several projects
Ocean fertilization	Global warming	Very high	Global	Modest	10–100 km	None	Preliminary experiments
Stratospheric sulfate	Global warming	Very high	Global	Poor	Global	None	None
Mirror in space	Global warming	Very high	Global	Poor	Extraplanetary	None	None

how to conceptualize problems and response on appropriate timescales, how to transform goals into actions, and how to measure progress. In these actions is truly contained the science of sustainability.

FURTHER READING

Ecological engineering:

Hobbs, R.J., and V.A. Cramer, Restoration ecology: Interventionist approaches for restoring and maintaining ecosystem function in the face of rapid environmental change, *Annual Review of Environment and Resources, 33*, 39–61, 2008.

Mitsch, W.J., and S.E. Jørgensen, Ecological engineering: A field whose time has come, *Ecological Engineering, 20*, 363–377, 2003.

NRC (U.S. National Research Council). 2007. *Progress Toward Restoring the Everglades: The First Biennial Review—2006.* Washington, DC: National Academy Press.

Carbon capture and sequestration:

Benson, S.M., and T. Surles, Carbon dioxide capture and storage: An overview with emphasis on capture and storage in deep geological formations, *Proceedings of the IEEE, 94*, 1795–1804, 2006.

Schrag, D.P., Preparing to capture carbon, *Science, 315*, 812–813, 2007.

Forest carbon storage:

Bonan, G.B., Forests and climate change: Forcings, feedbacks, and the climate benefit of forests, *Science, 320*, 1444–1449, 2008.

Iron fertilization:

Blain, S., Effect of natural iron fertilization on carbon sequestration in the Southern Ocean, *Nature, 446*, 1070–1074, 2007.

Woods Hole Oceanographic Institution, Fertilizing the ocean with iron, *Oceanus, 46* (1), 4–27, 2008.

Sulfate Aerosols:

Crutzen, P.J., Albedo enhancement by stratospheric sulfur injections: A contribution to resolve a policy dilemma? *Climatic Change, 77*, 211–219, 2006. Several other scientists comment on this article in the pages following.

Schneider, S.H., Earth systems engineering and management, *Nature, 409*, 417–421, 2001.

Space Mirrors:

Angel, R., Feasibility of cooling the Earth with a cloud of small spacecraft near the inner Lagrange point (L1), *Proceedings of the National Academy of Sciences of the U.S., 103*, 17184–17189, 2006.

Early, J.T., Space-based solar screen to offset the greenhouse effect, *Journal of the British Interplanetary Society, 42*, 567–569, 1989.

Earth System Engineering and Management:

Allenby, B.R. *Reconstructing Earth*, Washington, DC: Island Press, 2005.

Allenby, B.R., Earth systems engineering and management: A manifesto, *Environmental Science & Technology, 41*, 7960–7965, 2007.

Barrett, S., The incredible economics of geoengineering, *Environmental Resource Economics, 39,* 45–54, 2008.

Science, Special report: Human-dominated ecosystems, *267,* 485–525, 1997.

Smil, V. 1997. *Cycles of Life: Civilization and the Biosphere.* New York: Scientific American Library.

EXERCISES

27.1 What are the relationships among industrial ecology, ESEM, and sustainable development?

27.2 You have been selected as the technologist in charge of Everglades engineering and management, with the task of creating a South Florida regional policy that meets the requirements of all stakeholders involved in the Everglades.

(a) What stakeholders will you involve in your planning process, and why?

(b) How will you monitor the system to determine when relevant changes are occurring, and how will you respond to them in a principled way?

(c) An important stakeholder demands that the Everglades be returned to the state it was in before Europeans arrived in the New World. How will you respond?

27.3 Successful response to the challenge of stratospheric ozone depletion is often given as a model for policy formulation regarding global climate change. Do you think the comparison between the ozone depletion situation and global climate change is valid? Why or why not?

27.4 You are the technologist in charge of global climate change mitigation for the United Nations. Unexpectedly, global climate change appears to be getting much worse, and you have been asked to recommend one of three geoengineering responses:

1. major iron fertilization of the ocean;

2. installing mirrors in space to reduce incoming radiation; or

3. injection of sulfur particles into the atmosphere. Analyze your choices using ESEM principles, select the best option of the three, and defend your selection.

CHAPTER 28

Looking to the Future

28.1 A STATUS REPORT

All fields of science and technology are works in progress, and industrial ecology and sustainable engineering are no exceptions. Accordingly, it is worth taking stock as we end this book. What areas of these fields are in good shape—well understood and with high-quality theoretical and experimental analyses being generated? We would place four methodologies and their results in this category: Design for Environment and Sustainability (DfES), life cycle assessment (LCA), material flow analysis (MFA), and national material accounts. This is not to say that we know all we wish to know, or that improvements and new discoveries are not anticipated, but rather that these areas are where we can wave the banners of our profession and be proud of the work being done.

Other areas are in earlier stages, but show considerable promise. Here we would include industrial symbiosis, the energy implications of technology, and resource sustainability. These are important topics for both the present and the future. We know enough about them to anticipate that they will play important roles in industrial ecology and sustainable engineering, but at present they are in transition from conceptual to quantitative and have not yet realized their full potential.

A third group of topics relates to areas somewhere between infancy and childhood so far as their development is concerned. We hope for much from them, but cannot yet be very certain our hopes will be realized. In this category we would place physical input–output analysis, industrial ecology models, industrial ecology scenarios, the interplay of water and industry, urban ecology, the interactions of technology and society, the relationship of technology to sustainable development, and the role of developing countries in the industrial ecology and sustainable development enterprise. Each of these topics, in our view, is a prime area for research and development.

28.2   NO SIMPLE ANSWERS

In the complex systems with which industrial ecology and sustainable engineering deal, simple answers are generally incorrect, or at least poorly thought through. Consider the following example, devised by David Allen of the University of Texas (Figure 28.1). A chemical flow stream containing two reactive organic compounds in a solvent can, in principle, undergo a complete separation to yield each of the compounds. The solvent is recovered as a by-product. Assume that the energy required to carry out the separation comes from burning fossil fuels that contain sulfur and that generate nitrate and sulfate combustion products.

If our concern is about smog formation, we want to completely distill and capture the organic compounds, because they will assist in smog development if they escape. Conversely, if our concern is about acid rain, we want to perform no distillation at all, because the required energy production will intensify the acidity of rainfall. If the main concern is global climate change, we realize that fossil fuel combustion will release carbon dioxide (thus warming the atmosphere) but will also generate reflective sulfate aerosols (thus cooling the atmosphere), so the distillation may be possible to some degree. The point of this discussion is that there is no "right" answer; the choice depends on one's goals. Most situations related to sustainability are much more complex than this simple example, but the reality is that the society–technology–environment linkages are extensive and that no single approach, no matter how laudatory the cause, is likely to solve all the challenges.

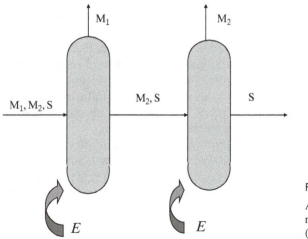

Figure 28.1

A two-stage distillation column. M_1 and M_2 are two materials of interest; S is the solvent containing them. (After a diagram by D.T. Allen, University of Texas.)

28.3 FOCI FOR RESEARCH

The status report for the tools and achievements of industrial ecology and sustainable development outlined above was primarily related to technology–environment interactions. We can also present topics that relate to the more sociological aspects of the field. Here is a suggestive, but far from complete, list:

At the *micro* level (the individual):

- What do people really need?
- Why do we want so much these days?
- How can people get what they need without so much stuff?

At the *meso* level (the firm):

- How can products be designed to minimize resource closure problems?
- What can firms do to produce authentic but dematerialized satisfaction?
- How can firms understand their roles in the holarchic systems within which they function?

At the *macro* level (society):

- How far can ecoefficiency take us before absolute limits set in?
- What are the regulatory, economic, and institutional barriers to improving humanity's path to sustainability?
- How can industrial ecology develop optimal solutions to multiobjective decision making in technologies related to sustainable development?
- Where are the most opportunistic leverage points for improving humanity's path to sustainability?

These are interdisciplinary topics, and ones in which progress will only be made by groups of colleagues willing to cross disciplinary boundaries. Ultimately, the quest for sustainability will require progress on questions such as these.

28.4 THEMES AND TRANSITIONS

It is clear that for engineering and technology our scope for thinking and analysis is broadening, and our temporal and spatial assessment scales are increasing. Among the environmentally related approaches of note are the following:

- Not looking to the past, but to the future
- Not fragmented, but systemic
- Emphasizing not gross insults, but microtoxicity
- From focusing on environmental improvement to focusing on sustainability

These approaches need to play out in the midst of a dramatic increase in global population and an equally vigorous reliance on technology. Our upcoming planetary population, the majority of which will live in urban areas, is likely to have increased income and thus increased demands for resources (or the services thereof). It will desire buffering from surprises. Five transitions capture most of this societal evolution.

- Not local, but global
- Not pastoral, but urban
- Not isolated, but connected
- Not more technology, but better technology
- Not emphasizing the developed world, but the developing world

Four themes characterize the approaches needed to succeed as we build upon this framework: We much approach the technology–society–environment web, linked and challenged as it is by its insatiable demand for resources of all kinds, in a *holistic* fashion, a *connected* perspective, a *parsimonious* use of resources, and a *metabolically benign* approach to design and use. With this philosophy we still "sin less"; perhaps we will even "change the story."

The twentieth century has turned out to be one of major anthropogenic change on the planet on which we live. The twenty-first century must be one in which we do better—measuring our every action against its impacts on the environment, on society, and on sustainability. It is the supreme challenge to the human species, and one in which industrial ecology and sustainable engineering must be the enablers.

FURTHER READING

Diamond, J., *Collapse: How Societies Choose to Fail or Succeed*, New York: Viking Press, 2005.

Ehrenfeld, J.R., *Sustainability by Design*, New Haven: Yale University Press, 2008.

Ostrom, E., J. Burger, C.B. Field, R.B. Norgaard, and D. Policansky, Revisiting the commons: Local lessons, global challenges, *Science, 284,* 278–282, 1999.

Speth, J.G., *The Bridge at the Edge of the World: Capitalism, the Environment, and Crossing from Crisis to Sustainability*, New Haven, CT: Yale University Press, 2008.

EXERCISE

28.1 It has been proposed that catastrophes related to the environment and sustainability may be likely in the next few decades. They could occur at a variety of levels, for example:
- Personal—$3/liter gasoline
- Corporate—unavailability of a principal raw material
- City/Country—uncertain water supplies
- Planet—ecosystem failures

Prepare a story line for a catastrophic challenge of your own choice at less than planetary level and devise a coherent industrial ecology–oriented set of response options.

Units of Measurement in Industrial Ecology

TABLE A1 Prefixes for Large and Small Numbers

Power of 10	Prefix	Symbol
+24	yotta	Y
+21	zetta	Z
+18	exa	E
+15	peta	P
+12	tera	T
+9	giga	G
+6	mega	M
+3	kilo	k
−3	milli	m
−6	micro	μ
−9	nano	n
−12	pico	p
−15	femto	f
−18	atto	a
−21	zepto	z
−24	yocto	y

The basic unit of energy is the joule ($J = 1 \times 10^7$ erg). One will often see the use of the British thermal unit (Btu), which is 1.55×10^3 J. For very large energy use, a unit named *quad* is common; it is a shorthand notation for one quadrillion Btu. Thus, 1 quad = 1×10^{15} Btu = 1.55×10^{18} J.

The units of mass in the environmental sciences and in this book are given in the metric system. Since many of the quantities are large, the prefixes given in Table A1 are common. Hence, we have such figures as $2 \text{ Pg} = 2 \times 10^{15}$ g. Where the word *tonne* is used, it refers to the metric ton $= 1 \times 10^6$ g.

The most common way of expressing the abundance of a gas phase atmospheric species is as a fraction of the number of molecules in a sample of air. The units in common use are *parts per million* (ppm), *parts per billion* (thousand million; ppb), and *parts per trillion* (million million; ppt), all expressed as volume fractions and therefore abbreviated ppmv, ppbv, and pptv to make it clear that one is not speaking of fractions in mass. Any of these units may be called the *volume mixing ratio* or *mole fraction*. Mass mixing ratios can be used as well (hence, ppmm, ppbm, pptm), a common example being that meteorologists use mass mixing ratios for water vapor. Since the pressure of the atmosphere changes with altitude and the partial pressures of all the gaseous constituents in a moving air parcel change in the same proportions, mixing ratios are preserved as long as mixing between air parcels is neglected.

For constituents present in aqueous solution, as in seawater, the convention is to express concentration in volume units of moles per liter (designated M) or some derivative thereof (one mole [abbreviated mol] is 6.02×10^{23} molecules). Common concentration expressions in environmental chemistry are millimoles per liter (mM), micromoles per liter (μM), and nanomoles per liter (nM). Sometimes one is concerned with the "combining concentration" of a species rather than the absolute concentration. A combining concentration, termed an *equivalent*, is that concentration which will react with eight grams of oxygen or its equivalent. For example, one mole of hydrogen ions is one equivalent of H^+, but one mole of calcium ions is two equivalents of Ca^{2+}. Combining concentrations have typical units of equivalents, milliequivalents, or microequivalents per liter, abbreviated eq/l, meq/l, and μeq/l. A third approach is to express concentration by weight, as mg/l or ppmw, for example. Concentration by weight can be converted to concentration by volume using the molecular weight as a conversion factor.

Acidity in solution is expressed in pH units, pH being defined as the negative of the logarithm of the hydrogen ion concentration in moles per liter. In aqueous solutions, an acidity of pH $= 7$ is neutral at $25°$C; lower pH values are characteristic of acidic solutions, and higher values are characteristic of basic solutions.

Glossary

Acid deposition — The deposition of acidic constituents to a surface. This occurs not only by precipitation but also by the deposition of atmospheric particulate matter and the incorporation of soluble gases.

Adaptive management — Management of complex technological–social systems that recognizes the importance of technological information, the inadequacy of predictive activity, and the need to act, nonetheless, to optimize human–natural systems over time.

Aggregate — Crushed stones, typically 1–3 cm in diameter, employed in concrete and for roadbed construction.

Anthropogenic — Derived from human activities.

Aquifer — Any water-bearing rock formation or group of formations, especially one that supplies ground water, wells, or springs.

Asphalt — A combination of small stones, mineral dust, and a few percent of bitumin (the lowest boiling fraction resulting from the refining of crude oil).

This glossary was compiled from various sources, including particularly: T.E. Graedel and P.J. Crutzen, *Atmospheric Change: An Earth System Perspective*, W.H. Freeman, New York, 446 pp., 1993. Organization for Economic Cooperation and Development, *Measuring Material Flows and Resource Productivity: Glossary of Terms*, Document ENV/EPOC/SE/RD(2005)2/REV1, Paris: OECD Environment Directorate, 2008.

Background consumption—Consumption that satisfies basic needs.

Bioaccumulation—The concentration of a substance by an organism above the levels at which that substance is present in the ambient environment. Some forms of heavy metals, and chlorinated pesticides such as DDT, are bioaccumulated.

Biomagnification—The increasing concentration of a substance as it passes into higher trophic levels of a food web. Many substances which are bioaccumulated are also biomagnified.

Budget—A balance sheet of the magnitudes of all of the sources and sinks for a particular species or group of species in a single reservoir.

By-product—A useful product that is not the primary product being produced. In life cycle analysis, by-products are treated as coproducts.

Carcinogen—A material that causes cancer.

Cascade recycling—See *Open-loop recycling*

Category—Derived from human activities.

Cement—A combination of limestone, shale, and gypsum used to bind the constituents of concrete.

CFCs—Chlorofluorocarbon compounds, that is, organic compounds that contain chlorine and/or fluorine atoms. CFCs are widely recognized as hazardous to stratospheric ozone.

Characterization—The process of quantitatively determining the impact resulting from the stress indicated by LCA inventory values.

Chronic—In toxicology, an exposure or effect of an exposure which becomes manifest only after a significant amount of time—weeks, months, or even years—has passed. Many carcinogens (substances causing cancer) are chronic toxins, and low level exposure to many heavy metals, such as lead, produces chronic, rather than acute, effects.

Classification—The process of assigning raw LCA data on flows of materials and energy to particular environmental concerns.

Closed-loop recycling— A recycling system in which a particular mass of material is remanufactured into the same product (e.g., glass bottle into glass bottle). Also known as "horizontal recycling."

Comet diagram—A diagram that details the stages of product lifetime and the opportunities for end-of-life reuse of products, components, and materials.

Commons—A geographical region or entity under common ownership or without ownership (e.g., a village green, the oceans, the atmosphere).

Consumption—The organism-induced transformation of materials and energy.

Cultural construct—A social idea or approach that has become absolute and unquestioned.

Cycle—A system consisting of two or more connected *reservoirs,* where a large part of the material of interest is transferred through the system in a cyclic manner.

Decoupling—Breaking the link between economic growth and the concomitant growth of environmental degradation and/or use of resources.

Dematerialization—An absolute or relative reduction in the use of materials per unit of value added or output.

Depletion time—The time required to exhaust a resource if the present rate of use remains unchanged.

Design for environment—An engineering perspective in which the environmentally related characteristics of a product, process, or facility design are optimized.

Discount rate—A rate applied to future financial returns to reflect the time value of money and inflation.

Disposal—Discarding of materials or products at the end of their useful life without making provision for *recycling* or *reuse.*

Dissipative product—A product that is irretrievably dispersed when it is used (e.g., paint, fertilizer).

Dose–response curve—A curve plotting the known dose of a material administered to organisms against the percentage response of the test population. If the material is not directly administered, but is present in the environment surrounding the organism (e.g., water, air, sediment), the resulting curve is know as a "concentration–response curve."

Ecological engineering—The design of sustainable ecosystems that integrate human society with its natural environment for the benefit of both.

Ecology (biological)—The study of the distribution and abundance of organisms and their interactions with the physical world.

Ecoprofile—A quick overview of a product or process design to ensure that no disastrous features are included or to determine whether additional assessment is needed.

Embodied energy—The energy employed to bring a particular material or product from its initial physical reservoir or reservoirs to a specific physical state.

Emergent behavior—The behavior of a system at a particular holonic level which is impossible to predict from detailed knowledge of adjacent holonic levels.

Emissions—Losses to the environment from any of a variety of human activities.

Energy audit—An accounting of input flows, output flows, and losses of energy within an industrial process, a facility, a corporation, or a geographical entity.

Enterprise resource planning (ERP)—The utilization of software to link the activities of a corporation with the associated requirements for labor, materials, financial systems, management, reporting, and so on.

Enzyme—A molecule that acts as a catalyst in a biological chemical reaction.

Expert system—A computer-based system combining knowledge in the form of facts and rules with a reasoning strategy specifying how the facts and rules are to be used to reach conclusions. The system is designed to emulate the performance of a human expert or a group of experts in arriving at solutions to complex and not completely specified problems.

Exposure—Contact between a hazard (see below) and the target of concern, which may be an organ, an individual, a population, a biological community, or some other system. The confluence of exposure and hazard gives rise to risk.

Extent—The size of a scale dimension.

Flux—The rate of emission, absorption, or deposition of a substance from one *reservoir* to another. Often expressed as the rate per unit area of surface.

Food chain—A sequence in which resources flow in linear fashion from one trophic level to the next.

Food web—A pattern in which resources flow largely from one trophic level to the next but may also flow across trophic levels in nonlinear fashion.

Fossil fuel—A general term for combustible geological deposits of carbon in reduced (organic) form and of biological origin, including coal, oil, natural gas, oil shales, and tar sands.

Fugitive emissions—Emissions from valves or leaks in process equipment or material storage areas that are difficult to measure and do not flow through pollution control devices.

Functional unit—The quantified function that a product or service will deliver and that will serve as a basis for comparing products, services, or product or service scenarios.

Global warming—The theory that elevated concentrations of certain anthropogenic atmospheric constituents are causing or will cause an increase in Earth's average temperature.

Green accounting—An informal term referring to management accounting systems that specifically delineate the environmental costs of business activities rather than include those costs in overhead accounts.

Green chemistry—Employing chemical techniques and methodologies that reduce or eliminate the use or generation of feedstocks, products, by-products, solvents, and reagents that are hazardous to the environment.

Green engineering—The design and implementation of engineering solutions that take environmental issues into account throughout the life cycle of the design.

Greenhouse gas—A gas with absorption bands in the infrared portion of the spectrum. The principal greenhouse gases in the Earth's atmosphere are H_2O, CO_2, O_3, CH_4, and N_2O.

Hazard— (as used in risk assessment) A material or condition that may cause damage, injury, or other harm, frequently established through standardized assays performed on biological systems or organisms. The confluence of hazard and exposure creates a risk.

Hidden flow—The indirect flows of materials such as resources, pollution, or waste that occur upstream in a production process but that are not physically embodied in the product itself.

Holarchy—A network of holons.

Holon—An individual entity in a system of systems.

Horizontal recycling—See *Closed-loop recycling.*

Impact analysis—The second stage of life cycle assessment, in which the environmental impacts of a process, product, or facility are determined.

Improvement analysis—The third stage of life cycle assessment, in which design for environment techniques are used in combination with the results of the first and second LCA stages to improve the environmental plan of a process, product, or facility.

Indicator—A nonquantitative measure of the status of a chosen parameter, environmental or otherwise.

Industrial ecology—An approach to the design of industrial products and processes that evaluates such activities through the dual perspectives of product competitiveness and environmental interactions.

Industrial enzyme—An industrial process or piece of equipment that results in a transformation of materials or energy.

Industrial symbiosis—See *Symbiosis.*

Infrastructure—The basic facilities, equipment, and installations needed for the functioning of a community or industrial operation.

Inventory analysis—The first stage of life cycle assessment, in which the inputs and outputs of materials and energy are determined for a process, product, or facility.

Inviolates list—A list of design decisions never allowed to be taken by product or process designers.

Level—A unit of analysis located in a particular position on a scale.

Life cycle—The stages of a product, process, or package's life, beginning with raw materials acquisition; continuing through processing, materials manufacture, product fabrication, and use; and concluding with any of a variety of waste management options.

Life cycle assessment—A concept and a methodology to evaluate the environmental effects of a product or activity holistically, by analyzing the entire life cycle of a particular material, process, product, technology, service, or activity. The life cycle assessment consists of three complementary components—inventory analysis, impact assessment, and improvement analysis—together with an integrative procedure known as scoping.

Lock-in—In which a particular process or product standard becomes embedded in product design, manufacturing, or consumption patterns, thus stifling innovation. The common example is the notably inefficient QWERTY keyboard.

Material flow analysis—An analysis of the flows of materials within and across the boundaries of a particular geographical region.

Material requirements planning (MRP)—The utilization of software to link industrial production schedules with the associated requirements for parts, components, and materials, as well as the status of current inventories.

Metabolic analysis—The analysis of the aggregate of physical and chemical processes taking place in an organism, biological or industrial.

Metabolism—The physical and chemical processes taking place in an organism, biological or industrial.

Metabolite—An intermediate product of chemical transformations within an organism.

Metric—A quantitative measure of the status of a chosen parameter, environmental or otherwise.

Mineral—A distinguishable solid phase that has a specific chemical composition, for example, quartz (SiO_2) or magnetite (Fe_3O_4).

Misconsumption—Consumption that undermines an individual's own well-being even if there are no aggregate effects on the population or species.

Molecular flow analysis—An analysis of the flows of a specific molecule within and across the boundaries of a particular geographical region.

Mutagen—A hazard that can cause inheritable changes in DNA.

Nonpoint source—See *Source*.

Normalization—In life cycle assessment, the process of relating environmental impact values derived at the characterization step to reference values in order to arrive at common indicator values.

NO_x—The sum of the common pollutant gases NO and NO_2.

Omnivory—The acquisition of resources from organisms at several different trophic levels.

Open-loop recycling—A recycling system in which a product from one type of material is recycled into a different type of product (e.g., plastic bottles into fence posts). The product receiving recycled material itself may or may not be recycled. Also known as "cascade recycling."

Ore — A natural rock assemblage containing an economically valuable resource.

Organism — An entity internally organized to maintain vital activities.

Overburden — The material to be removed or displaced that is overlying the ore or material to be mined.

Overconsumption — Consumption for which choice exists and that undermines a species' own life support system.

Ozone depletion — The reduction in concentration of stratospheric ozone as a consequence of efficient chemical reactions with molecular fragments derived from anthropogenic compounds, especially CFCs and other halocarbons.

Packaging, primary — The level of packaging that is in contact with the product. For certain beverages, an example is the aluminum can.

Packaging, secondary — The second level of packaging for a product that contains one or more primary packages. An example is the plastic rings that hold several beverage cans together.

Packaging, tertiary — The third level of packaging for a product that contains one or more secondary packages. An example is the stretch wrap over the pallet used to transport packs of beverage cans.

Pathway — The sequence of chemical reactions that connects a particular starting material with the final material that is produced.

Plating — The act of coating a surface with a thin layer of metal.

Point source — See *Source.*

Pollution prevention — The design or operation of a process or item of equipment so as to minimize environmental impacts.

Prompt scrap — Waste produced by users of semifinished products (turnings, trimmings, etc.). This scrap must generally be returned to the materials processor if it is to be recycled. (Also called *New scrap*).

Rebound effect — A consumption decision that appears beneficial, but which in practice stimulates behavior that nullifies the benefits of the original decision.

Recycling — The reclamation and reuse of output or discard material streams useful for application in products.

Remanufacture — The process of bringing large amounts of similar products together for purposes of disassembly, evaluation, renovation, and reuse.

Reserve — The total known amount of a resource that can be mined with today's technology at today's market prices.

Reserve base — The total known amount of a resource that can be mined, without regard for technology or market prices.

Reservoir—A receptacle defined by characteristic physical, chemical, or biological properties that are relatively uniformly distributed.

Residence time—The average time spent in a reservoir by a specific material.

Resolution—The precision used in measurement.

Resource productivity—The efficiency with which an economy or a production process uses material resources. It is often expressed as a ratio of materials used per monetary unit of production.

Reuse—Reemploying materials and products in the same use without the necessity for *recycling* or *remanufacture*.

Reverse fishbone diagram—A diagram detailing the steps involved in the disassembly of a product.

Risk—The confluence of exposure and hazard; a statistical concept reflecting the probability that an undesirable outcome will result from specified conditions (such as exposure to a certain substance for a certain time at a certain concentration).

Risk assessment—An evaluation of potential consequences to humans, wildlife, or the environment caused by a process, product, or activity, and including both the likelihood and the effects of an event.

Scale—A spatial, temporal, quantitative, or analytical dimension used to measure and study an object or process of interest.

Scenario—An alternative vision of how the future might unfold.

Self-Organizing Holarchic Open (SOHO) system—A conceptually or causally linked system of objects or processes grouped along an analytical scale, but with a structure that is self-imposed.

Servicizing—Satisfying a need by providing a service rather than by providing products that meet the need.

Sink—In environmental chemistry, the process or origin from which a substance is lost from a reservoir.

Slag—The fused residue that results from the separation of metals from their ores.

Smog—Classically, a mixture of smoke plus fog. Today the term "smog" has the more general meaning of any anthropogenic haze. Photochemical smog involves the production, in stagnant, sunlit atmospheres, of oxidants such as O_3 by the photolysis of NO_2 and other substances, generally in combination with haze-causing particles.

Social ecology—The study of societies and their evolutionary behavior from the perspective of societal use of energy and materials.

Solution—A mixture in which the components are uniformly distributed on an atomic or molecular scale. Although liquid, solid, and gaseous solutions exist, common nomenclature implies the liquid phase unless otherwise specified.

Solvent—A medium, usually liquid, in which other substances can be dissolved.

Source—In environmental chemistry, the process or origin from which a substance is injected into a reservoir. *Point sources* are those where an identifiable source, such as a smokestack, can be identified. *Nonpoint sources* are those resulting from diffuse emissions over a large geographical area, such as pesticides entering a river as runoff from agricultural lands.

Stock—The contents of a reservoir.

Stratosphere—The atmospheric shell lying just above the *troposphere* and characterized by a stable lapse rate. The temperature is approximately constant in the lower part of the stratosphere and increases from about 20 km to the top of the stratosphere at about 50 km.

Streamlined Life Cycle Assessment (SLCA)—A simplified methodology to evaluate the environmental effects of a product or activity holistically, by analyzing the most significant environmental impacts in the life cycle of a particular product, process, or activity.

Stressor—A set of conditions that may lead to an undesirable environmental impact.

Substance flow analysis—An analysis of the flows of a specific chemically identifiable substance within and across the boundaries of a particular geographical region. The word "substance" typically incorporates atoms and molecules of the entity of interest without regard for chemical form.

Sustainable engineering—The design and implementation of engineering solutions that take environmental and sustainability issues into account throughout the life cycle of the design.

Sustainability—In the context of industrial ecology, sustainability is the state in which humans living on Earth are able to meet their needs over time while nurturing planetary life-support systems.

Symbiosis—A relationship within which at least two willing participants exchange materials, energy, or information in a mutually beneficial manner.

Tailings—The residue left after ore has been ground and the target metallic minerals separated and retained.

Trophic level—A group of organisms that perform similar resource exchanges as part of natural food chains or food webs.

Troposphere—The lowest layer of the atmosphere, ranging from the ground to the base of the stratosphere at 10–15 km altitude, depending on latitude and weather conditions. About 85 percent of the mass of the atmosphere is in the troposphere, where most weather features occur. Because its temperature decreases with altitude, the troposphere is dynamically unstable.

Valuation—In life cycle assessment, the process of assigning weighting factors to different impact categories based on their perceived relative importance.

Visibility—The degree to which the atmosphere is transparent to light in the visible spectrum, or the degree to which the form, color, and texture of objects can be perceived. In the sense of visual range, visibility is the distance at which a large black object just disappears from view as a recognizable entity.

Waste—Material thought to be of no practical value. One of the goals of industrial ecology is the reuse of resources, and hence the minimization of material regarded as waste.

Waste audit—An accounting of output flows and losses of wastes within an industrial process, a facility, a corporation, or a geographical entity.

Water audit—An accounting of input flows, output flows, and losses of water within an industrial process, a facility, a corporation, or a geographical entity.

Weighting—In life cycle assessment, the process of assigning factors to different impact categories based on their perceived relative importance.

Index

Adaptability. *See* Adaptive management
Adaptive management, 219–221
   adaptability, 221
   resiliance, 221
   transformability, 221
Air conditioner design, 117
Alkali Act of 1863, 92
Allen, D., 108, 382
Anthrobiogeochemical, 250
Arizona-Phoenix, 298
Ashton, W., 214
Asphalt recycling, 157
Atom economy, 104
Ausubel, J., 57
Automobile
   streamlined life-cycle assessment, 197–207
   technology system, 3, 167–168
Ayres, R., 53

Basel Convention, 90
Bennett, E., 209
Biodiversity, 24, 27
Bioinformatics, 56–57

Biological ecology, 41
Biological ecosystem, 44
   type I system, 44
   type II system, 44
   type III system, 44
Biological organism, 42
Biotechnology, 8
Blue Angel, 137, 358
Bottom-up stock assessment. *See* Stock
Brent Spar, 77, 112
BRIC countries, 342
British East India Company, 94
Brooks, H., 2
Brownfield, 366, 378
Brunner, P., 296
Budyko, M., 373
Buildings, 152–158. *See also* Design for
       Environment and Sustainability

Carbon dioxide, 20, 26–27
Carnivores, 226
Carrying capacity, 5
Castro, M., 129–130

CH_4, *See* Methane
Chapagain, A., 286
Characterization. *See* Life Cycle Assessment
Chertow, M., 232
Chlorofluorocarbons (CFCs), 27, 31,
    112, 248, 369
Chromium, 276
Circular economy, 35, 348
Classification. *See* Life Cycle Assessment
Closed-loop recycling. *See* Recycling
CO_2. *See* Carbon dioxide
CO_2 sequestration, 370, 378
Coal, 337
Coca-Cola Company, 287
Cognitive science, 8
Comet diagram, 135–136
Command and control, 93
Common-pool resources, 2
Commons, 2
    tragedy of the, 2
Complex systems. *See* Systems analysis
Comprehensive policy support assessment
    (CPSA), 79
Computer-aided design (CAD), 117, 121
Computer-aided manufacturing
    (CAM), 121
Concrete recycling, 157
Connectance, 226
Conservation of energy, 32
Conservation of mass, 32, 242
Constable, D., 360
Consumption, 88–89
Copper
    global sustainability, 335
    Chinese cycle, 344
Coupled models. *See* Models
Cronon, W., 86
Cultural constructs, 83, 85
Cycle, 242, 250

Dahmus, J., 141
Daigo, I., 243
Decomposers, 223
Decoupling, 262
Dematerialization, 128–129, 168
Depletion time, 332

Design
    DfES guidelines, 143
    for assembly, 119
    for compliance, 119
    for disassembly, 119, 138, 140
    for environment and sustainability
        (DfES), 119, 126–159, 164, 195,
        353, 359, 381
    for manufacturability, 119
    for material logistics and component
        applicability, 119
    for recyclability, 135, 137
    for reliability, 120
    for safety and liability prevention, 120
    for serviceability, 120
    for sustainability, 31
    for testability, 120
    for X, 118, 119, 121
Dietz, T., 91
Discount rates, 95–96
Dissipative products, 135
Distribution system, 148
Domestic material consumption (DMC).
        See National material accounts
Domestic material input (DMI). *See*
        National material accounts
Domestic processed output (DMO).
        See National material accounts

Earth system engineering and
        management, 367
    governance principles, 375–376
    theoretical principles, 375
Easter Island, 13
Eckelman, M., 243
Ecological economics, 93
Ecological engineering, 365–366
Ecological rucksack, 260
Economics and industrial ecology, 93–97
Ecoprofile, 192
Ecosystem engineering, 47
Ehrenfeld, J., 29
Electric power infrastructure. *See*
        Infrastructure
Elemental substance analyses, 245–246
Embodied energy, 260

Embodied flows, 260–261
Emergent behavior. *See* Systems
    analysis
Endogenous variables, 308
Energy, 269–280
    energy analysis, 52
    embodied energy, 260
    energy Star, 358
    refrigerators, energy use by, 275
Enterprise resource planning
    (ERP), 64
Environmental footprint, 36
Environmental policy objectives. *See* Grand
    environmental objectives
Environmentally-responsible product
    assessment matrix, 194
Enzymes
    biological, 56
    industrial, 59, 243
Ethical concerns, 91–92
Evolution, 49
Exogenous variables, 308
Experienced resources, 33
Extent. *See* Models
Extractors, 223
Exxon, 77

Fertilizer, 31
Fischer-Kowalski, M., 87
Fish stocks, 2
Florida Everglades, 367, 377
Flux, 242
Food chain
    biological, 223–224
    industrial, 225
Food web
    biological, 226–228
    industrial, 229–232
Free rider problem, 3
Frosch, R., 41

Gallopoulos, N., 41
Germanium, sustainable supplies of, 19
Glaxosmithkline, 360
Global climate change, 27, 370
Goal and scope. *See* Life cycle assessment

Gossage, W., 92
Governance and industrial ecology, 89–91
Grand environmental objectives, 22
Green accounting, 96–97
Green buildings, design principles,
    152–157, 158
Green chemistry, principles of, 103–105
Green engineering, principles of, 33,
    105–106
Greenhouse gases, 20–21, 26
Green marketing, 360
Green Star, 152
Greenpeace, 112
Green taxes, 91
Gross national product, 96
Guilds, 226
Gutowski, T., 141

Hammond, A., 319
Hardin, G., 2
Herbivores, 226
Hertwich, E., 88
Hidden flows, 128, 254–255, 260–262,
    268, 278, 285, 295
Hockstra, A., 286
Holarchy, 216
Holon, 216
Hong Kong, 300
House of Quality, 118
Human organism damage, 27
Human demographics
    expected life span, 10
    global population growth, 4
    urban population growth, 296
Hybrid input-output analysis, 265

Impact analysis. *See* Life cycle
    assessment
Industrial ecology
    characteristics, 38–39
    concepts, 30–36
    definition, 32, 41
    template, 34
    type II system, 46
Industrial ecosystem, 46,
    223–237

Industrial enzyme. *See* Enzyme
Industrial metabolite. *See* Metabolite
Industrial organism. *See* Organism
Industrial pathway, 59
Industrial process. *See* Process
Industrial symbiosis, 232–233, 381
Information and communications
    technology (ICT), 8–9
Infrastructure, 146–151. *See also* Design
    for environment and sustainability
  electric power infrastructure
    distribution, 148–149
    generation, 148–149
    transmission, 148–149
  green design principles, 158
  telecommunications infrastructure, 151
  transportation infrastructure, 150
  water infrastructure, 149
Input-output tables, 262
Intergenerational equity, 92
Intragenerational equity, 92
Interpretation. *See* Life cycle assessment
Inventory analysis. *See* Life cycle assessment
IPAT equation, 6, 88
Iron, 276
Island biogeography, 52, 236–237
Island industrogeography, 236–237

Jones, C., 47

Kay, J., 215
Kalundborg, Denmark, 233–234
Kondratiev waves, 8
Kyoto Protocol, 90

Land use, 27
Landfilling, 243
LCA software. *See* Life cycle assessment
Leadership in energy and environmental
    design (LEED), 152–155
Legal concerns and industrial ecology, 92–93
Leontief inverse, 264
Level. *See* Models
Life cycle assessment, 161–189, 353, 381
  boundaries, 164–167
  characterization, 177

classification, 176
  five percent rule, 165
  goal and scope, 162, 164
  impact analysis, 162, 175–181
  interpretation, 162, 181–182
  inventory analysis, 162–173
  limitations of, 188
  microchip, 168–170
  normalization, 178–180
  palm oil in Malaysia, 181
  prioritization, 183–187
  software, 182
  streamlining of, 191–209
  valuation, 94, 180–181
  women's shoes, 178–180
Life cycle of products, 120
Linked. *See* Models
Lock-in, 93
London. *See* Metabolic analysis
Lovel, R., 286
Ludwig, D., 221

Maier, M., 217
Markov chain modeling, 243
Master equation, 5
Material flow analysis (MFA), 52,
    240–252, 353, 381
Material requirements planning
    (MRP), 59
Maximum acceptable risk, 78
Maximum economic yield, 2
Maximum sustainable yield, 2
McDonough, W., 36, 112–113
Mercedes-Benz, 168, 209
Metabolic analysis
  biological, 55–57
  definition, 52
  industrial, 57–65
  of London, 297–298
Metabolic cycle, 243
Metabolic flows, 295
Metabolic pathways
  biological, 57
  industrial, 59, 63–64
Metabolic structure of society, 87
Metabolism, 55

Metabolite
  biologial, 56
  industrial, 59, 243
Methane, 26, 249
Midgley, T., 369
Mirrors in space, 373, 378
Mitsch, W., 365
Models, 304–316, 381
  classification, 306
  coupled, 311
  extent, 308
  level, 308
  resolution, 308
  scale, 308
  validation, 310
Molina, M., 248
Montreal Protocol, 3, 90, 369
Morgan, G., 78

N_2O. *See* Nitrous oxide
Nanotechnology, 8
National material accounts, 254–267, 381
  domestic material consumption (DMC), 257
  domestic material input (DMI), 255
  domestic processed output (DMO), 259
  total domestic output (TDO), 255
  total material requirements (TMR), 255, 260
Natural Step, 113
NBIRC technologies, 9
Network analysis, 213–214
Nickel, 244, 276
Norgate, T., 286
Normalization. *See* Life Cycle Assessment

Ocean fertilization, 378
Oil, 337
Omnivory, 223
1.7 kilogram microchip, 168
Open-loop recycling. *See* Recycling
Ore grade, 279, 286
Organism
  biological, 42
  industrial, 43
Oreskes, N., 311

Our Common Future, 4
Ozone hole, 249, 370

Packaging, 131–134
Philips Electronics, 362
Physical input-output analysis, 381
Pollution prevention, 107–108, 164
Princen, T., 88
Prioritization. *See* Life cycle assessment
Processes, 329
  process design, 107
  process life cycle, 110
Products
  energy use, 134
  gates, 123–124
  integrated product development, 122
  life cycle, 120
  product delivery, 131
  product development, 115–124
  product realization process (PRP), 122
Pugh selection matrix, 117

REACH, 97, 128
Rebound effect, 89
Recycling
  closed-loop, 138
  open-loop, 138
Refining, 243
Refrigerators, energy use by
  See Energy
Remanufacturing, 134
Reserve base, 333
Residence time, 243
Residues (other than wastes), 33
Resilience. *See* Adaptive management
Resolution. *See* Models
Resources
  depletion, 27
  human/natural dominance, 250
  resource productivity, 261
  scarcity, 329–340
  sustainability, 381
Reverse fishbone diagram, 138, 140

Risk, 71
  dimensions of, 71
  quantification, 72
  assessment, 75
  management, 78
Robotics, 8
Rowland, F., 248
Rose, C.M., 137
Royal Dutch Shell, 77, 112

Sankey diagram, 271
Saxton, E., 209
Scale. *See* Models
Scenarios, 319–326, 381
  definition, 319
  class, 321
  typology, 321
Self-organizing holarchic open systems
    (SOHO systems), 217, 365
Semiconductors
  manufacturing capacity, 346, 348
  water use in manufacturing, 283–284
Service sector, 354
Silicon Valley, 94
Simple systems. *See* Models
Skinner, B., 333
Slag, 243
Smelting, 243
Social ecology, 86–87
Solder, 112, 177–178
Species richness, 226
Stainless steel, 276
Stern, P., 88
Stevels, A., 362
Stock
  bottom-up determination, 245
  definition, 241
  top-down determination, 245,
    297–299
Stratospheric ozone depletion, 27
Stratospheric sulfate, 378
Streamlined life cycle assessment (SLCA),
    192–209
  assets and liabilities, 207
  matrix, 194
  weighting in, 201

Strong sustainability. *See* Sustainability
Substance analysis, 244–249
Supply chain, 359
Sustainable engineering
  concepts, 33–36
  characteristics, 38
  industrial practice, 101–114
  tools, 353
Sustainable consumption,
    88–89
Sustainability, 13–29, 85–86
  strong, 16
  weak, 16
Sydney, Australia, 299
Symbiosis, 52
System of National Accounts, 96
Systems analysis, 211–221
  complex systems, 212
  emergent behavior, 218–219
  holarchies, 215–217
  simple systems, 212

Tailings, 243
Target plots, 195–197
Technology
Accelerating, 1, 9, 68
  characteristics, 34, 38, 50, 68, 71
  emerging, 9
  evolution, 7, 68
  historical impacts, 67
  scale dependence, 82
  transformative, 8
Technological evolution, 7, 68
Telecommunications infrastructure.
    See Infrastructure
Tetraethyl lead, 31
Tilton, J., 333
Top-down. *See* Stock
Total domestic output. *See* National
    material accounts
Total material requirements. *See* National
    material accounts
Toyota Motor Company, 58
Tragedy of the commons, 2
Transformability. *See* Adaptive
    management

Transportation infrastructure.
    See Infrastructure
Tree of life, 49–51
Tricarboxylic acid cycle, 56
Triple bottom line, 363
Trophic levels, 223
Type I system. *See* Biological ecosystem
Type II system. *See* Biological ecosystem
    and Industrial ecosystem
Type III system. *See* Biological ecosystem

Uranium, 337
Urban industrial ecology, 294–303, 381
Urbanization, 296
Urban mining, 301

Valuation
    in life-cycle assessment. *See* Life
        Cycle Assessment
    in economics, 95
Virtual flows, 260–261

Virtual impacts, 357–358
Virtual water, 287–288

Walker, B., 221
WalMart, 357
Waste management, 243
Water, 282–293
    availability and quality, 27, 290
    efficiency, 287
    water footprint, 288
    water infrastructure.
        See Infrastructure
Weak sustainability. *See* Sustainability
Wei, J., 31
Wernick, I., 57
Wetlands, 378
Wulf, W., 118

Xerox, 131

Zinc, sustainable supplies of, 18